Medical Microbiology

BIOS INSTANT NOTES

Series Editor: B.D. Hames, School of Biochemistry and Molecular Biology, University of Leeds, Leeds, UK

Biology
Animal Biology, Second Edition
Biochemistry, Third Edition
Bioinformatics
Chemistry for Biologists, Second Edition
Developmental Biology
Ecology, Second Edition
Genetics, Second Edition
Immunology, Second Edition
Mathematics & Statistics for Life Scientists
Medical Microbiology
Microbiology, Second Edition
Molecular Biology, Third Edition
Neuroscience, Second Edition
Plant Biology, Second Edition
Sport & Exercise Biomechanics
Sport & Exercise Physiology

Chemistry
Consulting Editor: Howard Stanbury
Analytical Chemistry
Inorganic Chemistry, Second Edition
Medicinal Chemistry
Organic Chemistry, Second Edition
Physical Chemistry

Psychology
Sub-series Editor: Hugh Wagner, Dept of Psychology, University of Central Lancashire, Preston, UK
Cognitive Psychology
Physiological Psychology
Psychology
Sport & Exercise Psychology

Medical Microbiology

Professor Will Irving

Department of Microbiology, University Hospital, Queens Medical Centre, Nottingham, UK

Professor Dlawer Ala'Aldeen

Department of Microbiology, University Hospital, Queens Medical Centre, Nottingham, UK

Dr Tim Boswell

Nottingham City Hospital NHS Trust, Nottingham, UK

Taylor & Francis
Taylor & Francis Group

Published by:

Taylor & Francis Group

In US: 270 Madison Avenue
          New York, NY 10016

In UK: 4 Park Square, Milton Park
          Abingdon, OX14 4RN

© 2005 by Taylor & Francis Group

First published 2005

ISBN: 1-8599-6254-8

Library of Congress Cataloging-in-Publication Data

Irving, William L.
   Medical microbiology / Will Irving, Dlawer Ala'Aldeen, Tim Boswell.
      p. ; cm. –– (BIOS instant notes)
   Includes index.
   ISBN 1-85996-254-8 (alk. paper)
   1.   Medical microbiology –– Outlines, syllabal, etc.
   [DLM: 1.   Communicable Diseases –– microbiology –– Outlines. 2.   Infection –– microbiology –– Outlines. 3.   Bacterial Infections –– microbiology –– Outlines. 4.   Mycoses –– microbiology –– Outlines. 5.   Parasitic Diseases –– microbiology –– Outlines. 6.   Virus Diseases –– microbiology –– Outlines. QW 18.2 I72m 2005] I.   Ala'Aldeen, Dlawer A. A. II.   Boswell, Tim. III.   Title. IV.   Series.
QP46.I78 2005
616.9'041 –– dc22   2005022853

Editor: Elizabeth Owen
Editorial Assistant: Chris Dixon
Production Editor: Georgina Lucas/Simon Hill
Typeset by: Phoenix Photosetting, Chatham, Kent, UK
Printed by: TJ International, Padstow, Cornwall

Printed on acid-free paper

10 9 8 7 6 5 4 3 2 1

Taylor & Francis Group
is the Academic Division of T&F Informa plc.

Visit our web site at http://www.garlandscience.com

CONTENTS

ABBREVIATIONS

AAFB	acid–alcohol-fast bacilli	GRE	glycopeptide-resistant enterococci
Ad	adenovirus	GVHD	graft-versus-host disease
AIDS	acquired immunodeficiency syndrome	HACEK	*Haemophilus, Actinobacillus, Cardiobacterium, Eikenella, Kingella* group of organisms
AME	aminoglycoside-modifying enzymes	HAI	hospital-acquired infection
APC	antigen-presenting cells	HAP	hospital-acquired pneumonia
ART	automated reagin test	HAV	hepatitis A virus
BCG	Bacille Calmette–Guérin vaccine	HBIg	hepatitis B immunoglobulin
BCYE	buffered charcoal yeast extract agar	HBV	hepatitis B virus
		HCV	hepatitis C virus
BSE	bovine spongiform encephalopathy	HDV	hepatitis D virus
		HEPA	high-efficiency particulate air
CAP	community-acquired pneumonia	HEV	hepatitis E virus
CAPD	continuous ambulatory peritoneal dialysis	HHV	Human herpesviruses
		Hib	*Haemophilus influenzae* type b
CF	cystic fibrosis	HIV	human immunodeficiency virus
CFT	complement fixation test	HLA	human leukocyte antigens
CGD	chronic granulomatous disease	hMPV	human metapneumovirus
CJD	Creutzfeldt–Jakob disease	HPV	human papillomavirus
CMV	cytomegalovirus	HSE	Herpes simplex encephalitis
CNS	central nervous system	HSV	herpes simplex virus
CoNS	coagulase-negative staphylococci	HTLV	human T-cell lymphotropic virus
COPD	chronic obstructive pulmonary disease	HUS	hemolytic–uremic syndrome
		IE	infective endocarditis
CPE	cytopathic effect	IFN	interferon
CRS	congenital rubella syndrome	IL	interleukin
CSU	catheter specimen of urine	IUCD	intrauterine contraceptive device
CT	computed tomography	KDO	ketodeoxyoctonate
DEAFF	detection of early antigen fluorescent foci test	KSHV	Kaposi's-sarcoma associated human herpesvirus
DHF	dengue hemorrhagic fever	LCMV	lymphochoriomeningitis virus
ds	double-stranded (DNA/RNA)	LJ	Löwenstein–Jensen medium
EBV	Epstein–Barr virus	LOS	lipo-oligosaccharide
EIA	enzyme immunoassay	LP	lumbar puncture
EIEC	enteroinvasive *E. coli*	LPS	lipopolysaccharides
ELISA	enzyme-linked immunosorbent assay	LRTI	lower respiratory tract infection
		MAI	*Mycobacterium avium–intracellulare* complex
ENT	ear, nose and throat		
EPEC	enteropathogenic *E. coli*	MBC	minimum bactericidal concentration
ESBL	extended spectrum β-lactamases		
ETEC	enterotoxigenic *E. coli*	MHC	major histocompatibility complex
FI	fusion inhibitors	MIC	minimum inhibitory concentration
FTA-ABS	fluorescent treponemal antibody-adsorption test	MMR	measles-mumps-rubella vaccine
		MRI	magnetic resonance imaging
GE	gastroenteritis	MRSA	methicillin-resistant *Staph. aureus*
GNAB	Gram-negative anaerobic bacteria	MSU	mid-stream urine

NA	neuraminidase	SPA	suprapubic aspirate
NNRTI	non-nucleoside analogue reverse transcriptase inhibitors	ss	single-stranded (DNA/RNA)
		SSPE	subacute sclerosing panencephalitis
NRTI	nucleoside analogue reverse transcriptase inhibitors	TB	tuberculosis
OMP	outer membrane proteins	Tc	cytotoxic T cells
PAE	post-antibiotic effect	TCBS	thiosulfate–citrate bile sucrose
PBP	penicillin binding protein	TCR	T-cell receptor
PCP	*Pneumocystis* pneumonia	Th	helper T cells
PFGE	pulsed field gel electrophoresis	TPHA	*T. pallidum* hemagglutination test
PI	protease inhibitors	TPN	total parenteral nutrition
PID	Pelvic inflammatory disease	TPPA	*Treponema pallidum* particle agglutination
PMLE	progressive multifocal leukoencephalopathy	TSE	transmissible spongiform encephalopathies
PMN	polymorphonuclear leukocytes		
PTLD	post-transplant lymphoproliferative disorder	TTP	thrombotic thrombocytopenic purpura
PUO	pyrexia of unknown origin	URTI	upper respiratory tract infection
PVE	prosthetic valve endocarditis	UTI	urinary tract infection
RAPD	random amplified polymorphic determinants	VAP	ventilator-associated pneumonia
		VDRL	venereal disease reference laboratory test
RPR	rapid plasma reagin		
RSV	respiratory syncytial virus	VRE	vancomycin-resistant enterococci
RT–PCR	reverse transcriptase–polymerase chain reaction	VRSA	vancomycin-resistant *Staph. aureus*
		VZV	varicella-zoster virus
SARS CoV	severe acute respiratory syndrome coronavirus	WHO	World Health Organization
		ZN	Ziehl–Neelson stain
SBE	subacute endocarditis		

PREFACE

Medical microbiology is potentially an intimidating subject for new students. Not only does it appear to have a language all of its own, including plenty of obscure Latin terminology, but it covers a bewilderingly wide range of material, from the molecular biology of the infectious agents themselves (of which there is an alarmingly large – and ever-increasing – number) right through to the clinical management of the infected patient, passing disease pathogenesis, diagnosis, and the use of antimicrobial therapy on the way. This book seeks to identify, explain and expound upon the essentials of each of these aspects of the subject, in suitable 'bite-sized' chunks (a philosophy inherent in all of the books in the Instant Notes series).

We do not expect students to start this book at page 1 and work their way meticulously through to the last page (although to do so would undoubtedly assist them in gaining excellent examination marks!). Many courses concerned with medical microbiology will have their own particular emphasis, and will provide scant, if any, cover of the other aspects of the subject. Also, certainly at undergraduate level, it is unlikely that any one course would intend to cover each and every microbe, at least not in the same depth of detail. It would be appropriate therefore for students undertaking those courses to concentrate on the subject areas and microorganisms relevant to their own course. Nevertheless, as practising medical microbiologists, we felt it entirely appropriate to aim to cover the many facets of our subject in a single text, and also to be comprehensive in dealing with all of the infectious agents with which our patients present on a routine basis.

The book is divided into a number of sections. In section A we provide introductory background information on the nature of the infectious agents, the host response to infection, and how micro-organisms give rise to disease. Sections B, C, and D deal with the organisms themselves in more detail, the sections being ordered from the least (viruses) to the most (eukaryotes) structurally complicated. Within these sections, each group of related organisms is dealt with in a separate topic, which are ordered, very roughly, according to their clinical importance, and, for the bacteria, their Gram stain characteristics. Section E is concerned with general principles of the laboratory diagnosis and management of infection, including the mechanisms of action of antimicrobial agents, and possible strategies for prevention of infection. The final section, F, deals with infection from the point of view of the patient, i.e. clinical microbiology.

We hope that students find this approach to be logical and useful. We have benefited enormously from our own experiences of teaching students, both medical and non-medical, and would welcome any constructive feedback which students might wish to raise. Most of all we hope that at least some students will be as fascinated by this subject as we are, and will see this text as their first step on the road to a medical microbiologically related career.

William Irving, Tim Boswell and Del Ala'Aldeen

A1 INTRODUCTION

Key Notes

Microbiology	Microbiology is the 'biology of microscopic organisms'.
Medical microbiology	Medical microbiology is the study of microbes that cause disease in humans. Microbes are everywhere, both within and outside our body. Most are harmless.
Types of microbes	Microbes vary in shape, size and structure and are categorized into eukaryotic (fungi, worms and protozoa), prokaryotic (bacteria, rickettsia and chlamydia) or noncellular (viruses and prions). Eukaryotic organisms are uni- or multicellular; prokaryotes are unicellular. Viruses and prions are incapable of independent life.
Epidemiology of infection	Epidemiology is the study of spread of infection, including the source, transmission, distribution and prevalence of infection in the community. Sources of infection are either endogenous or exogenous. Acquisition of pathogens may occur via many routes, including direct contact, inhalation, ingestion, injection or vertical transmission. Epidemiologists assess infection in a community by using various measurements, including incubation period of disease, incidence, prevalence, attack rate and mortality rate.
Definitions and terms	Infection is a generic term used to indicate invasion of the host by a microorganism. Infection may be subclinical or asymptomatic when the patient is unaware of the infection. Clinical infection is associated with the presence of overt signs and symptoms of disease. The term 'colonization' should be restricted to the presence of a microbe at an expected site. A pathogen is a microbe that potentially can cause harm, i.e. tissue damage. An opportunistic pathogen is a microbe that causes infection in patients with impaired immunity, e.g. fungal infections in cancer patients.

Introduction

Microbiology is the study of the 'biology of microscopic organisms' whereas medical microbiology is the study of microbes (infectious agents) that cause disease in humans. Microorganisms vary tremendously in terms of shape, size, structure and importance.

Medical microbiology is of increasing importance in health and disease. For example:

- approximately half of all patient visits to general practitioners are for infections;

- there are increasing numbers of patients with impaired immune systems in hospital, who are susceptible to a wide range of life-threatening infections;
- infectious diseases are associated with major public health implications, e.g. in the control of infection with human immunodeficiency virus (which leads to the acquired immunodeficiency syndrome – AIDS), tuberculosis and food-poisoning;
- costs of antimicrobial agents are increasing, as is the problem of microbial resistance to these agents;
- infection is by far the most common cause of morbidity and mortality in developing countries.

Microbes are ubiquitous. We are surrounded by microorganisms, the vast majority of which are harmless. Microorganisms live inside and outside the human body. They colonize the skin and mucous membranes of the mouth, nose, eyes, ears, sinuses, throat, gastrointestinal tract and vagina. They also live in the surrounding environment, including water, food, vegetables, animals and birds.

There are 10^{12} bacteria in each gram of feces. The vast majority of these are harmless, and some possibly useful.

Types of microbes   Microbes can be divided into three main categories (*Fig. 1*).

Eukaryotic organisms
These include fungi (e.g. molds and yeasts) and parasites (helminths and protozoans). They have a complex cellular structure, similar to those of humans and animals. Their cells have nuclei and mitochondria and they are largely self-sufficient and capable of independent life.

Prokaryotic organisms
These are simple and largely self-sufficient unicellular organisms which have no nuclei or internal dividing membranes but are usually capable of independent life (*Table 1*). Cell walls contain mucopeptides (peptidoglycans). They are collectively named bacteria. Some genera, e.g. rickettsiae and chlamydiae, are not capable of independent life and are therefore named 'atypical bacteria'. The latter group of organisms are obligate intracellular pathogens which require the presence of viable eukaryotic host cells for growth and reproduction.

Noncellular organisms
Viruses consist simply of DNA or RNA plus a few other components such as proteins. They are not capable of independent life, and therefore must infect cells of higher organisms (eukaryotic or prokaryotic) for their growth and reproduction. Some viruses infect bacterial cells – these are called bacteriophage or simply phage. Also included among the noncellular infectious agents are the recently characterized 'prions' which cause bovine spongiform encephalopathy (BSE) in animals and Creutzfeldt–Jakob disease (CJD) in humans.

Fig. 1(a).   Cell structure of a typical eukaryotic organism.

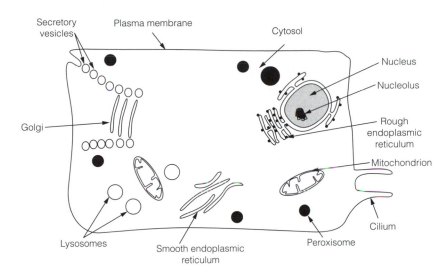

Fig. 1(b).   Cell structure of a typical (Gram negative) prokaryotic organism (bacterium).

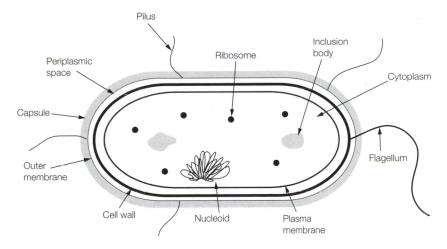

Fig. 1(c).   A non-cellular organism (virus)

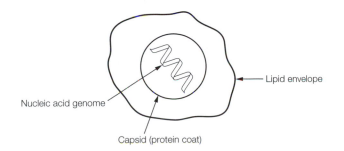

Fig. 1.   Types of microbes.

Table 1:   Some basic differences between eukaryotic and prokaryotic cells

	Eukaroytic cell	Bacterial cell
Size	>5 µm	1–3 µm
DNA and RNA	+	+
Paired chromosomes	+	–
Mitotic division	$+_m$	–
Binary fission	$–_m$	+
Structured nucleus	+	–
Golgi apparatus	+	–
Endoplasmic reticulum	+	–
Mitochondria	+	–
Ribosomes	+	+
Cell membrane	+	+

m = majority

Size and characteristics of microbes

Most microbial cells cannot be seen except by microscopy (light or electron).

Fungi are larger organisms with thick walls containing chitin. They exist either in the form of **molds** which grow by tubular branching filaments, or **yeasts** which are oval or spherical and grow by budding (*Fig. 2a*).

Protozoa are larger than bacteria. **Helminths** vary in size, ranging from a few millimetres to meters long (*Fig. 2b*).

Bacteria can be **seen under light microscopes**. They are 0.5–1 µm broad, 0.5–8 µm long and vary in shape and size. They are mostly either spherical (**cocci**) or cylindrical (**bacilli**) (*Fig. 2c*). Some bacteria assume different cellular morphologies, e.g. rickettsiae are pleomorphic (different shapes) and spirochaetes are spiral. Different bacteria have different growth characteristics. **Strictly aerobic** bacteria cannot grow in the absence of oxygen, **strictly anaerobic** bacteria cannot grow in the presence of oxygen and **facultatively anaerobic** bacteria can grow in the presence or absence of oxygen. Some bacteria are **fastidious** and have specific nutritional or other environmental requirements for growth.

Viruses are too small (fractions of a micrometer) to be seen by an ordinary light microscope, but can be seen under an **electron microscope** (EM). They vary in shape and structure (*Fig. 2d*). Many species have distinctive morphological characteristics that can be spot-diagnosed under EM.

Epidemiology of infection

Epidemiology is the study of spread of infection. This includes the source, transmission, distribution and prevalence of infection in the community.

Infecting organisms commonly originate from **endogenous sources** (the patient's own normal human flora, *Fig. 3*) or **exogenous** sources (e.g. other infected patients, animals, plants or contaminated objects, food or water *Fig. 4*). Endogenous sources are by far the most common. The source and **reservoirs** of infection are often (but not always) the same.

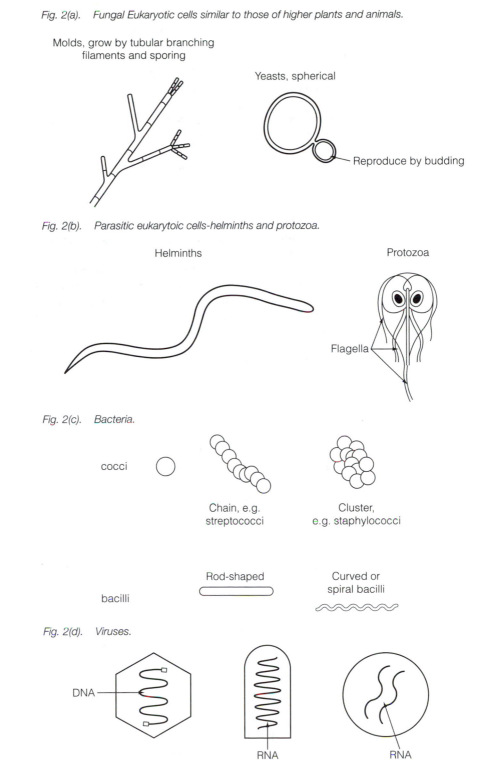

Fig. 2(a). Fungal Eukaryotic cells similar to those of higher plants and animals.

Molds, grow by tubular branching
filaments and sporing

Yeasts, spherical

Reproduce by budding

Fig. 2(b). Parasitic eukarytoic cells-helminths and protozoa.

Helminths

Protozoa

Flagella

Fig. 2(c). Bacteria.

cocci

Chain, e.g.
streptococci

Cluster,
e.g. staphylococci

Rod-shaped

Curved or
spiral bacilli

bacilli

Fig. 2(d). Viruses.

DNA

RNA

RNA

Fig. 2.   Differing shapes of microbs (not drawn to scale).

Routes of infection

Acquisition of pathogens may occur via many routes, including:

* **direct skin or mucosal contact** with contaminated hands, body fluids, animals or objects; this includes sexual transmission of infection;
* **inhalation** of contaminated aerosols or droplets generated by sneezing, coughing, talking;
* **ingestion** of contaminated food or drink;
* **inoculation, penetration or injection** of contaminated fluids (e.g. during surgery or drug administration);
* **vertical transmission** from mother to baby (e.g. transplacental);
* **vector-borne** transmission.

The control and prevention of **outbreaks** of infections depend mainly (not entirely) on understanding the epidemiology of the disease and identification of the reservoirs, sources and routes of transmission. Important factors in control and prevention of infectious diseases include the number of **susceptible** humans, pathogenicity (virulence capability) of the organism, route of spread, identification of carriers, **incubation period** of disease and other environmental contributors. **Surveillance** (i.e. constant monitoring) of important infectious diseases is very important for implementing measures for the control of outbreaks.

Epidemiologists assess infection in a community by using various measurements. These include:

* incubation period of disease;
* **incidence and incidence rate**: number of cases of disease in the community per given population (e.g. 100 000);
* **prevalence**: incidence of disease within a given time;
* **attack rate**: incidence within a defined population group;
* **secondary attack rate**: number of cases of disease among contacts of the primary (index) case;
* **mortality rate**: number of deaths from a particular infection within a given population (e.g. 100 000).

Definitions and terms

Infection is a generic term used to indicate invasion of the host by a microorganism. This invasion is usually associated with a host response (e.g. inflammation) with or without clinical manifestations. Thus, infection may be **subclinical (asymptomatic, nonapparent)** when there is no apparent disease, and the patient is unaware of the infection. Alternatively, a **clinical** infection is one associated with the presence of overt signs and symptoms of disease.

It is important to remember that the presence of microorganisms in the host does not always mean disease. Healthy individuals may be 'colonized' without disease. The term **colonization** should be restricted to the presence of a microbe at an expected site, e.g. *Escherichia coli* in the large bowel, or *Staphylococcus epidermidis* on the skin, but occasionally it is also used to describe the presence of a potential pathogen in an unexpected site without causing symptomatic infection, e.g. *Staph. aureus* in the upper respiratory tract. Colonized patients are often also described as **carriers**.

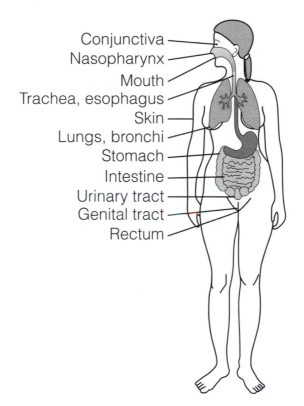

Conjunctiva
Nasopharynx
Mouth
Trachea, esophagus
Skin
Lungs, bronchi
Stomach
Intestine
Urinary tract
Genital tract
Rectum

Fig. 4.   Endogenous sources of infection.

Sex

Needles, sharps, blood transfusion

Animals

Vertical transmission

Food

Fig. 5.   Exogenous sources of infection.

A **pathogen** is a microbe that causes harm, i.e. tissue damage, and potentially may therefore cause disease when in contact with a host. An **opportunistic pathogen** is a microbe that is not usually pathogenic in an immunocompetent or otherwise normal host, but may cause life-threatening infection in patients with impaired host defenses, e.g. fungal infections in cancer patients.

A2 HOST DEFENSE MECHANISMS TO INFECTION

Key Notes

Definition of terms

Innate immunity is mediated by nonspecific mechanisms which operate without the need for previous contact with the infectious agent. **Adaptive immunity** comprises a number of effector mechanisms triggered by specific recognition of an infectious agent by the host immune system. **Humoral** immune mechanisms are present within the plasma, whereas **cellular** immunity is mediated by cellular components.

Innate immunity

Prevention of entry of organisms into the host is prevented by intact skin, secreted mucus which can trap particulate matter, gastric acid and commensal flora.

Nonspecific humoral mechanisms include:
- the acute phase response – a rapid increase in a number of plasma proteins with a variety of protective functions;
- the complement system, a set of circulating proteins which can be activated in cascade fashion by a number of different pathways, resulting in the generation of opsonizing, anaphylactic, chemotactic, and lytic components;
- the interferons (IFN), which have antiproliferative, immunomodulatory and antiviral properties.

Nonspecific cellular mechanisms are mediated by:
- neutrophils, which have phagocytic activity, and can kill ingested organisms by both oxygen-dependent and independent mechanisms;
- monocytes/macrophages, which are also phagocytic, and can release soluble monokines with inflammatory activity;
- natural killer cells, with spontaneous cytotoxicity for a variety of target cells, greatly enhanced by IFN.

Adaptive Immunity

Cellular interactions in adaptive immunity
Foreign material (antigen) is expressed on a cell surface bound to one of two classes of human leukocyte antigens (HLA). Class I HLA molecules (HLA-A, -B, -C) are expressed on all nucleated cells; class II HLA molecules (HLA-DP, -DQ, -DR) are expressed only on certain cells of the immune system. Cells of the adaptive immune response include: antigen-presenting cells derived from the monocyte/macrophage lineage. These take up and express foreign antigens bound to HLA class II molecules. T lymphocytes all express CD3; helper T cells (Th) also express CD4; and cytotoxic T cells CD8. The T-cell receptor (TCR) of Th cells recognizes antigen bound to HLA class II, resulting in cytokine release. Two subtypes of Th are recognized: Th1 cells secrete interleukin (IL-2) and

IFN-γ, which stimulate Tc cells, whereas Th2 cells secrete IL-4 and IL-10, which stimulate B cells. Tc cells recognize antigen bound to HLA class I molecules. B lymphocytes are the precursors of antibody-secreting plasma cells.

The first response to an antigen is referred to as the primary immune response. This may take several days to develop, but it also creates immunological memory, such that later challenge with the same antigen results in a much faster, secondary immune response.

Antigen-specific humoral defense mechanisms

Antigen-specific humoral defense is mediated by antibodies (immunoglobulins). These act by enhancing phagocytosis, activating complement, neutralizing viruses and secreted toxins, preventing bacterial and viral adherence, and causing degranulation of mast and basophil cells. The humoral response thereby protects against extracellular pathogens.

Antigen-specific cellular defense mechanisms

These are mediated by activated Tc, which, on recognition of foreign antigen on the cell surface, cause the death of the cell by apoptosis and release of perforins and granzyme B. The Tc response thereby provides protection against intracellular pathogens.

Related topics   Vaccines and immunoprophylaxis (E10)

The structure and function of the human immune system is a complex matter. Many excellent – and several lengthy – textbooks are devoted exclusively to this subject. Here, only the essence of the host defense mechanisms to infection are elucidated in the briefest of fashions. Students are referred to the aforementioned texts for further information and explanation.

Definition of terms   **Innate immunity:** this refers to a variety of nonspecific mechanisms which operate automatically to protect the host against infection without the need for previous contact with any infectious agent.

Adaptive immunity: this refers to a number of effector mechanisms initiated and stimulated by specific recognition of an infectious agent by the host immune system.

Innate immunity provides the first line of defense and operates immediately, hopefully containing the invasion of an infectious agent until the adaptive mechanisms can come into play.

Humoral immune mechanisms are those mediated by proteins circulating in the plasma, as opposed to **cellular** immune mechanisms which are mediated by cellular components of the peripheral blood. There are humoral and cellular components of both innate and adaptive immunity.

Innate Immunity   *Prevention of entry*
The most obvious way to protect a host against infection is to prevent entry of the infectious agent into the body. Thus, intact skin is impermeable to most infectious agents, and breaches by trauma or surgical wounds are a major risk factor for infection. Within the respiratory tract, secreted mucus can trap inhaled

particulate matter (e.g. bacteria), which is subsequently eliminated by the co-ordinated upward beating of small hairlike structures (cilia) lining the respiratory tract, coughing or sneezing. In the alimentary tract, the mucous membrane acts as a physical barrier and gastric acid has powerful microbicidal properties. In addition, the normal commensal flora (e.g. on the skin, most mucosal membranes, and in the gut) compete out potential pathogens (a phenomenon referred to as **colonization resistance**) by preventing access to cellular binding sites or essential nutrients, or by production of inhibitory substances such as bactericidins. Lachrymation, nasal secretions, salivation and urination can be considered as 'washing' processes whereby internal access of microorganisms is impeded.

Nonspecific humoral defense mechanisms
Most forms of tissue injury or inflammation, including infection, result in a rapid increase in a number of plasma proteins, including coagulation proteins, protease inhibitors, transport proteins, C-reactive protein, and complement components (see below). This is referred to as the **acute phase response.** Collectively, these proteins aid in anatomical limitation of the inflammatory process, and in recognition and clearance of toxic products released whenever cell death or tissue necrosis occurs.

The **complement system** comprises a large number of plasma proteins that mediate several functions of the inflammatory process. These circulate as **proenzymes**, i.e. in an inactive form. Complement activation arises through a cascade process, whereby activation of one proenzyme results in activation of the next proenzyme in the pathway, and so on. The two most important mechanisms for activation of the early complement components are the classical pathway, initiated by antigen–antibody complexes, and the alternate pathway, initiated by microbial surface molecules. These pathways result in the generation of enzymes capable of splitting component C3. Once C3 is split into C3a and C3b, the subsequent pathway of activation of the terminal components is identical, regardless of the initiating stimulus. The end result is the generation of several active molecules which mediate distinct biological properties:

- *Opsonization*: adherence of C3b to the surface of foreign particles results in enhanced phagocytosis (see below) through binding to C3b receptors on macrophages and polymorphonuclear leukocytes.
- *Anaphylaxis*: C3a and C5a bind to mast cells, basophils, and platelets causing release of potent mediators such as histamine and serotonin, causing vasodilatation and increased vascular permeability.
- *Chemotaxis*: C5a and C5b67 are chemotactic for polymorphs and eosinophils, i.e. these cells are attracted to the site of inflammation.
- *Cell lysis:* The lytic attack complex C5b6789 formed on a cell (e.g. bacterial) surface can generate a stable channel through the lipid bilayer of the cell membrane resulting in cell lysis.

A simplified diagram of complement activation is provided in *Fig. 1*.

The **interferons** (IFN) are a family of glycoproteins that have a multiplicity of effects on cells. IFN-α and -β are produced by a wide variety of cells in response to stimuli such as viral or bacterial infection. IFN-γ is produced by T lymphocytes in response to stimulation by a specific antigen. Different IFN have different properties. These include inhibition of cell division, modulation (both suppressive and enhancing effects) of the immune system, and inhibition of viral replication. Binding of an IFN molecule to a cell receptor results in the

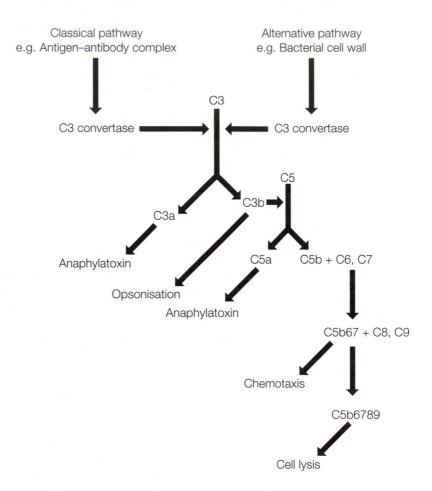

Fig. 1.   Complement activation.

switching on of several cellular genes. The protein products of these genes act to inhibit the synthesis of viral proteins within the cell.

Nonspecific cellular defense mechanisms

Neutrophils (also known as polymorphonuclear leukocytes, PMN) are the main inflammatory cell induced by bacterial infection. They are mature, nondividing, short-lived cells rich in structures required for cell migration and antimicrobial activity. Various biologically active molecules, such as myeloperoxidase, lysozyme, lactoferrin and vitamin B_{12} binding protein are contained within **cytoplasmic granules**. Neutrophils are able to change shape and move towards an inflammatory stimulus (chemotaxis). They adhere to the vascular endothelium, and then migrate across the endothelial barrier. Once at the site of inflammation, neutrophils can *phagocytose* (i.e. engulf) particulate matter, including bacteria. This is enhanced if the matter to be engulfed is coated by **opsonins** such as C3b (see above) or antibodies, as the cells have surface receptors for these molecules. Lysosomes and intracellular granules then fuse with the **phagosome** (the engulfed particle) in the cell cytoplasm. **Phagolysosome** formation activates a number of antimicrobial mechanisms:

oxygen-dependent mechanisms include generation of superoxide anions, hydrogen peroxide, hydroxyl and other organic free radicals, processes enhanced by the presence of lactoferrin and myeloperoxidase;

oxygen-independent mechanisms include deprivation of essential nutrients, cell wall digestion by lysozyme, the microbicidal action of cationic proteins, and lysis induced by the low pH.

Mononuclear cells include both circulating **monocytes** and fixed tissue **macrophages** (e.g. Kupffer cells in the liver). As with neutrophils, these cells have surface receptors for complement components and immunoglobulins, and contain a number of cytoplasmic granules, and thereby are efficient **phagocytes**. When appropriately activated, these cells secrete a variety of soluble factors known as **monokines** (e.g. IL-1, endogenous pyrogen, collagenase, elastase, IFN-β) that mediate inflammation and regulate the function of other cell types. Monocytes and macrophages also act as antigen-presenting cells, an important step in the adaptive immune response (see below).

Natural killer (NK) cells are a subset of lymphocytes, identified morphologically as large granular lymphocytes, that have spontaneous cytotoxicity for a variety of target cells in the absence of prior sensitization. Such target cells include tumor cells and virally infected cells. The killing activity of these cells is markedly enhanced by IFN.

Adaptive immunity *Cellular interactions*

The adaptive, or antigen-specific, immune response includes both humoral and cellular mechanisms stimulated by recognition of foreign antigens by cells of the host immune system. The cells involved must be able to recognize both foreign antigens and also each other. These recognition functions are subserved by the gene products of the **major histocompatibility complex** (MHC), also known in man as the **human leukocyte antigen** (HLA) system. There are two important types of HLA molecules, known as class I (HLA-A, -B, and -C) and class II (HLA-DQ, -DP, -DR) antigens. Class I HLA antigens are expressed on all nucleated cells, whereas class II HLA antigens are only expressed on certain cells of the immune system, mostly the antigen-presenting cells. Both classes of HLA molecules can bind foreign antigens. The cells involved in the adaptive immune response are:

(i)   **Antigen-presenting cells (APC)**. These are derived from the monocyte/macrophage series. They are able to take up foreign antigens, and present these antigens on their cell surface, bound to HLA class II molecules.

(ii) **T lymphocytes**. These cells undergo differentiation and maturation in the thymus (hence T cells), and can be identified as they all express a molecule known as CD3. There are two important subsets of T cells: **helper T cells** (Th), which also express CD4, and **cytotoxic T cells** (Tc) which express CD8. CD4-positive Th cells recognize antigen bound to HLA class II molecules presented to them by APC. Th cells are further subdivided into two differentiated subtypes. Th1 cells release cytokines such as IFN-γ and IL-2, which act to stimulate Tc cells, whereas Th2 cells release cytokines such as IL-4 and IL-10, which stimulate B cells. CD8-positive Tc cells recognize antigen bound to HLA class I molecules. As the latter are present on all nucleated cells of the body, Tc cells can be stimulated by any nucleated cell invaded by a foreign antigen.

(iii) **B cells**. These are the precursors of antibody-secreting plasma cells. They carry antibody on their cell surface, which acts as a specific receptor for antigen.

The essence of the adaptive immune response is shown in *Fig. 2*. Th cells are stimulated by antigen presented to them by APC; this results in the release of cytokines which can activate either B cells (generating a humoral response), or Tc cells (mediating the cellular response) to the initiating antigen. Note that this **primary immune response** takes several days to develop. However, once stimulated, the adaptive immune system has the property of **immunological memory**, such that re-stimulation with antigen at a later date (even years later) results in a much more rapid, or **secondary**, immune response.

Antigen-specific humoral defense mechanisms
B cells are stimulated to proliferate and differentiate into antibody-secreting plasma cells by the binding of antigen to their surface antibody in the presence of appropriate cytokine signals released by antigen-stimulated Th1 cells (e.g. IL-4, IL-10). Each **antibody**, also known as an **immunoglobulin** (Ig), comprises two heavy (H) and two light (L) chains (*Fig. 3*). Both H and L chains consist of a constant region and a variable region. There are five classes of antibody, defined by differences in the constant regions of their H chains, known as IgA, IgD, IgE,

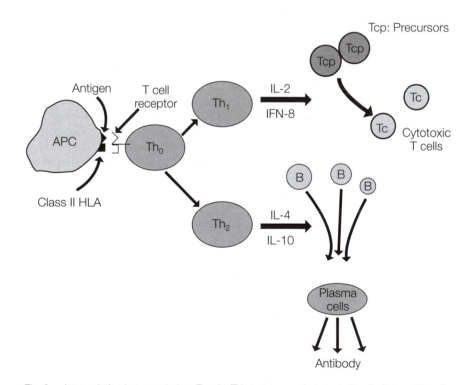

Fig. 2.   *Inter-relation between helper T-cells (T$_h$), suppressor/cytotoxic T-cells (T$_c$) and B-cells. The T$_h$-cell recognises antigen in association with a class II MHC molecule on the surface of a macrophage. It then releases a number of soluble factors which help both the antibody and the cell-mediated responses to antigen. The T$_h$ cell therefore has a pivotal role in the immune response.*

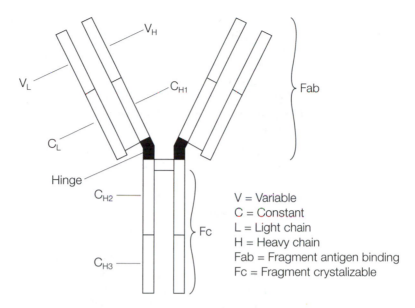

Fig. 3. *Structure of an immunoglobulin molecule.*

IgG and IgM, which have different properties (*Table 1*). An antibody molecule contains two identical antigen-binding sites derived from the variable parts of the H and L chains, known as the Fab (fragment antigen binding) sites.

Antibodies mediate the humoral component of the adaptive immune response, and protect the host against infection via a number of mechanisms:

(i)   enhancing phagocytosis: by coating infectious agents, such as encapsulated bacteria, and then binding to neutrophils and monocytes via their Fc region;

(ii) activating complement: Ag–Ab complexes trigger the classical pathway of complement activation, leading to the generation of a variety of defense mechanisms (see above);

Table 1.   Biological properties and functions of different immunoglobulin classes

	Heavy chain	No. of Ig units	Complement Fixation	Crosses placenta	Comment
IgG	γ	1	++	Yes	Antigen-specific IgG is marker of past infection
IgA	α	1 or 2	+/–	No	Major Ig in seromucous secretions where it defends mucous membranes
IgM	μ	5	+++	No	Transient production early in immune response (i.e. marker of recent infection)
IgD	δ	1	–	No	Mostly present on B-cell surface
IgE	ε	1	–	No	Binds to mast cells, basophils; responsible for symptoms of allergy

(iii) neutralization of viruses: by binding to the surface of the virus, and thereby preventing the virus from binding to its cellular receptor;

(iv) coating of bacteria and viruses by IgA, thus preventing their adherence to mucosal surfaces;

(v) neutralization of toxins released by bacteria;

(vi) IgE-mediated release of vasoactive and chemotactic factors from mast cells, basophils and eosinophils, which leads to recruitment of all the agents of an immune response.

Note that the effector mechanisms of the humoral arm of the adaptive immune response are directed against the control of extracellular pathogens.

Antigen-specific cellular defense mechanisms

Tc cells are stimulated to proliferate and become activated by IFN-γ and IL-2 released by antigen-stimulated Th cells. Recognition by the Tc cells of antigen on the surface of host cells in association with class I HLA molecules then results in killing of the target cell through induction of apoptosis (cellular suicide) and release of effector molecules (e.g. perforin, granzyme B) from granules within the Tc cytoplasm. Thus, the Tc response provides a defense mechanism against organisms which have developed the capacity for living and multiplying within the cells of the host, thereby evading the humoral immune response (e.g. intra-cellular bacteria, viruses).

A3 VIRUSES AND VIRUS REPLICATION

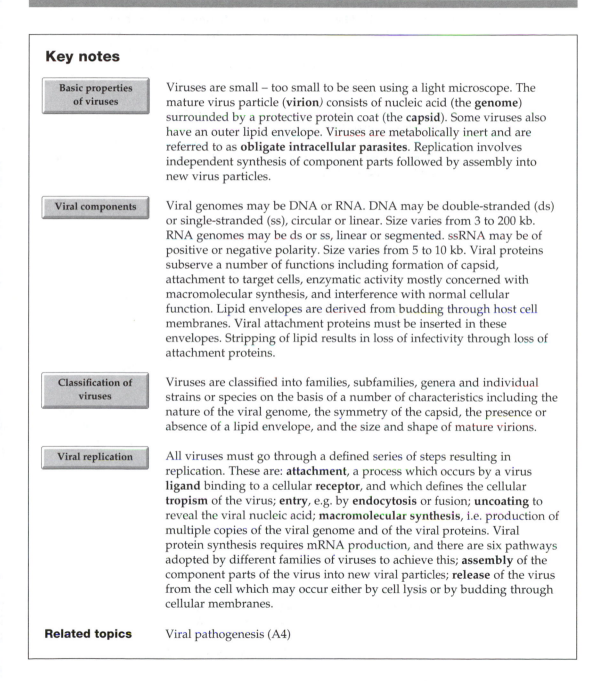

Key notes

Basic properties of viruses

Viruses are small – too small to be seen using a light microscope. The mature virus particle (**virion**) consists of nucleic acid (the **genome**) surrounded by a protective protein coat (the **capsid**). Some viruses also have an outer lipid envelope. Viruses are metabolically inert and are referred to as **obligate intracellular parasites**. Replication involves independent synthesis of component parts followed by assembly into new virus particles.

Viral components

Viral genomes may be DNA or RNA. DNA may be double-stranded (ds) or single-stranded (ss), circular or linear. Size varies from 3 to 200 kb. RNA genomes may be ds or ss, linear or segmented. ssRNA may be of positive or negative polarity. Size varies from 5 to 10 kb. Viral proteins subserve a number of functions including formation of capsid, attachment to target cells, enzymatic activity mostly concerned with macromolecular synthesis, and interference with normal cellular function. Lipid envelopes are derived from budding through host cell membranes. Viral attachment proteins must be inserted in these envelopes. Stripping of lipid results in loss of infectivity through loss of attachment proteins.

Classification of viruses

Viruses are classified into families, subfamilies, genera and individual strains or species on the basis of a number of characteristics including the nature of the viral genome, the symmetry of the capsid, the presence or absence of a lipid envelope, and the size and shape of mature virions.

Viral replication

All viruses must go through a defined series of steps resulting in replication. These are: **attachment**, a process which occurs by a virus **ligand** binding to a cellular **receptor**, and which defines the cellular **tropism** of the virus; **entry**, e.g. by **endocytosis** or fusion; **uncoating** to reveal the viral nucleic acid; **macromolecular synthesis**, i.e. production of multiple copies of the viral genome and of the viral proteins. Viral protein synthesis requires mRNA production, and there are six pathways adopted by different families of viruses to achieve this; **assembly** of the component parts of the virus into new viral particles; **release** of the virus from the cell which may occur either by cell lysis or by budding through cellular membranes.

Related topics

Viral pathogenesis (A4)

Basic properties of viruses

Size. Viruses are much smaller than bacteria. Viruses were first known as **'filterable agents'**, as material containing viruses was still infectious after passing through a filter able to restrict the passage of bacteria. Sizes range from ~ **20 to 250 nm**; as light microscopes can only resolve down to ~ 300 nm, even the largest viruses can only be visualized using **electron microscopy**.

Structure (Fig. 1). Mature virus particles (**virions**), comprise nucleic acid (the **genome** of the virus), and a protein shell (the **capsid** of the virus), and some viruses in addition have an outer lipid coat, the **envelope**. The nucleic acid surrounded by its capsid is referred to as the viral **nucleocapsid**.

Metabolism. Virus particles have no means of energy generation or of biosynthesis, and are therefore metabolically inert. In order to replicate, viruses must utilize the machinery and energy-generating properties of host cells. They are thus **obligate intracellular parasites**.

Replication. Viruses are replicated by the independent synthesis of their component parts (genome and proteins) within a host cell, with a subsequent assembly of these subcomponents into new virus particles.

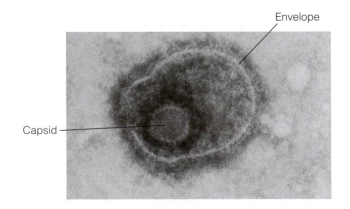

Fig. 1. A herpesvirus.

Viral components

Viral genomes. Viral genomes may be carried in the form of DNA *or* RNA; viruses never contain both DNA *and* RNA.

(i) *DNA viruses.* The DNA genome may be double-stranded (ds) (as in eukaryotic cells), or single-stranded (ss). An important family of viruses (the **Hepadnaviridae**) carry their genome in the form of partially dsDNA, with a complete circular strand and an incomplete complementary strand. The DNA may be linear or circular. The size of DNA viral genomes is quite variable, from as few as ~ 3000 bases (3 kilobases, or 3kb), up to 200 000 bases. As a rough guide, an average protein molecule is encoded by ~ 1000 bases. Thus, the small DNA viruses have enough genetic information to encode three or four proteins, whereas the larger ones are considerably more complex, as they may encode a couple of hundred proteins.

(ii) *RNA viruses.* Numerically, these form the majority of viruses. The RNA may be ss or ds (note that ds RNA is a molecule completely foreign to eukaryotic cells). ssRNA can exist in either positive (+ve) or negative (–ve) **polarity**. RNA of +ve polarity is defined as RNA which can be directly processed by a ribosome, e.g. messenger RNA (mRNA). Thus, –ve ssRNA must first be copied into its complementary, +ve, strand before being translated on the ribosome. RNA genomes may be continuous or segmented. Most RNA viruses are of the order of 5–10 kb, although there are exceptions to this. Thus, they would contain enough genetic information to encode ~ 5–10 proteins.

Viral proteins. These are encoded by the viral genome, and subserve a number of functions:

(i)   *Capsid formation.* This acts as a protective coat to prevent digestion of the viral genome, and is essential for virus survival. Enzymes able to digest nucleic acids (DNases and RNases) are widespread in nature, and if viruses existed as naked nucleic acid, they would not survive long enough extracellularly to be transmitted. Capsids consist of multiple copies of a small number of proteins which spontaneously self-assemble into structures of defined geometric shape, into which the viral genome can be inserted. The geometric **symmetry** of capsids usually takes one of two forms, icosahedral (i.e. essentially spherical) or helical (wound around the helical structure of the nucleic acid).

(ii) *Attachment.* The first step in the viral replication cycle involves attachment of the virus to its target cell surface. This process is mediated by a protein molecule (referred to as a viral **ligand**) on the outer surface of the virus (see below).

(iii) *Enzymes.* Some viruses encode and carry proteins which have enzymatic function. The vast majority of these are involved in macromolecular synthesis (e.g. DNA or RNA polymerase enzymes), or are protein-modifying enzymes such as proteases.

(iv) *Interference with host cell function.* Many viruses subvert host cell function in order to divert the activity of the cell towards production of virus particles. Viruses must also prevent the natural defense mechanisms of the host cell and the host immune system from eliminating the cell by **apoptosis** (cellular suicide) or by T-cell cytotoxicity, as this would result in prevention of virus replication. Thus, there are many regulatory proteins encoded by viruses which can switch off host cell macromolecular synthesis, prevent host cells from undergoing apoptosis, and prevent recognition of infected cells by circulating T cells (e.g. by reducing the expression of HLA molecules on the cell surface).

Viral envelopes. These are composed of lipid, and derived from membranes of the host cell, acquired as the virus buds through when exiting the cell. As the lipid envelope is on the outside of the virus, the viral attachment proteins must be inserted into the envelope. If the envelope is stripped off, the attachment proteins are also removed, rendering the virus non-infective. Stripping lipid envelopes is a relatively easy thing to do (e.g. with 70% alcohol, or detergents). Enveloped viruses are therefore generally more fragile and more easily inactivated than non-enveloped ones.

Classification of viruses

More than 4000 different viruses have been described. These are grouped together into 69 large families (designated by the suffix **-viridae**), in turn subdivided into subfamilies (designated by the suffix **-virinae**) and 243 genera (designated by the suffix **-virus**) on the basis of a number of characteristics, as defined by The International Committee on Taxonomy of Viruses. Genera are comprised of strains or species.

The characteristics used to assign a given virus to a given family include the nature of the viral genome, the geometric symmetry of the capsid, the presence or absence of a lipid envelope, and, if known, the size and shape of virus particles as seen by electron microscopy. A list of virus families containing important human pathogens is provided in *Table 1*.

Viral replication   The replication process of viruses involves the independent synthesis of their individual components, followed by assembly of these into new viral particles. The precise details of the replication cycle will vary between viruses, but all viruses must complete a series of distinct steps (*Fig. 2*), which can be broadly summarized as follows:

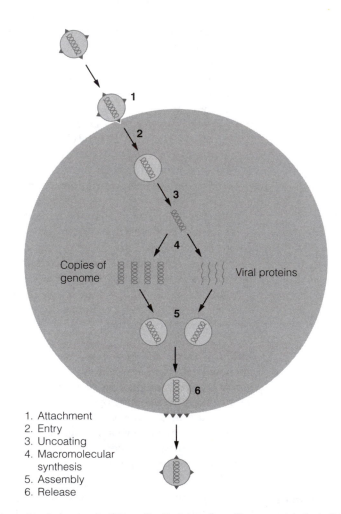

1. Attachment
2. Entry
3. Uncoating
4. Macromolecular synthesis
5. Assembly
6. Release

Fig. 2.   Schematic of virus replication cycle. Redrawn from Greenwood (ed), Antimicrobial chemotherapy, 1995, with permission from Oxford University Press.

Table 1.   Classification of important human virus pathogens

Family	Viruses	Nucleic acid	Lipid envelope
DNA viruses			
Herpesviridae	Herpes simplex 1 and 2 (HSV) Varicella–zoster (VZV) Cytomegalovirus (CMV) Epstein–Barr virus (EBV) Human herpesviruses 6 and 7 (HHV6, HHV7) Human herpesvirus 8 = Kaposi's sarcoma-associated herpesvirus (HHV8)	ds	yes
Poxviridae	Variola (smallpox), Vaccinia, Molluscum contagiosum, Orf	ds	yes
Adenoviridae	Adenoviruses (> 50 serotypes)	ds	no
Papovaviridae	Human papillomaviruses (HPV, > 80 genome types) Polyomaviruses (BK and JC)	ds	no
Hepadnaviridae	Hepatitis B virus (HBV)	partial ds	no
Parvoviridae	Parvovirus B19	ss	no
RNA viruses			
Orthomyxoviridae	Influenza viruses A,B,C	-ss, segmented	yes
Paramyxoviridae	Measles, mumps, parainfluenza viruses, respiratory syncytial virus (RSV)	-ss	yes
Rhabdoviridae	Rabies virus	-ss	yes
Picornaviridae	Enteroviruses (including polio, Coxsackie A and B, Echo, Hepatitis A virus (HAV) Rhinoviruses (> 100 serotypes)	+ss	no
Togaviridae	Rubella	+ss	yes
Flaviviridae	Yellow fever virus, dengue viruses Hepatitis C and G viruses (HCV, HGV)	+ss	yes
Reoviridae	Rotavirus	ds	no
Retroviridae	Human immunodeficiency viruses (HIV), human T-cell lymphotropic viruses (HTLV)	+ss	yes

ds = double stranded   ss = single stranded   + or – refers to polarity of ss RNA

1. *Attachment.* Attachment of a virus particle to the surface of its target cell involves a specific interaction between proteins (**ligands**) on the outer surface of the virus, with **receptors** on the outer surface of the cell. The nature of the viral ligands and cellular receptors are known in great detail for some viruses (e.g. the gp120 of human immunodeficiency virus (HIV) and the CD4 molecule on T lymphocytes), but not at all for others. Cells have not evolved to carry these proteins in order that they may be infectable by a virus – quite the converse is true. Viruses have evolved to bind to molecules already present on cell surfaces.

   Thus a given virus can only infect cells which carry the appropriate receptor for that virus. This interaction determines the cell **tropism** of the virus. Viral tropism explains to a large extent the kind of disease the virus will cause. For example, HIV is tropic for cells of the host immune system, and therefore causes immunodeficiency, whereas hepatitis A virus is tropic for the liver and causes hepatitis.

2. *Entry.* Viral entry into the host cell is a separate and distinct process which follows attachment. It may occur through a number of different mechanisms. Viral particles may invaginate into the host cell membrane, eventually pinching off a membrane-bound vesicle (the **endosome**) into the cytoplasm of the cell. This process is known as **endocytosis**, and is another example of viruses evolving to take advantage of normal cellular functions – for example, cells may use this process for acquisition of certain nutrients. Some viruses enter cells by **fusion** of their own outer lipid membrane with the plasma membrane of the cell, resulting in release of viral nucleocapsid into the cell cytoplasm. Such viruses possess an F (for fusion) protein on their surface.

3. *Uncoating.* Once the virus has entered a cell, it must release its nucleic acid from its capsid before replication can take place. This process is known as *uncoating.* For viruses that have entered via an endosome, fusion of the endosome with a lysosome results in a sharp decrease in the pH of their environment. This results in spontaneous disaggregation of the viral capsid.

4. *Macromolecular synthesis.* This process encompasses the production of multiple copies of the viral genome, and multiple copies of the viral proteins. Viruses utilize the relevant host cell organelles to perform these functions. In order for viral proteins to be synthesized, all viruses have to generate positive ssRNA (i.e. mRNA), which can be appropriately processed by the host cell ribosomes. The various strategies by which viruses generate mRNA have been categorized in the **Baltimore classification** (after Dr D. Baltimore, co-discoverer of the enzyme reverse transcriptase). These are illustrated schematically in *Fig. 3*, and comprise:

    (i)   Viruses with +ve ssRNA genomes: the genome needs no further processing, as it is already in a form recognizable by ribosomes.

    (ii) Viruses with −ve ssRNA genomes: these must copy their genomic RNA into the complementary +ve strand. The synthesis of RNA from an RNA template is not a normal cellular process, and therefore the mature virus particles must carry the relevant enzyme, an RNA-dependent RNA polymerase, with them into the cell, in order to expedite this biosynthetic pathway. The protein must also be encoded within their own genome.

    (iii) Viruses with dsRNA genomes: these viruses unwind their ds genome, and copy the −ve strand into the complementary +ve sequence. Thus, viral particles from this group of viruses also need to contain the relevant enzyme to enable this process to happen.

    (iv) Viruses with dsDNA genomes: mRNA can be transcribed from the dsDNA genome, a process which occurs naturally in all eukaryotic cells. Some dsDNA viruses therefore rely entirely on the host cell to do this process for them. Others (e.g. the pox viruses) carry their own enzymes for this.

    (v)   Viruses with ssDNA genomes: firstly the ssDNA is converted into dsDNA, and then transcription into mRNA takes place as above for the dsDNA viruses. Again, these are processes which the cell is perfectly able to carry out, therefore these viruses do not need to burden themselves with any of the necessary enzymes.

    (vi) Retroviruses: these viruses have a +ve ssRNA genome, but rather than use this as their mRNA, their genome is copied firstly into ssDNA, and

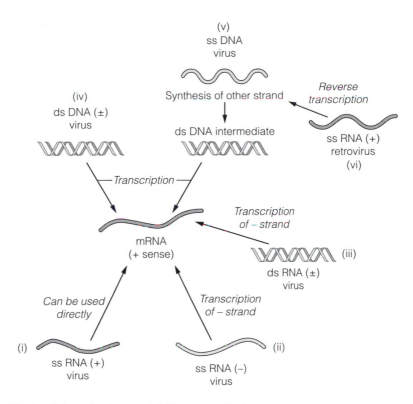

(v)
ss DNA
virus

(iv)
ds DNA (±)
virus

Synthesis of other strand

Reverse transcription

ds DNA intermediate

ss RNA (+)
retrovirus
(vi)

— *Transcription* —

Transcription of – strand

mRNA
(+ sense)

ds RNA (±)
virus (iii)

Can be used directly

Transcription of – strand

(i) ss RNA (+)
virus

ss RNA (–)
virus (ii)

Fig. 3.   Schematic diagram of Baltimore classification.

then into dsDNA. Transcription into mRNA by the host cell then follows. The conversion of an RNA template into DNA was originally viewed as a backwards step (hence the name 'retroviruses'), as the central dogma of molecular biology held that genetic information flows unidirectionally from DNA into RNA into protein. Host cells are not able to perform this step, so the necessary enzyme, called reverse transcriptase, must be encoded, and carried into the cell, by the virus.

5. *Assembly.* Once the individual components of the virus have been replicated within a cell, they must come together to form new viral particles, a process referred to as assembly. Before final assembly into new viral particles, viral proteins may be modified by host cell processes such as glycosylation, or phosphorylation.

6. *Release.* The final stage in the replication cycle involves the release of new virus particles from the infected cell. This may be achieved by loss of the integrity of the cell membranes and subsequent death and lysis of the cell, with an explosive release of several thousand viral particles from the dying cell. Alternatively, virus particles may bud through the cell membranes in a process similar to endocytosis in reverse. Budding viruses will pinch off part of the cell membrane through which they are budding, thereby becoming enveloped. Note that for enveloped viruses, the viral attachment proteins must be present in the envelope itself, as this is the outermost part of the virus. Thus, prior to budding, these proteins will have been inserted into the relevant cell membrane.

A4 VIRAL PATHOGENESIS

Key notes

Acute cytolytic infection	Many virus infections result in death of the infected cell. This may arise for a number of reasons, e.g. virus-induced inhibition of host cell protein synthesis, or disturbance of cell architecture. Cell death results in the release of new virus particles, which can infect neighboring cells. The host immune system may also kill virally infected cells. The end result for the patient will vary from no disease through to organ failure and death, depending on which organs and tissues are infected, the extent of virally-induced cell death and the host inflammatory response to infection.
Chronic infection	In a chronic infection, virus replicates within the cell without killing it. Virus production continues for the normal lifespan of the cell. The patient thereby becomes a chronic carrier of the virus. Disease may arise through recognition of infected cells by the host immune system, resulting in chronic inflammation, or through virus-induced interference with the specialized or luxury function of the cell.
Latent infection	In a latent infection, virus is present within a cell, but there is no replication of the viral genome, and very little or no production of viral proteins. Latent infection does not damage the cell. However, virus may be reactivated from latency, start to replicate, and cause disease. Once infected with a latent virus, the host then carries that virus for the rest of his/her life. All herpesviruses exhibit latency.
Transformation	Virus infection of a cell may result in the uncontrolled proliferation of the cell, a process known as transformation. There are many examples of transforming viruses, and infection with such viruses is a major cause of malignant disease in humans.
Related topics	Viruses and virus replication (A3)

The presence of a virus within a cell may have a number of profound effects on the cell.

Acute cytolytic infection

The most obvious potential outcome of virus infection of a cell is death of the infected cell. This is referred to as a **cytolytic** or **cytocidal** infection. Many viruses shut off host cell macromolecular synthesis in order to maximize production of virus components. Cells need to replenish damaged proteins in order to maintain their integrity, and cessation of this process will severely damage the cell. The accumulation of thousands of viral particles within a cell may have a considerable effect on the architecture of the cell. The ultimate result of these and other processes is death of the cell. Lysis of the cell membranes results in the explosive release of huge numbers of viral particles, which can then infect neighboring cells. This cycle may be complete within several hours of initial infection.

Evidence of cellular damage due to virus infection is not hard to find. Viruses can be grown in cell culture *in vitro*. Light microscopy of these cell cultures reveals morphological changes within the cells as the virus replicates. These changes are referred to as a **cytopathic effect** (CPE). Thus, although viruses themselves cannot be visualized using a light microscope, their effect on cells can be, and diagnostic laboratories take advantage of this when using virus isolation in cell culture to prove that a patient is infected with a particular virus. Different viruses produce different CPE. With experience, it is possible to make a highly educated guess as to which virus is present within a particular culture showing a CPE.

Virus-induced CPE also occurs *in vivo*. Histological sections of virus-infected tissues may reveal **inclusion bodies** within cells (consisting of rafts of viral particles), or the presence of multinucleate giant cells.

The dependence of the virus on the host cell for replication creates an interesting dilemma. Although the virus is ultimately released from the cell once the cell is killed, it is not in the interests of the virus to kill the cell too quickly, before adequate virus replication has taken place. Within a normal host, there is the added complication of the host immune response, one of whose primary roles is to destroy virally infected cells as soon as possible in order to minimize production of new viral particles. The end result for the patient will depend on the amount of cell death induced by the infection, and also on the degree of inflammatory response to the infection. This will depend on a number of factors including the dose of infecting virus, the speed of virus replication and virus-induced cell death, and the speed and quality of the host immune response. In many virus infections, the amount of cell death within a target organ is insufficient to impair the function of the organ or to cause the patient any symptoms. For most infections in immunocompetent hosts, the immune system succeeds in eliminating replicating virus, and the patient survives. Sometimes the amount of virus-induced or immune-mediated cell death is so great that the function of the organ is impaired and the patient suffers an acute symptomatic infection, e.g. in acute hepatitis. At its extreme, this can result in acute organ failure, e.g. fulminant hepatic failure may arise from an overwhelming infection with hepatitis B virus.

Chronic infection

Another form of virus-cell interaction is **chronic** or **persistent** infection, in which there is steady production and release of virus from the infected cell, but the cell is able to survive. Virus production continues for the natural lifespan of the host cell. A patient with a chronic virus infection may have no symptoms at all from their infection – such an individual is known as an asymptomatic (or healthy) carrier of the virus. However, disease may arise from chronic virus infection in one of two ways.

(i)   As viruses replicate within cells, viral antigens are expressed on the cell surface. The **host immune system**, in particular the T lymphocytes, is able to recognize these antigens as foreign, and mount an **inflammatory response** against the infected cells. If this response, for whatever reason, is unsuccessful in eliminating the virally infected cells, then the patient will suffer from a chronic inflammatory state within the organ concerned. Thus, in chronic hepatitis B virus infection of the liver, much of the liver damage is actually caused by the host immune response to the infected hepatocytes, rather than any virus-induced cell death.

(ii) All cells must perform certain functions simply in order to remain alive. Collectively, these are termed **housekeeping** functions. In addition, many cells have specialized functions, also known as **luxury** functions. In a chronic infection, by definition the virus must not be interfering with the housekeeping functions of the cell (otherwise the cell would die). However, the presence of the virus may affect the luxury functions of the cell, and this may give rise to clinical disease. When newborn mice are infected with lymphochoriomeningitis virus (LCMV), and then followed over the next several months, the mice fail to grow. LCMV infects the pituitary gland in these animals but does not cause cell death, or an inflammatory response. However, the virus is present within the growth-hormone-producing cells, and interferes with the production and secretion of the hormone – hence the failure of the mice to develop. As yet, there is no human disease whose pathogenesis fits this model.

Latent infection

In a **latent** infection, the viral genome is present within the cell, but there is no replication of the genome, and few, if any, viral proteins are synthesized. The presence of latent virus within a cell causes the cell no harm. However, a latent virus may, in response to certain stimuli, become **reactivated**. The virus then does replicate, and this may have consequences for the cell and for the whole organism.

All herpesviruses undergo latency. Following a **primary** infection, the virus will remain within that individual for the rest of their life. Mostly, the virus is in a latent state, but reactivation and consequent disease may arise. Such reactivated, or **secondary** infections are usually relatively trivial and highly localized,

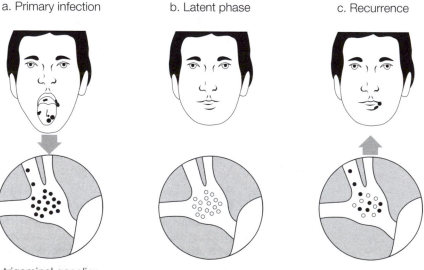

a. Primary infection   b. Latent phase   c. Recurrence

trigeminal ganglion

Fig. 1.   Schematic diagram of HSV latency and reactivation.

as, in contrast to the very first or primary infection, which can be quite severe, the virus is reactivating in a host whose immune system has seen it before.

Thus, herpes simplex virus (HSV) may infect the oral cavity to cause orolabial herpes, or gingivostomatitis (*Fig. 1*), which may or may not be symptomatic. Virus enters nerve terminals (from the trigeminal nerve), and travels up the axons to reach the cell body (or ganglion), which is the site of latency for HSV. Subsequent reactivation of virus from the trigeminal ganglion results in virus travelling back down to the periphery, where it causes one or a small cluster of vesicles known as a cold sore.

Transformation **Transformation** is the process whereby virus infection of a cell leads to uncontrolled cell division, resulting in an immortal cell line. For example, Epstein–Barr virus infection of human B lymphocytes *in vitro* results in stimulation of cell division, and the establishment of a continuous **lymphoblastoid cell line**. The fact that viruses can transform cells raised the obvious question as to whether viruses cause cancer. The answer is an unequivocal 'yes'. A number of virus infections have been shown to be intimately involved in the pathogenesis of a variety of malignant diseases (*Table 1*). These will be discussed in more detail in the chapters dealing with the relevant viruses.

Table 1. Viruses as a cause of human cancer

Virus	Malignant disease
Human T-cell lymphotropic virus (HTLV)-1	Adult T-cell leukemia/lymphoma
Epstein–Barr virus	B-cell lymphomas; Burkitt's lymphoma; nasopharyngeal carcinoma; Hodgkin's lymphoma (some forms)
Human herpesvirus 8	Kaposi's sarcoma; body cavity lymphoma
Hepatitis B virus	Hepatocellular carcinoma
Hepatitis C virus	Hepatocellular carcinoma
Human papilloma virus (types 16, 18)	Cancer of the uterine cervix; cancer of the penis
Simian virus 40 (SV40)	Non-Hodgkin's lymphoma

A5 BACTERIAL ANATOMY

Key Notes

Structure	Bacteria vary in size, shape and structure.
Cell contents	Bacterial cells consist of a fluid cytoplasm surrounded by a plasma membrane and multiple other layers, depending on the species. The cytoplasm contains DNA, ribosomes, RNA, proteins, carbohydrates and all the elements required for survival, growth and pathogenesis. The bacterial genome (entire DNA content) consists mainly of the chromosome and partly of other genetic material including plasmids.
Bacterial cell layers and their constituents	The **plasma membrane** is composed primarily of protein and phospholipid. It performs many functions, including transport, biosynthesis, and energy transduction. The **outer membrane** is another protein-rich layer present in Gram-negative bacteria. The space between the plasma and outer membranes of Gram-negative bacteria is called the **periplasmic space**. **Teichoic acids** are negatively charged polymers found in Gram-positive bacteria. **Lipoteichoic acids** are antigenic polymers anchored in the cytoplasmic membrane. They are cytotoxic and adhesins. **Cell wall peptidoglycan** is a protective layer that also determines the shape of bacterial cells. In Gram-positive bacteria this layer is thick and outermost, in Gram-negative bacteria it is thin and situated between the inner and outer membranes (in the periplasmic space). **Lipopolysaccharides (LPS)** are complex molecules consisting of a lipid A anchor, a polysaccharide core, and a chain of carbohydrate molecules. It is present on the outer leaflet of the outer membrane of Gram-negative bacteria. LPS is responsible for symptoms of Gram-negative sepsis, hence described as endotoxin. **Capsules**. Some bacteria form a thick outer capsule, usually made of a polysaccharide gel. Capsules confer resistance to phagocytosis. **Surface Appendages**. Flagella are long and narrow cylindrical appendages that are used for locomotion. Pili (fimbriae) are hair-like, appendages on the surface of mainly Gram-negative bacteria. Pili serve many functions including adherence and DNA uptake.
Related topics	Bacterial survival mechanisms (A6) Bacterial pathogenicity (A7)

Structure

Bacteria are unicellular organisms that reproduce by binary fission. With the exception of the obligate intracellular pathogens, *Chlamydia* and *Rickettsia*, all other bacteria are capable of independent metabolic existence and growth.

Size

Bacterial cells are extremely small but vary in length and width. *Bacillus* species are among the largest known, up to 10 μm long and > 1 μm wide. In contrast, parvobacteria may be < 0.5 μm long and 0.2 μm wide.

Shape

Bacteria may be round (cocci), e.g. *Staphylococcus* or *Streptococcus* species, or cylindrical (rod-shaped, bacilli), e.g. Enterobacteriaceae and *Clostridium* species (*Fig. 3(c), Topic A1*). Some of the rod-shaped bacteria are filamentous with branching cells, e.g. *Actinomyces* species. Vibrios are comma-shaped and spirochaetes (e.g. *Treponema pallidum*) are spiral.

Bacteria grow two-dimensionally along one axis, producing chains (e.g. streptococci) or in all dimensions, producing clusters (e.g. staphylococci).

Cell contents

Bacterial cells consist of a fluid **cytoplasm** surrounded by multiple protective layers (*Fig. 2*). The chromosome (DNA), ribosomes, RNA, proteins, carbohydrates and all the essential elements required for homeostasis, growth and pathogenesis are found in the cytoplasm.

Unlike eukaryotic cells, prokaryotic bacteria lack nuclei, nuclear membranes and various complicated cellular compartments such as endoplasmic reticulum. The DNA is free in the cytoplasm and often becomes closely associated with the cell membrane.

All the properties of a bacterial cell, including its structural and functional characteristics and virulence capabilities, are determined by the **genome**, the genetic information (DNA) contained within the cell. Bacterial genome consists mainly of a single, circular, dsDNA **chromosome** (1–5 million nucleotides long), and partly of additional genetic material such as **plasmids** (tens of thousands of nucleotides) and **bacteriophages** which may be present in some bacteria.

The genome of the cell acts as a template for transcription by RNA polymerase, leading to subsequent protein production by the cell. A segment of DNA that specifies the production of a particular protein is called a gene. The chromosome contains between several hundred to a few thousand genes. Plasmids are small, circular, extrachromosomal genetic elements that may encode a variety of supplementary genetic information, including antibiotic resistance, toxin production, enhanced pathogenicity, and the information necessary to allow transfer to other cells. Bacteriophages are viruses that infect bacteria. Occasionally they will survive in a dormant state (lysogeny) within the bacterial cell and may bestow additional properties on the cell.

Genes express proteins which may be modified before they are released into the cytoplasm, integrated into the cell membranes or secreted. Proteins required for cell maintenance, referred to as house-keeping enzymes, remain in the cytoplasm. Proteins functioning as respiratory elements, receptors, sensor elements or adhesins integrate into the plasma membranes or outer membranes. Toxins and many other virulence factors are often secreted outside the bacterial cell.

Bacterial cell layers and their constituents

All bacteria contain their cytoplasm within a complex protein-rich layer, the plasma membrane. The structures outside this membrane differentiate bacteria into two main groups, those that retain the Gram stain (Gram positive) and those that do not (Gram negative); see *Box 1* and *Fig. 3*.

The plasma membrane of Gram-positive bacteria is covered with a thick layer of peptidoglycan which may be surrounded by layers of usually

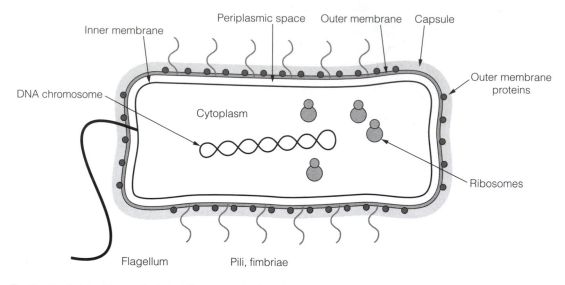

Fig. 2.   Bacterial cell layers (typical of Gram-negative bacilli).

polysaccharide-based capsule and/or loose slime. In Gram-negative bacteria, the plasma membrane is covered with a thin layer of peptidoglycan followed by a complex and protein-rich 'outer membrane' which may be covered by a poly-saccharide capsule. The space between the inner and outer membranes of Gram-negative bacteria is called the periplasmic space. The outer membrane structure is anchored non-covalently to lipoprotein molecules, which, in turn, are covalently linked to the peptidoglycan. In some bacteria, the outer membrane forms 'Bayer's patches' which break the continuity of the peptidoglycan and bind directly to the inner membrane. The lipopolysaccharides (LPS) of the Gram-negative cell envelope form part of the outer leaflet of the outer membrane structure.

Plasma membrane
The plasma membrane, which is also known as the inner (in Gram negatives) or cytoplasmic (in Gram positives) membrane, shares some properties with the eukaryotic mitochondrial plasma membrane. The bacterial plasma membrane is composed primarily of proteins and phospholipids at ratios of ~ 3:1, and performs multiple functions. It contains the tools of active transport, respiratory chain components, energy-transducing systems and the membrane-associated stages of the biosynthesis of phospholipids, peptidoglycan, LPS, and capsular polysaccharides. The membrane also contains components of protein export (secretion) systems that enable outer membrane proteins or secreted

Box 1. Gram staining. Bacteria take up Crystal Violet and iodine which form a complex inside the cell. The thin and often discontinuous peptidoglycan layer of Gram-negative bacteria does not impede solvent extraction of this complex. However, the reduced porosity of the thick and continuous peptidoglycan cell wall traps the complex inside Gram-positive cells.

Fig. 3(a). Gram staining.

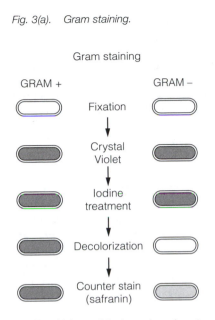

Fig. 3(b). Cell wall location: The thick peptidoglycan layer is outermost in Gram-positive bacteria. This layer is much thinner in Gram negatives and is trapped between the outer and inner membranes.

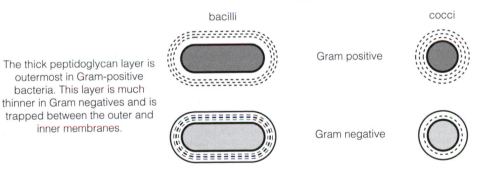

The thick peptidoglycan layer is outermost in Gram-positive bacteria. This layer is much thinner in Gram negatives and is trapped between the outer and inner membranes.

Fig. 3. Gram staining of bacteria.

proteins to exit the cell. The plasma membrane is also the anchoring site for DNA.

Mesosomes are plasma membrane invaginations that are seen in thin sections of mainly Gram-positive bacteria and their function remains unknown.

Peptidoglycan
Except for mycoplasmas, all pathogenic bacteria possess cell wall peptidogly-cans (sometimes called **mucopeptide or murein**), which are synthesized by specific enzymes. Inhibition of these enzymes prevents peptidoglycan biosyn-thesis and results in cell death. Peptidoglycans form the **skeleton of bacteria** and without them cylindrical organisms become spherical protoplasts (sphero-plasts).

Teichoic acids and lipoteichoic acids

Teichoic acids are phosphate polymers bearing a strong negative charge. They are covalently linked to the peptidoglycan in some Gram-positive bacteria. They are strongly antigenic, but are generally absent in Gram-negative bacteria.

Lipoteichoic acids are plasma membrane-bound teichoic acids which are polymers of glycophosphates and lipophilic glycolipid. They are **antigenic, cytotoxic** and many act as **adhesins** (e.g. in *Streptococcus pyogenes*).

Outer membrane and outer membrane proteins of Gram-negative bacteria

The outer membrane of Gram-negative bacteria is a bilayered molecular sieve (permeability barrier), with the lipopolysaccharide and phospholipids on the outer leaflet, whereas the inner leaflet consists of phospholipids and the lipid portion of lipoproteins. The bilayer is interrupted by numerous proteins, known as **outer membrane proteins** (OMP). These are either integrated in monomers or multimers to form surface-exposed **sensors, receptors and adhesins**, or to form barrel-shaped '**porins**' that allow nutrition into the cell. Without these porins, hydrophilic molecules are not able to cross the hydrophobic barrier of the outer membrane. Most OMP link the periplasmic space with the outside environment, but a few may traverse both inner and outer membranes.

Lipopolysaccharides

A characteristic feature of Gram-negative bacteria is their possession of lipopolysaccharide (LPS), also called **endotoxin**. This is integrated within the outer layer of the outer membrane and is shed by live organisms as part of **outer membrane vesicles (blebs)**. LPS is **biologically highly active** and plays an important role in the pathogenesis of many Gram-negative bacterial infections. It strongly induces the release of a vast array of proinflammatory cytokines and the activation of complement and coagulation pathways, leading to **endotoxic shock**. The LPS endotoxic properties reside largely in the **lipid A** components embedded within the outer membrane and covalently attached to the core part (*Fig. 4*). Covalently linked to lipid A is an eight-carbon sugar, **ketodeoxyoctonate** (KDO), in turn linked to the chains of sugar molecules (saccharides) which form the **antigenically variable O antigen** structures of Gram-negative bacteria. The latter carbohydrate component can be long (LPS) or short (lipo-oligosaccharide, LOS), forming hydrophilic smooth and hydrophobic rough bacterial colonies, respectively.

Capsule and slime

Some bacteria form capsules, which constitute the outermost layer of the bacterial cell and may be too thin to be detected or up to 10 μm thick. Bacterial capsules enable bacteria to survive inside the host and avoid phagocytosis. Some organisms lack a well-defined capsule but have loosely bound layers of polysaccharide-based slime which covers the organism. Overproduction of slime may help to form a protective **biofilm**.

Surface Appendages

Flagella and pili (also called **fimbriae**) are the only two surface appendages known. The former is the major organ of locomotion, expressed by both Gram-positive and -negative bacilli, but rarely cocci.

Flagella are ≤ 12 μm long (i.e. longer than the bacterium) and ≤30 nm in diameter. They are made of proteins assembled into a long hollow cylinder anchored to the cell membranes via a hook and a basal body. The flagellum rotates fast in

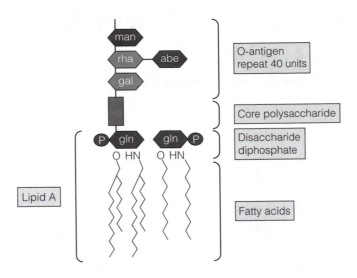

Fig. 4.   Structure of lipopolysaccharide.

helical twists which create movement towards nutrition (**chemotaxis**) or air (**airotaxis**) or away from noxious stimuli. The amount and distribution of flagella on the bacterial surface are characteristic for any given species and hence are useful in identifying and classifying bacteria.

The terms pili and fimbriae are used interchangeably. These are hair-like appendages of varying length, but shorter than flagella, and are distributed in large numbers all over the cell surface. They are expressed almost exclusively by Gram-negative bacteria (bacilli or cocci). Pili are key **adhesion molecules** that mediate colonization, play important roles in the **acquisition of external DNA** and are required for **twitching motility** (different from flagella-mediated loco-motion). Many Gram-negative bacteria have both flagella and pili.

A6 BACTERIAL SURVIVAL MECHANISMS

Key Notes

Microorganisms express numerous survival and virulence factors to overcome hostile *in vivo* conditions.

Overcoming physical barriers of the host

Some bacteria secrete potent toxins that kill host cells and connective tissues of the skin and mucous membranes.

Overcoming physiological barriers of the host

Bacteria express adhesins to enable binding to host cells. Some also express flagella. Both adhesions and flagella help to resist the normal physiological processes designed to eliminate foreign material from the host.

Avoiding phagocytosis

Many bacteria produce a hydrophilic extracellular capsule which inhibits phagocytosis, and which also prevents complement deposition. The *Streptococcus pyogenes* protein M performs similar functions.

Avoiding recognition by antigenic mimicry

Mimicry of self-antigens is an important mechanism by which the bacterium attempts to avoid recognition by the immune system.

Avoiding detection by hiding inside cells

Many organisms avoid exposure to immune attacks by invading host cells and living intracellularly.

Antigenic variation

By changing surface antigen composition, bacteria may avoid adaptive immune responses. Almost all bacteria are capable of changing their antigenic structures using several highly efficient mechanisms, including mutation, phase variation and horizontal transfer of DNA. Gene transfer can occur via transformation, conjugation and transduction. Bacteria can also move genes within their own genomes.

Immunoglobulin A proteases

Bacteria produce proteases that cleave host molecules, including antibodies. Several species of pathogenic bacteria produce a protease that specifically cleaves immunoglobulin A (IgA).

Serum resistance

Some bacteria resist complement-mediated lysis and thereby survive in the bloodstream. Lipopolysaccharide (LPS) may hinder deposition of complement components on the cell surface, and bacterial capsules may prevent efficient complement activation.

Biofilms

Some bacteria produce profuse amounts of extracellular slime which forms a protective matrix around the bacterial populations.

Related topics

Host defense mechanisms to infection (A2) Bacterial anatomy (A5) Bacterial pathogenicity (A7)

The human immune system consists of a complex and highly efficient array of structural and functional barriers that are designed to prevent colonization and invasion by pathogenic organisms. There is also a range of cellular and non-cellular (humoral) mechanisms designed to remove pathogens after they become established inside the body. Conditions inside the human body are therefore very harsh for bacterial survival, growth and virulence. Pathogenic bacteria have evolved ways of avoiding or neutralizing these highly efficient clearance systems. Since most of the interactions between the bacterium and the host immune effector mechanisms involve the bacterial surface, bacterial resistance to host immune responses is related to the molecular architecture of the bacterial surface layers or secreted virulence factors which attack host molecules.

Overcoming physical barriers of the host

Some bacteria secrete potent enzymes (**toxins**) that kill host cells and destroy connective tissues of the skin and mucous membranes (e.g. toxins produced by *Bacillus anthracis*, the anthrax organism). Others can penetrate intact skin by combining secretion of enzymes and powerful spiral movement mediated by flagella, e.g. *Treponema pallidum*.

Overcoming physiological barriers of the host

Bacteria express **adhesins**, such as pili and outer membrane proteins (Topics A5, A7) to bind to host cells via specific or nonspecific interactions with receptors. These will enable the organisms to resist the body's physiological forces that help to expel microbes, such as coughing, sneezing, peristalsis of the intestine and the flow of urine and tears. Motile bacteria use their **flagella** to enable them to swim against the tide of urine flow or peristalsis (e.g. *Neisseria gonorrhoea*).

Avoiding phagocytosis

Production of an **extracellular capsule** is the most common mechanism by which bacteria avoid phagocytosis. Many bacterial pathogens which are associated with meningitis and pneumonia express capsules. These include *Neisseria meningitidis*, *Haemophilus influenzae* and *Streptococcus pneumoniae*. Noncapsulate variants of these bacterial species are usually much less virulent. Most capsules are polysaccharides composed of sugar monomers that vary between different bacteria. Capsules are normally hydrophilic, thereby hindering uptake by phagocytes. They may also prevent or impair complement deposition on the bacterial surface, a process that normally facilitates phagocytosis and/or complement-mediated killing.

The **M protein** present on the surface of *Streptococcus pyogenes* is not a capsule but functions in a similar manner to prevent complement deposition at the bacterial surface. The M protein binds both fibrinogen and fibrin, and deposition of these on the streptococcal surface partly hinders the access of complement activated by the alternative pathway.

Avoiding recognition (antigen mimicry)

This is an important mechanism by which bacteria attempt to avoid recognition by the immune system. A classical example is the **sialic-acid-containing capsules** of serogroup B *N. meningitidis* and serotype K1 *E. coli*, which mimic brain sialic acid moieties. These are seen by the immune system as self-antigens, and therefore fail to generate antibody responses.

Avoiding detection by hiding inside cells

Many organisms avoid exposure to humoral and cellular attacks by invading host cells (including immune cells such as phagocytes) and **living intracellularly**. However, survival within phagocytes, such as macrophages, is usually extremely difficult due to the presence of very low pH, free radical and enzymatic conditions. Bacteria are normally enclosed in a 'phagosome' which then fuses with **lysosomal granules** present in the cell cytoplasm. These granules contain enzymes and cationic peptides involved in oxygen-dependent and oxygen-independent bacterial killing mechanisms. Some bacteria, e.g. *Mycobacterium tuberculosis,* may **prevent the phagosome–lysosome fusion**. Others are able to resist the killing action of lysosomal components. Many bacteria, e.g. *Listeria monocytogenes,* lyse the phagosome membrane and **escape into the cytoplasm** where they multiply in the presence of plenty of nutrition and no immune attack. (Related Topic: A2.)

Antigenic variation

Changing surface antigen composition during the course of infection provides a powerful mechanism by which bacteria avoid adaptive immune responses directed at vital antigens. Almost all bacteria are capable of changing their antigenic structures using several highly efficient mechanisms, including **mutation** of individual amino acids, switching genes on and off (**phase variation**) and **horizontal exchange of DNA** material (gene transfer). These mechanisms can generate an unlimited number of antigenic determinants that can change very rapidly. The most variable antigens are usually those that are vital for the survival or virulence of the organism and these are largely surface-exposed molecules. Antigenic variation also creates problems for effective vaccine development, e.g. against *Helicobacter pylori* or *Neisseria meningitidis*.

Mutations occur at fairly constant low rates, normally between 10^{-4} and 10^{-10} per cell division; however, since a large bacterial colony contains $\geq 10^9$ cells, even a 'pure' culture will contain thousands of mutations affecting many of the genes of the cell. Such mutations may enhance the ability of an organism to grow in the body by (e.g.) conferring antibiotic resistance, enhancing virulence, or altering surface antigens. In a clinical situation, cells with favorable mutations will rapidly outgrow cells without the mutation, and will therefore be **selected** to become the predominant type.

Gene transfer in bacteria occurs by one of three mechanisms: (i) **Transformation,** i.e. the acquisition of free DNA from the growth medium followed by integration into the genome. Only short random pieces of DNA can be transferred in this way. (ii) **Conjugation** involves the movement of large fragments of DNA (a plasmid or even a whole genome) from one bacterium to another. Pili play important roles in mediating both transformation and conjugation. (iii) **Transduction** is transportation of DNA from one bacterium to another via a bacteriophage which acts as vehicle. A small fragment of bacterial DNA is packaged inside the bacteriophage and later becomes integrated within the genome of the recipient organism.

Finally, bacteria can rearrange their genetic structure by **moving genes around within their chromosome**. Also, there are large number of recognizable mobile genetic elements, **transposons,** which have specific nucleotide sequences and are designed to help DNA fragments (e.g. virulence or antibiotic resistance genes) move within and between genomes of bacteria.

Immunoglobulin A proteases

Several species of pathogenic bacteria that cause disease on mucosal surfaces produce a protease that specifically cleaves immunoglobulin A (IgA), the principal antibody type produced at these sites. These proteases are specific for human IgA isotype I. Nearly all the pathogens causing meningitis possess an IgA protease in addition to a polysaccharide capsule (see above), enabling them to persist on the mucosal surface and resist phagocytosis during the invasive phase of the disease.

Serum resistance

To survive in the bloodstream, bacteria must be able to resist lysis as a result of deposition of complement on the bacterial surface. In the Enterobacteriaceae, resistance is primarily due to the composition of the LPS present in the bacterial outer membrane. **Smooth** colonial variants which possess polysaccharide 'O' side-chains in their LPS are more resistant than **rough** colonial variants that lack such side-chains. The side-chains sterically hinder deposition of complement components on the bacterial surface. In *Neisseria meningitidis* group B and *Escherichia coli* K1, sialic acid capsules prevent efficient complement activation and, in *N. gonorrhoeae*, complement binds but forms an aberrant configuration in the bacterial outer membrane so that it is unable to effect lysis. (Related topic: A5.)

Biofilms

Some bacterial species are capable of producing profuse amounts of slime, made up of **extracellular polysaccharide** molecules (also known as extracellular polymeric substances), which form a tough and highly protective matrix around the bacterial populations. Bacteria usually form biofilms around foreign bodies such as implanted medical devices. Penetration of biofilms by host immune cells, antibodies or antibiotics is extremely difficult if not impossible. Thus, eradication of infection in the presence of established biofilms is not easily achievable.

A7 BACTERIAL PATHOGENICITY

Key notes

Pathogenicity

Pathogenicity means the capacity to initiate disease. This will require the properties of **transmissibility** between hosts, **survival** in the new host, **infectivity** and **virulence**. Pathogenicity and the immune status of the host together determine the outcome of disease.

Types of bacterial pathogens

Pathogenic bacteria can be 'primary' pathogens, i.e. capable of causing disease in immunocompetent individuals, or they are 'opportunistic', causing disease mainly in the immunocompromised or otherwise damaged host.

Transmission

Many opportunistic pathogens are carried as part of the normal human flora. Primary pathogens are usually transmitted between hosts via various routes, including the respiratory, gastrointestinal, urinary or genital tracts. Some are inoculated into body tissue through insect bites or accidental or surgical trauma.

Colonization

The establishment of a stable population of bacteria in the host is called colonization. This is initiated by bacterial 'adhesins' which help the organism adhere to host cell receptors.

Invasion

Invasion is penetration of host cells and tissues, and is mediated by a complex array of molecules, often described as 'invasins'. These can be in the form of bacterial surface or secreted proteins which target host cell molecules (receptors).

Virulence determinants

Virulence is the ability to cause damage to host cells and tissues. Pathogenic bacteria produce large numbers of virulence-related molecules (virulence factors) such as toxins.

Toxins

Pathogenic bacteria produce toxic protein molecules, named toxins, which damage host cells directly or indirectly. Two major types of toxin are described: **endotoxin** and **exotoxins**. The former (also called LPS, lipopolysaccharide) is a component of the outer membrane of Gram-negative bacteria and also released from the bacterial surface via blebs or following bacterial lysis. Exotoxins are diffusible proteins secreted into the external medium by the pathogen. These facilitate adhesion, invasion or host cell damage. Many bacteria also secrete other extracellular aggressins including enzymes such as urease and metalloprotease.

Related topics　　Bacterial anatomy (A5)　　　　Bacterial survival mechanisms (A6)

Pathogenicity　　Pathogenicity is the capacity to initiate disease. It requires the attributes of **transmissibility** or communicability from one host or reservoir to a fresh host, survival in the new host, **infectivity** or the ability to breach the new host's defenses, and **virulence**, a variable that is multifactorial and denotes the capacity of a pathogen to harm the host. Virulence in the clinical sense is a manifestation of a **complex bacterial–host relationship** in which the capacity of the organism to cause disease is considered in relation to the resistance of the host.

Types of bacterial pathogens

Bacterial pathogens can be classified into two broad groups, primary and opportunistic pathogens.

Primary pathogens are capable of establishing infection and causing disease in previously healthy individuals with intact immunological defenses. However, these bacteria may more readily cause disease in individuals with impaired defenses. **Opportunistic pathogens** rarely cause disease in individuals with intact immunological and anatomical defenses. Only when such defenses are impaired or compromised, as a result of congenital or acquired disease or by the use of immunosuppressive therapy or surgical techniques, are these bacteria able to cause disease. Many opportunistic pathogens, e.g. coagulase-negative staphylococci and *Escherichia coli*, are part of the **normal human flora** and are carried on the skin or mucosal surfaces where they cause no harm and may actually have beneficial effects, by preventing colonization by other potential pathogens. However, introduction of these organisms into anatomical sites in which they are not normally found, or removal of competing bacteria by the use of broad-spectrum antibiotics, may allow their localized multiplication and subsequent development of disease.

The above classification is applicable to the vast majority of pathogens; however, there are exceptions and variations within both categories of bacterial pathogens. Different strains of any individual bacterial species can vary in their genetic make up and virulence capacity. For example, the majority of *Neisseria meningitidis* strains are harmless **commensals** and considered opportunistic bacteria, however, some **hypervirulent clones** of the organism can cause disease in a previously healthy individual. Conversely, people vary in their genetic make-up and susceptibility to invading bacteria. For example, *Mycobacterium tuberculosis* is a primary pathogen but does not cause disease in every host it invades.

Transmission

Potential pathogens may enter the body by various routes, including the respiratory, gastrointestinal, urinary or genital tracts. Alternatively, they may directly enter tissues through insect bites or by accidental or surgical trauma to the skin. Many opportunistic pathogens are carried as part of the normal human flora, and this acts as a ready source of infection in the compromised host (e.g. in cases of AIDS or when the skin barrier is breached). For many primary pathogens, however, transmission to a new host and establishment of infection are more complex processes.

Colonization

The establishment of a stable population of bacteria on the host's skin or mucous membranes is called colonization (*Fig. 1*). For many pathogenic bacteria, the initial interaction with host tissues occurs at a mucosal surface and colonization normally requires adhesion to the mucosal cell surface. This allows the establishment of a focus of infection that may remain localized or may subsequently spread to other tissues. Adhesion is necessary to avoid innate host defense mechanisms such as peristalsis in the gut and the flushing action of mucus, saliva and urine, which remove nonadherent bacteria. For bacteria, **adhesion** is

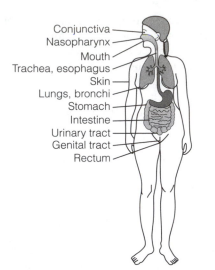

Conjunctiva
Nasopharynx
Mouth
Trachea, esophagus
Skin
Lungs, bronchi
Stomach
Intestine
Urinary tract
Genital tract
Rectum

Fig. 1.   Endogenous source of infection where the organisms are from patients' own flora.

an essential preliminary to colonization and then penetration through tissues. Successful colonization also requires that bacteria are able to acquire essential nutrients – in particular iron – for growth.

At the molecular level, **adhesion involves surface interactions** between specific **receptors** on the mammalian cell membrane (usually carbohydrates) and **ligands** (usually proteins) on the bacterial surface. The presence or absence of specific receptors on mammalian cells contributes significantly to tissue specificity of infection. Nonspecific surface properties of the bacterium, including surface charge and hydrophobicity, also contribute to the initial stages of the adhesion process. Several different mechanisms of bacterial adherence have evolved, all utilizing specialized cell surface organelles or macromolecules, that help to overcome the natural forces of repulsion that exist between the pathogen and its target cell. Many bacteria express pili (or fimbriae) which are involved in mediating attachment to mammalian cell surfaces. Different strains or species of bacteria produce different types of pili which can be identified on the basis of antigenic composition, morphology and receptor specificity

Invasion

Invasion is penetration of host cells and tissues (beyond the skin and mucous surfaces), and is mediated by a complex array of molecules, often described as **'invasins'**. These can be in the form of bacterial surface or secreted proteins which target host cell molecules (receptors).

Once attached to a mucosal surface, some bacteria, e.g. *Corynebacterium diphtheriae* or *Clostridium tetani,* exert their pathogenic effects without penetrating the tissues of the host. These produce biologically active molecules such as toxins, which mediate tissue damage at local or distant sites.

For a number of pathogenic bacteria, however, adherence to the mucosal surface represents only the first stage of the invasion of tissues. Examples of organisms that are able to invade and survive within host cells include mycobacteria, *Salmonella, Shigella* and others. The initial phase of cellular invasion involves penetration of the mammalian cell membrane and many intracellular pathogens use normal phagocytic entry mechanisms to gain access. Inside the cell, they become surrounded by host cell-derived membrane vesicles. Many intracellular pathogens escape from these vesicles into the cell cytoplasm where they multiply rapidly before spreading to adjacent cells and repeating the process of invasion.

The availability of specific receptors on host cells defines the type of host cells that are involved. As a result, some pathogens can invade a wide range of cell types whilst others have a much more restricted invasive potential. The receptors for some of the invasive pathogens have been identified.

Virulence determinants

Both primary and opportunistic pathogens possess **virulence determinants or aggressins** that **facilitate pathogenesis**. Possession of a single virulence determinant is rarely sufficient to allow the initiation of infection and production of pathology. Many bacteria possess several virulence determinants, all of which play some part at various stages of the disease process. In addition, not all strains of a particular bacterial species are equally pathogenic. For example, although six separate serotypes of encapsulated *Haemophilus influenzae* are recognized, serious infection is almost exclusively associated with isolates of serotype b (hence Hib vaccine). Moreover, even within serotype b isolates, 80% of serious infections are caused by six out of > 100 clonal types.

Different strains of a pathogenic species may cause distinct types of infection, each associated with possession of a particular complement of virulence determinants. Different strains of *E. coli*, for example, cause several distinct gastro-intestinal diseases, urinary tract infections, septicemia, meningitis and a range of other minor infections.

Many pathogens produce an impressive armory of virulence determinants; however, their **expression is coordinated or regulated** by several nutritional and environmental factors. Among virulence regulators are the availability of nutrition (e.g. iron), oxygen, suitable temperature or other growth requirements. Importantly, differences in virulence between similar organisms may be due to additional cryptic phenotypic or genotypic variations. For example, some virulence factors are only expressed when in direct contact with host cells.

Virulence genes can move between bacteria via special genetic vehicles e.g. plasmids, bacteriophage and transposons. The horizontally transferred virulence factors (e.g. toxins) may or may not transform the recipient bacteria into better-adapted or more virulent pathogens.

Toxins

In many bacterial infections the characteristic pathology of the disease is caused by toxic protein molecules (or enzymes), named toxins. Toxins may exert their pathogenic effects directly on a target cell or may interact with cells of the immune system resulting in the release of immunological mediators (cytokines) that cause pathophysiological effects. Such effects may not always lead to the death of the target cell but may selectively impair specific functions. Two major types of toxin have been described: **endotoxin and exotoxins.**

Endotoxin, also called **LPS** (lipopolysaccharide) or LOS (lipo-oligosaccharide), is a component of the outer membrane of Gram-negative bacteria, and is released from the bacterial surface via outer membrane vesicles (blebs), following natural lysis of the bacterium or by disintegration of the organism *in vitro*. The term endotoxin was originally introduced to describe the component of Gram-negative bacteria responsible **for the pathophysiology of endotoxic shock**, a syndrome with high mortality, particularly in immunocompromised or otherwise debilitated individuals.

Exotoxins, in contrast to endotoxin, are diffusible proteins secreted into the external medium by the pathogen. Most pathogens secrete various protein molecules that **facilitate adhesion to, or invasion of, the host**. Many others cause **damage to host cells**. The **damage may be physiological**, e.g. cholera toxin promotes electrolyte (and fluid) excretion from enterocytes without killing the cells. Alternatively, the **damage may be pathological**, e.g. diphtheria toxin, inhibits protein synthesis and induces cell death. Some toxins cause damage to distant internal organs following absorption from the focus of infection. Exotoxins vary in their molecular structure, biological function, mechanism of secretion and immunological properties. The list of bacterial exotoxins is now long and increasing (examples are given in *Table 1*).

Many bacteria secrete a range of **enzymes** that may be involved in pathogenic processes. *Proteus* spp. and some other Enterobacteriaceae that cause urinary tract infections produce **ureases** that break down urea in the urine, and the subsequent release of ammonia may contribute to the pathology. The urease produced by the gastric and duodenal pathogen *Helicobacter pylori* is similarly implicated in the virulence of the organism. *Legionella pneumophila* produces a metalloprotease thought to contribute to the characteristic pathology seen in legionella pneumonia. Many other degradative enzymes, including **mucinases**, **phospholipases**, **collagenases** and **hyaluronidases**, are produced by pathogenic bacteria.

Table 1.   Some examples of bacterial exotoxins and their clinical effects

Toxic effect	Examples
Lethal action (e.g. by inhibition of metabolic activity)	
Effect on neuromuscular junction	*Clostridium botulinum* toxin A
Effect on voluntary muscle	Tetanus toxin
Damage to heart, lungs, kidneys.	Diphtheria toxin
Pyrogenic effect	
Increase in body temperature	Exotoxins of *Staphylococcus aureus* and *Streptococcus pyogenes* Staphylococcal toxic shock syndrome toxin 1
Action on gastrointestinal tract	
Secretion of water and electrolytes	Cholera and *Escherichia coli* enterotoxins
Pseudomembranous colitis	*Clostridium difficile* toxins A and B
Bacillary dysentery	Shigella toxin
Vomiting	*Staphylococcus aureus* enterotoxins A–E
Action on skin	
Necrosis	Clostridial toxins; staphylococcal α-toxin
Erythema	Diphtheria toxin; streptococcal erythrogenic toxin
Permeability of skin capillaries	Cholera enterotoxin; *Esch. coli* heat-labile toxin
Nikolsky sign*	*Staphylococcus aureus* epidermolytic toxin
Cytolytic effects	
Lysis of blood cells	*Staphylococcus aureus* α-, β- and δ-lysins, leukocidin Streptolysin O and S *Clostridium perfringens* α and θ toxins

* Separation of epidermis from dermis.

A8 PRIONS

Key Notes

Prions	'Prion' is a term originally derived from 'proteinaceous infectious particles'. Prions are believed to be composed entirely of protein. A normal host cellular gene encodes the prion protein, which exists as a cellular isoform, PrP^C, of Mr 33-35kDa, and as an abnormal isoform, PrP^{Res}. PrP^C is completely digestible by proteinase K but PrP^{Res} is digested only as far as a 27–30kDa product. PrP 27–30 therefore accumulates in tissues containing PrP^{Res}.
Prion diseases	These are also referred to as slow infections, transmissible spongiform encephalopathies (TSE), and, in humans, transmissible dementias. Animal diseases include scrapie (sheep), and bovine spongiform encephalopathy (BSE, cows); human diseases include Kuru and Creutzfeldt–Jakob disease (CJD).
Prion replication	Current theory suggests that spontaneous conversion of PrP^C into PrP^{Res} is catalyzed by the presence of PrP 27–30. Thus, once PrP^{Res} has appeared, there will be an inevitable accumulation of PrP^{Res} and PrP 27–30.
Creutzfeldt–Jakob disease	CJD is the commonest prion disease in humans. It usually presents after the age of 60 years, as a progressive dementia, but cases in younger individuals arise by iatrogenic transmission (e.g. from contaminated neurosurgical instruments, or human pituitary-derived hormones). Familial CJD (15% of all cases) is associated with mutations in the prion gene.
Bovine spongiform encephalopathy and variant CJD	An epidemic of BSE emerged in the UK in the late 1980s. The BSE agent thus entered the food chain, and has resulted in the appearance of a new variant of CJD in humans. The ultimate size of the resultant vCJD epidemic is as yet unknown. Theoretically, person-to-person spread of vCJD may occur through contaminated surgical instruments and transfusion of contaminated blood.
Related Topics	Other CNS infections (F4)

Prions

The term **'prion'** was originally derived from the phrase **'proteinaceous infectious particle'**. It describes a class of agents with unique properties responsible for a number of transmissible diseases of humans and animals. The infectious agent is believed to consist **entirely of protein**, based on the following evidence:

(i) Infectivity of scrapie brain in experimental animals is resistant to a wide range of physical and chemical treatments which destroy nucleic acids, but is, however, affected by treatments which destroy or alter protein molecules.

(ii) Despite intensive efforts, no-one has succeeded in identifying any nucleic acid in the scrapie agent.

(iii) The scrapie agent consists of an **isoform** of a normal cellular protein derived from a normal cellular gene. The normal isoform of the protein is referred to as prion protein^C, or **PrP^C**, and has a molecular weight of **33–35 kDa**. The **scrapie isoform** is referred to as **PrP^{Sc}** or **PrP^{Res}** (for 'resistant'). The fundamental difference between these two isoforms is that the former is completely sensitive to protease digestion, whereas the latter is only broken down by proteases as far as a molecule with M_r 27–30 kDa, referred to as **PrP 27–30**. This molecule therefore accumulates in affected brain, as it cannot be digested any further.

Prion diseases   Diseases caused by prions (*Table 1*) share a number of clinical features. They are all experimentally **transmissible** from an infected animal to an uninfected recipient, the most efficient way of transmitting infection being direct **intracerebral inoculation** of contaminated material. The infections have a prolonged incubation period, often many years, but once disease presents, a progressive and fatal clinical course inevitably follows – hence these diseases are also known as **slow infections**. The pathology in naturally-acquired infection is largely confined to the **central nervous system**. Macroscopically, the brain of affected animals has a **spongy appearance**, and histologically this is shown to be because of extensive vacuolation within neurones. This gives rise to another synonym for these diseases – **transmissible spongiform encephalopathies (TSE)**. In humans, dementia is the pre-eminent clinical feature of disease – hence yet another name, **transmissible dementias**.

Table 1.   Prion diseases

Disease	Host species
Scrapie	Sheep, goats
Bovine spongiform encephalopathy ('mad cow disease')	Cows
Kuru	Humans (cannibals)
Creutzfeldt–Jakob disease	Humans
Variant Creutzfeldt–Jakob disease	Humans

Prion replication   How can a protein molecule, PrP 27–30, not shown to contain any nucleic acid, cause increased production of itself when transferred from an infected animal to a noninfected recipient? The current hypothesis is that **PrP 27–30** causes the **normal PrP^C** present within neurons to undergo a **conformational change** into **PrP^{Res}**. This will eventually be digested by cellular proteases – but only as far as PrP 27–30, which therefore causes more production of PrP^{Res} from PrP^C, and hence accumulation of more PrP 27–30, and so on. This is illustrated diagramatically in (*Fig. 1*).

Thus experimental inoculation of material containing PrP 27–30 will result in transmission of disease, as the abnormal isoform of PrP is generated in the recipient brain, leading to the accumulation of ever more PrP 27–30. The physical difference between PrP^C and PrP^{Res} is not known. In scrapie, the gene encoding PrP is identical to that in nonaffected animals, indicating that the difference does not simply reside in the amino acid sequence of the two isoforms of the protein. It must therefore arise as a post-translational event, the nature of which has not yet been identified.

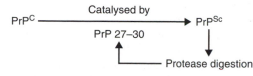

Fig. 1.   Mechanisms of accumulation of PrP 27–30.

Creutzfeldt–Jakob disease (CJD)

Sporadic CJD is the commonest prion disease of humans. It occurs worldwide, but is rare – ~ 1 case per million population annually. The disease presumably arises from a rare spontaneous conversion of PrP^C into PrP^{Res}. Most patients are aged > 60 years, and present with deficits in higher cortical function, progressing to profound dementia and death within 12 months of presentation.

CJD brain is infectious, and prions are unusually resistant to inactivation by conventional means (e.g. autoclaving at 121°C). Thus anything contaminated by CJD tissue can inadvertently transmit the disease if it is subsequently inoculated into another patient. Unfortunately, such transmission has now been well described, giving rise to CJD with some atypical features, e.g. in much younger patients, with a shorter incubation period. This form of CJD, originating from accidents of medical intervention, is known as **iatrogenic** CJD. Contaminated neurosurgical instruments, corneal grafts, and human pituitary-derived hormones (e.g. growth hormone) have all been shown to cause iatrogenic CJD.

Fifteen per cent of CJD patients have other family members with the same disease, and in these kindred, the disease is hereditary (as opposed to sporadic). Analysis of the *PrP* gene in families thus afflicted reveals mutations within the prion gene. The effect of these mutations is to increase considerably the likelihood of spontaneous conversion of PrP^C into PrP^{Sc}, and hence the emergence of disease.

Bovine spongiform encephalopathy (BSE) and variant CJD (vCJD)

BSE or 'mad cow disease' is a prion disease of cattle, first identified in the UK in 1986. A huge BSE epidemic occurred due to the practise (now banned) of feeding ruminant offal to cows, demonstrating that the infectious agent is transmissible through the oral route.

In 1996, a new variant of CJD was described in the UK, now known as vCJD. Several features allowed recognition of vCJD as a distinct entity, including the unusually young age of presentation (average age < 30 years), and the particular distribution of abnormal prion protein present in the brain of sufferers. There is now no doubt that vCJD has arisen through oral ingestion by humans of the BSE agent. How many cases of vCJD in humans will arise from this disaster remains uncertain. Prion diseases have a long incubation period, and mathematical models of the outbreak of vCJD give results varying from an optimistic few hundred to a pessimistic hundreds of thousands. Thus far, < 200 individuals are known to have died from this new disease.

The BSE epidemic in cows has all but subsided to zero, following the ban on using ruminant offal to feed cows in the late 1980s. Several additional safeguards have also been put in place to minimize the risk of transmission of disease to humans, including a ban on the use of bovine offal for human consumption, and steps taken to prevent nervous tissue contamination of meat destined for human consumption.

Now that vCJD has become established in the human population, the possibility of human-to-human spread of the disease must be considered. In patients incubating the disease (i.e. who do not yet have symptoms), the prion is present in lymphoid tissue (e.g. tonsils, appendix), and therefore transmission through contaminated surgical instruments is at least a theoretical possibility. There is also concern that the prion may be present in peripheral blood, raising the specter of transmission through blood transfusion. In response to the latter threat, all blood in the UK is now routinely depleted of white blood cells (leukodepletion) before transfusion.

B1 HUMAN IMMUNODEFICIENCY VIRUSES

Key Notes

Virology	Human immunodeficiency viruses (HIV) are retroviruses. They have a positive ssRNA genome, which is reverse-transcribed in the host cell into DNA.
Epidemiology	HIV infection arose in humans from direct transfer from monkeys. There were an estimated 40 million people living with HIV infection in 2004. Subsequent person-to-person spread is by sexual transmission, mother-to-baby (vertical), or through exposure to contaminated blood or blood products.
Consequences of infection	Initial infection may present with an HIV-seroconverting illness, which has many features of a glandular fever-like syndrome. Patients then enter a stage of asymptomatic carriage, during which there is an inexorable decline in the number of circulating CD4-positive T cells, resulting in severe dysfunction of the immune system, and rendering the individual susceptible to a wide range of infectious and other life-threatening complications. There are a large number of illnesses, which, once diagnosed, define the patient as having the acquired immunodeficiency syndrome (AIDS).
Diagnosis	This is by demonstration of anti-HIV antibodies in the patient's serum, and/or the presence of HIV RNA.
Treatment	There are now a number of antiretroviral drugs available for the management of patients with HIV infection, discussed in detail in Topic E5.
Prevention	There is no vaccine available. Prevention therefore depends on screening of donated blood and organs, and avoidance of risk behavior. Post-exposure prophylaxis may be effective after a single known exposure incident. Mother-to-baby transmission can be prevented by antenatal antiretroviral therapy, delivery by Cesarean section, and nonbreast-feeding.
Related topics	Antiretroviral agents (E5)

Virology   Thus far, two human immunodeficiency viruses have been identified: HIV-1 and HIV-2. These are **retroviruses** (Topic A3). Thus, they possess a positive ssRNA genome, which is reverse-transcribed within an infected cell into DNA. The DNA then becomes integrated into the host cell chromosome. Their genomes encode a number of proteins, including **reverse transcriptase**, a **DNA integrase**, and a protease enzyme. The virological properties and clinical consequences of infection with HIV-1 and HIV-2 are more or less identical, so it is convenient to refer to them simply as HIV. The only other known human retroviruses are human T-cell lymphotropic viruses 1 and 2 (HTLV-1 and -2). HTLV-1 causes a rare form of leukemia and a rare demyelinating disease.

The primary receptor for HIV on target cells is the **CD4 molecule**. This is only present on a subset of T lymphocytes and also monocytes, hence virus replicates within these cells of the immune system. CD4 binding is followed by subsequent binding to a co-receptor molecule, one of the chemokine receptors **CCR5** or **CXCR4**.

Epidemiology

The human immunodeficiency viruses originated from monkeys. There have been several transfers of simian immunodeficiency viruses to humans in the past 100 years in Africa. Acute HIV infection inevitably becomes chronic – spontaneous clearance of virus from an infected individual has not been reported. Most virus replication occurs in lymph nodes, from where virus may gain access to the bloodstream, and thence every bodily fluid. Person-to-person spread therefore arises through exposure to contaminated blood or other bodily fluids. The three most important routes of transmission are via **sexual intercourse**, from **mother to baby**, and through **exposure to blood** or **blood products** (e.g. blood transfusion, or sharing of contaminated needles).

Global estimates of the number of individuals currently infected with HIV are of the order of 40 million, including 2 million children. The bulk (26 million) of infections are in Africa, with 1, 2, 0.5 and 10 million in North America, Latin America, Western Europe and Asia respectively. In 2004, there were an estimated 5 million new infections, and 3 million deaths from AIDS.

Consequences of infection

Acute infection with HIV is often followed some 6–8 weeks later by an '**HIV seroconversion illness**', which has many features of glandular fever (fever, pharyngitis, generalized lymphadenopathy). Following this, patients then enter a stage of asymptomatic carriage, where they are infected (and infectious), but the damage to their immune system is not yet sufficient to cause clinical disease.

During this stage, HIV-induced cell death of CD4-positive T-helper cells is compensated for by increased production of these cells. However, eventually (over several years), cell loss outstrips cell replenishment, and there is an inexorable decline in the peripheral blood CD4 count. The loss of CD4 cells results in impaired functioning of the host immune system, rendering the individual susceptible to a wide range of infectious and other life-threatening complications. Once these occur, the patient is said to have **AIDS**. Note that AIDS is a very carefully defined diagnosis. In order to fulfil the criteria for a diagnosis of AIDS, an HIV-infected patient must either have one of a number of specific illnesses – thereby known as AIDS-defining illnesses – or have a CD4 count of $< 200/\text{mm}^3$. The official categorization of stages of HIV infection (A, B or C, followed by 1, 2 or 3) is shown in *Table 1*. Common clinical manifestations of AIDS, including those which are AIDS-defining illnesses, are listed in *Table 2*.

Diagnosis

The primary diagnostic test for HIV infection is to look for the presence of anti-**HIV antibodies** in a blood sample. Antibodies are, strictly speaking, only a marker of past infection, but, given that an individual infected with HIV will

Table 1.   Classification system for human immunodeficiency virus (HIV)-infected patients

Clinical categories:
A.   Asymptomatic, acute HIV infection, or persistent generalized lymphadenopathy
B.   Symptomatic disease, not A or C conditions
C.   AIDS indicator conditions (*Table 2*)

CD4-positive cell counts:
1. $\geq 500/mm^3$
2. $200–499/mm^3$
3. $< 200/mm^3$

always be infected with HIV, a positive anti-HIV test indicates current HIV infection.

It is now possible to measure the amount of virus circulating in the peripheral blood – the **viral load** – using **genome amplification techniques**. This has become a very important tool in the management of HIV-infected patients. Decisions as to when to start antiretroviral therapy (Topic E5) are based, in part, on the magnitude of the viral load, and once therapy has begun, the effectiveness of the drug regimen must be monitored by serial viral load measurements.

Treatment

There is now a range of **antiretroviral drugs** available for the treatment of HIV-infected individuals, discussed in detail in Topic E5. These include nucleoside analogue and nonnucleoside analogue reverse transcriptase inhibitors, protease inhibitors and fusion inhibitors. Antimicrobial management of immunosuppressed individuals is also discussed in Topic F14.

Prevention

There is an urgent need for an effective HIV vaccine. Several candidate vaccines are at various stages of clinical trials, but none thus far has shown any evidence of inducing useful protection against infection. Prevention strategies are therefore currently focused on:

(i)   screening of donated blood and organs;
(ii)   behavioral modification to reduce the risks of person-to-person spread (e.g. 'safe sex' practises, needle exchange schemes for injecting drug users);
(iii) the use of post-exposure prophylaxis: a 4-week course of a cocktail of antiretroviral agents taken immediately after a known exposure incident. This is of relevance to healthcare workers after a needlestick injury, or following sexual exposure, e.g. after rape;
(iv)   prevention of mother-to-baby transmission of HIV; this can be achieved by appropriate use of antiretroviral drugs in pregnancy, delivery by elective Cesarean section, and avoidance of breast-feeding.

Table 2.   Common clinical manifestations of AIDS

(a)   Opportunistic infections
Pulmonary

Pneumonia	***Pneumocystis jirovecii (formerly carinii)***
	Cytomegalovirus (CMV)
	Cryptococcus neoformans
Tuberculosis	***Mycobacterium tuberculosis***
Gastrointestinal	
Oral and **esophageal thrush**	***Candida albicans***
Oral hairy leukoplakia	Epstein–Barr virus (EBV)
Oral ulcers, **esophagitis**	**Herpes simplex virus** (HSV)**, CMV**
Diarrhea	*Salmonella* species
	Shigella
	Giardia lamblia
	Cryptosporidium (chronic, >1 month)
	***Mycobacterium avium* complex**
Colitis	**CMV**
Central nervous system	
Multiple brain abscess	***Toxoplasma gondii***
Meningitis	***Cryptococcus neoformans***
Retinitis, encephalitis	CMV
Progressive multifocal leukoencephalopathy (PMLE)	**JC polyomavirus**
Skin and mucous membranes	
Prolonged (>1 month), recurrent ulceration	**HSV**
Herpes-zoster	Varicella-zoster virus
Multiple skin lesions	***Cryptococcus neoformans***
Reticuloendothelial system	
Lymphadenopathy	*Toxoplasma gondii*
	***Mycobacterium avium* complex**
	EBV

(b)   Malignant disease
Kaposi's sarcoma
Non-Hodgkin's lymphoma, cerebral or peripheral
EBV-associated lymphomas, e.g. polyclonal lymphoproliferation, Burkitt's lymphoma
Carcinoma of the uterine cervix
Hodgkin's lymphoma
Squamous carcinoma
Testicular cancers
Basal cell cancer
Melanoma

(c)   Neurologic disease

Aseptic meningitis	May be HIV-related
HIV encephalopathy	**HIV**
Intracranial mass lesions	**Cerebral toxoplasmosis**
	Primary CNS lymphoma
	Metastatic lymphoma
	Kaposi's sarcoma
	PMLE
	CMV or HSV encephalitis
	Cerebral tuberculosis

(d)   Other
Wasting syndrome of HIV infection

Those which are AIDS-defining are shown in bold.

B2 HEPATITIS A AND E VIRUSES

Key Notes

Viral hepatitis	Hepatitis means inflammation of the liver. The illness presents non-specifically with malaise, fever, anorexia and right upper quadrant abdominal pain. Impairment of normal liver function results in jaundice, with dark urine and pale stools. Viruses whose major tropism is the liver are referred to as hepatitis viruses, and are identified with different letters of the alphabet.
Hepatitis A virus (HAV)	*Virology:* HAV has a positive ssRNA genome and belongs to the Picornaviridae. *Epidemiology:* spread is via the fecal–oral route. Risk groups include travellers to countries where infection is endemic, sewage workers, and people who eat shellfish. *Consequences of infection:* asymptomatic infection, acute hepatitis, fulminant hepatitis leading to acute liver failure (rare). *Diagnosis:* demonstration of HAV-specific IgM antibodies in serum. *Treatment:* supportive only. *Prevention:* by use of a heat-inactivated whole virus vaccine; good hygienic practises (water supplies, sewage disposal).
Hepatitis E virus (HEV)	*Virology:* HEV has a positive ssRNA genome, but is currently unclassified in terms of virus families *Epidemiology:* spread is via the fecal–oral route. Massive outbreaks arise through faecal contamination of water supplies. *Consequences of infection:* as for HAV, but mortality from HEV infection is higher than with HAV, particularly in pregnant women. *Diagnosis:* demonstration of HEV-specific IgM antibodies in serum. *Treatment:* supportive only. *Prevention:* no vaccine available; good hygienic practises (as for HAV)
Related topics	Hepatitis B and D viruses (B3) Hepatitis C virus (B4)

Viral Hepatitis

Hepatitis simply means **inflammation of the liver**, of which there are many causes. The commonest cause in the UK is probably toxin-mediated, where the toxin is alcohol. This is followed by viral infections of the liver (i.e. viral hepatitis). For some viruses, their effect on the liver is the dominant feature of infection, and collectively these are known as the hepatitis viruses, identified by different letters of the alphabet. However, hepatitis may be an unusual feature of a wide range of virus infections including Epstein–Barr virus (Topic B8) and Cytomegalovirus (Topic B7).

Clinically, acute hepatitis presents nonspecifically. Symptoms and signs include:

- lethargy;
- loss of appetite (anorexia);
- nausea;
- fever;
- alcohol and cigarette intolerance (adults);
- right upper quadrant abdominal pain (arises as the inflamed liver swells and stretches its innervated capsule).

The liver serves a number of functions in the body, one of which is the conjugation of pigments derived from hemoglobin in the blood, and their excretion in the bile. As this process is impaired, the pigments (e.g. bilirubin) remain in the bloodstream, giving rise to **jaundice** – seen as a yellowing of the skin and sclera of the eyes. Instead of being excreted in bile and emerging in feces, the pigments are filtered through the kidneys, and the patient may notice very **dark urine**, and correspondingly **pale stools**. Once the patient has become jaundiced, he/she often feels much better.

The **hepatitis viruses** all share a **tropism** (Topic A3) for the liver, but otherwise they are quite distinct, belonging to different virus families and having different epidemiological and clinical characteristics.

Hepatitis A virus (HAV)

Virology
HAV has a positive ssRNA genome, and belongs to the family **Picornaviridae**. Only one serotype has been described.

Epidemiology
Transmission of virus is via the **fecal–oral route**. An infected host excretes virus in feces, and a new host becomes infected through ingestion of contaminated food or water, the virus being highly resistant to the low pH found in the stomach. Initial replication takes place in the small bowel wall, leading to excretion in the feces, but some virus enters the bloodstream (i.e. there is a **viremia**) and thereby gains access to the liver.

In the developing world, with overcrowding, and less-than-optimal disposal of feces and clean water supplies, the vast **majority** of the population acquire **infection in childhood**. In more developed parts of the world, a higher percentage of the adult population is nonimmune, and these individuals are at **risk when travelling abroad** in highly endemic areas. Infection may be acquired from any type of foodstuff, but the risk is greatest for food which may be washed in contaminated water and then eaten uncooked (e.g. salads). **Shellfish** are also a potent source of HAV infection, as they have a mechanism for concentrating any virus present in the water in which they are grown, which is then difficult to inactivate. Transmission may also occur **person-to-person** (e.g. within a household), if strict standards of hygiene are not maintained. As there is a viremic phase in the infection, which occurs before the patient is clinically ill, hepatitis A may rarely be transmitted through blood transfusion.

Consequences of infection
(i) **Asymptomatic** infection. This is **age-related**: children (who represent the vast majority of infected individuals worldwide) are very unlikely to present with clinical illness, whereas over half of adults infected will be jaundiced.

(ii) **Acute hepatitis**. In some individuals, the damage to the liver is sufficient to impair function to a level where clinical illness ensues. The **incubation period** is 2–6 weeks, at which time the patient presents with signs and symptoms as outlined above. The illness may last for several days, and it may be weeks or even months before the individual returns to full fitness.

(iii) **Fulminant hepatitis**. In a very small minority of patients, the degree of liver damage is so overwhelming that the patient goes into acute liver failure or **fulminant hepatitis**. The **mortality** from this is high: 70% in the absence of a liver transplant.

Diagnosis

HAV infection is diagnosed by demonstration of **IgM anti-HAV** in a serum sample (Topic E1). This is an extremely useful diagnostic test; it can be performed within a couple of hours of receipt of the serum sample, and a positive result means that the clinician has a clear-cut diagnosis.

Treatment

The management of a patient with acute HAV infection is with bed rest and appropriate support. Only those severely affected will require hospitalisation.

Prevention

This used to be by administration of normal human immunoglobulin, i.e. passive immunisation. However, there is now an effective **vaccine**, comprised of **whole killed virus**. A single dose induces protective levels of antibody after ~ 2 weeks, which last for several months. A booster dose one year later will protect for several years. The vaccine is recommended for people who travel abroad to endemic countries, and others at high risk (e.g. sewage workers).

Hepatitis E virus (HEV)

Virology

HEV carries a positive ssRNA genome. The genome organisation is unusual, and it is currently an unclassified virus in terms of which virus family it should belong to.

Epidemiology

Although HEV is entirely distinct virologically from HAV, it shares many epidemiological and clinical features. Thus, it is also transmitted via the fecal-oral route. Huge outbreaks of HEV infection have been described in a number of countries (e.g. India, Mexico) through fecal contamination of water supplies. In the UK, most cases arise in travellers returning from parts of the world where HEV is endemic, but there is increasing evidence that the virus may also be present in the UK.

Consequences of infection

These are the same as for HAV infection (i.e. asymptomatic infection, acute hepatitis, and fulminant hepatitis). The mortality from HEV infection is higher than with HAV, especially in **pregnant women**, where mortality rates are 10–20%, although the reasons for this are not clear.

Diagnosis

Acute HEV infection is diagnosed by demonstration of an IgM anti-HEV response, but this test is only performed in reference laboratories in the UK.

Treatment and Prevention

Treatment is supportive only – there are no effective antiviral agents. There is also no vaccine at present.

B3 HEPATITIS B AND D VIRUSES

Key Notes

Hepatitis B virus

Virology: HBV belongs to the Hepadnaviridae family, and contains a small, partially dsDNA genome, which encodes surface (HBsAg) and core (HBcAg) proteins, and a DNA polymerase enzyme with reverse transcriptase activity. HBeAg is also derived from the core gene.

Consequences of infection: Asymptomatic seroconversion, acute hepatitis, fulminant hepatitis (in 1% of HBV infections). In addition, following acute infection, 5% of adults and 90% of neonates become chronically infected. Chronic carriers may be asymptomatic, or may suffer from chronic inflammatory hepatitis, leading eventually to cirrhosis of the liver, and development of primary hepatocellular carcinoma.

Epidemiology: HBV is a blood-borne virus. Routes of transmission are: (i) vertical (i.e. from mother-to-baby); (ii) sexual; (iii) exposure to contaminated blood. There are estimated to be 350 million carriers of HBV.

Diagnosis: The presence of HBsAg in serum is diagnostic of infection. The presence of IgM anti-HBc indicates an acute infection. Chronic infection is defined as the persistence of HBsAg for > 6 months. The HBeAg/anti-HBe system is used to distinguish highly infectious carriers at increased risk of liver disease (HBeAg positive) from less infectious carriers at lower risk of liver disease (anti-HBe positive). Recovery from infection is associated with the appearance of anti-HBc, anti-HBs, and anti-HBe. The viral load of HBV DNA in serum is also an important measure of infectivity and the risk of liver disease progression.

Treatment: Interferon-α acts as an immunomodulatory agent in some chronic carriers. Suppression of HBV DNA synthesis can be achieved using reverse transcriptase inhibitors such as lamivudine or adefovir.

Prevention: Active vaccination is with multiple doses of a subunit HBsAg vaccine. Protection is directly proportional to the amount of anti-HBs induced. Passive immunization is with hepatitis B immunoglobulin derived from blood donors.
   All blood and organ donors should be screened for evidence of current HBV infection.

Hepatitis D virus

Virology: An incomplete RNA virus that requires HBsAg in order to form viral particles.

Epidemiology: Can only infect patients at the same time as HBV (co-infection) or those who are already carriers of HBV (super-infection). Thus routes of transmission and risk factors for acquisition of infection are the same as for HBV.

Consequences of infection: Co-infection increases the risk of chronic carriage; super-infection increases the risk of serious liver disease.

Treatment and prevention: as for HBV.

Related topics   Hepatitis A and E viruses (B2)   Antiviral agents (E4)
                     Hepatitis C virus (B4)   Vaccines and
                                                              immunoprophylaxis (E10)

Hepatitis B virus   Hepatitis B virus (HBV) is an extremely important pathogen, especially in hospital medicine. It is easier to discuss the clinical consequences of infection *before* describing the epidemiology, as the latter is very much easier to understand once the salient points of the former are understood.

Virology
HBV has an unusual, partially dsDNA genome. It belongs to the hepadna (hepa–DNA) virus family, which contains a number of viruses each of which infects the liver of its natural host.

   The compact genome (only 3.2 kb in length) encodes a small number of proteins, or antigens:

- **core antigen (HBcAg)**, derived from the core gene, encapsidates the viral genome
- **'e' antigen (HBeAg)**, also derived from the core gene
- **surface antigen (HBsAg)**, derived from the surface gene, surrounds the nucleoprotein core of the virus
- **'X' antigen**, derived from the X gene, function unknown
- **polymerase** enzyme, derived from the Pol gene, has both DNA polymerase and reverse transcriptase activity.

Consequences of infection
These are outlined in *Fig. 1*
The majority of infections are **asymptomatic**. Alternatively, patients may present with an **acute hepatitis** (Topic B2), which is clinically indistinguishable from acute hepatitis A. There may be such overwhelming liver damage that the patient goes into acute liver failure – **fulminant hepatitis**. This occurs in ~ 1% of all HBV infections, and is more common than in HAV infection. However, the major difference between HBV and HAV is that some individ-

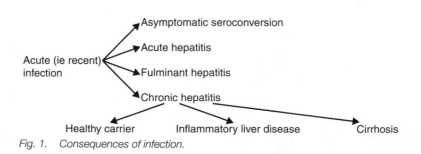

Fig. 1.   Consequences of infection.

uals may fail to clear HBV from their livers, and become chronically infected. In a **chronic infection** (Topic A4), virus actively replicates within cells, without interfering with the normal lifespan of those cells. As a result, the infected cell, in this case a hepatocyte, will continually excrete mature virus particles into the bloodstream. From here, the virus gains entry into every bodily compartment and fluid.

In many chronic carriers, the presence of replicating virus does not impact on liver function, and they are not ill. However, in others, the virus stimulates a **chronic inflammatory** response, which may compromise liver function and cause symptomatic **chronic liver disease**. Over a period of 20–30 years, this may result in extensive replacement of liver tissue by **fibrosis**, and considerable **distortion of the normal liver architecture**, a condition known as **cirrhosis**. The complications of cirrhosis and liver failure are potentially life-threatening. In addition, HBV is a transforming virus (Topic A4), and a chronic HBV carrier is ~ 400 times more likely than a non-carrier to develop a **primary tumor** of the liver cell, known as **hepatocellular carcinoma (HCC)** or **hepatoma**.

About **5% of adults** infected with HBV become **chronic carriers**. In contrast, up to **90% of neonates** who acquire virus from their carrier mothers will become chronically infected, reflecting the immaturity of their immune system at birth. This has two important consequences. Firstly, chronic carrier baby girls will pass the infection on to their offspring years later when they reach childbearing age. Secondly, the long-term complications of chronic carriage such as cirrhosis and cancer development, which become manifest 20–30 years after infection, will be a major cause of death in these carriers at a relatively young age.

Epidemiology

As explained above, in chronic carriers, HBV is a blood-borne virus. There are three routes of transmission of HBV:

(i) **Vertical** i.e. from **mother to baby**. Globally, this is the most important route. Most commonly, this occurs **perinatally** (i.e. at the time of birth), through exposure to a contaminated birth canal and maternal blood. This results in a high rate of carriage in the offspring, and in countries where vertical transmission is the main route, up to 20% of the population may be HBV carriers. In those countries (e.g. South East Asia, China, sub-Saharan Africa), for reasons explained above, hepatoma is one of the commonest tumors.

(ii) **Sexual** due to the presence of virus in genital tract secretions of chronic carriers. In the UK, HBV is predominantly a sexually transmitted disease. As this happens in adults, the resultant rate of chronic carriage is low. However, this route of transmission means that sexually promiscuous individuals, including sex workers, are at increased risk of infection, as are male homosexuals.

(iii) **Exposure to contaminated blood**. This can occur via a number of mechanisms, most obviously by blood transfusion. This is preventable by screening all blood donors for evidence of HBV infection. However, it can occur in much more subtle ways such as sharing of contaminated needles by injecting drug users, or in a domestic context by sharing toothbrushes or razors with a chronic carrier. In hospitals, accidental needlestick transmission of HBV is an important occupational health hazard for healthcare workers.

The World Health Organization (WHO) estimates that there are 350 million HBV carriers on the planet. These are not evenly distributed around the globe – there are some countries with very high carriage rates (e.g. 20%), and others with very low ones (e.g. the UK carriage rate is ~ 0.1%).

Diagnosis

Infection with HBV presents the host immune system with a number of antigens to which an antibody response may or may not be made. The important laboratory markers of HBV infection are HBsAg/anti-HBs, anti-HBc, and HBeAg/anti-HBe. The significance of these various markers is as follows:

(i) **HBsAg**. The primary diagnostic test for HBV infection is detection of hepatitis B surface antigen (HBsAg) in serum. If a serum is HBsAg positive, then the patient is infected with HBV – there is no alternative explanation. The infection may be acute or chronic – the latter is defined as persistence of HBsAg for > 6 months.

(ii) **Anti-HBc**. All patients infected with HBV will make antibodies to the core antigen (anti-HBc). This marker can be used to distinguish acute from chronic infection, as in the former, but not the latter, antibodies of the IgM class will be present (Topic E1).

(iii) **HBeAg** and **anti-HBe**. These markers allow a distinction between two groups of chronic carriers. HBeAg is a marker of the amount of virus replication occurring in the liver – the presence of HBeAg in a serum sample indicates a high degree of viral replication, and an HBeAg-positive carrier is highly infectious and at increased risk of serious liver disease. Over time, some HBV carriers lose HBeAg, and shortly thereafter become anti-HBe positive. This reflects a sharp downturn in virus replication, and, although such an individual is still HBsAg positive, and is still chronically infected, he/she is much less infectious, and is at much lower risk of serious liver disease.

(iv) **Anti-HBs**. This appears shortly after patients clear HBsAg, although in addition, anti-HBs is also a marker of response to hepatitis B vaccine (see below).

Patients who clear virus will have a full set of antibodies in their sera, i.e. anti-HBc, anti-HBs, and anti-HBe.

Common patterns of HBV diagnostic markers are shown in *Table 1*. There are additional complications which arise because of the occurrence of various HBV mutants, but these are beyond the scope of this text.

It is also possible to detect the presence of HBV DNA in a serum sample using molecular genome detection techniques such as PCR (Topic E1). This can be

Table 1. Diagnostic markers of hepatitis B virus (HBV) infection

	HBsAg	IgM anti-HBc	IgG anti-HBc	HBeAg	anti-HBe	anti-HBs	HBV DNA
Acute HBV infection	+	+	+	+	–	–	+++
Cleared HBV infection	–	–	+	–	+	+	–
Chronic HBV infection, high risk	+	–	+	+	–	–	+++
Chronic HBV infection, low risk	+	–	+	–	+	–	+/–
Responder to HBV vaccine	–	–	–	–	–	+	–

quantified – known as **viral load** testing. Infectivity and risk of liver disease progression are proportional to viral load in individual patients. Viral load is also an important parameter to measure when monitoring patients on anti-HBV therapy.

Treatment

The first successful therapy for chronic HBV carriers was interferon-α (IFN-α, see Topic E4), which in this context acts as an immunomodulatory agent, empowering the host immune system to eliminate the majority of HBV-infected hepatocytes and resulting in seroconversion from HBeAg to anti-HBe. As this mode of action of IFN is dependent on a functioning immune system, IFN therapy is not effective in carriers who are co-infected with HIV, nor in patients who acquired HBV infection at birth (i.e. the majority of patients world-wide), whose immune system is to a large extent tolerant of the virus.

There is a step in the replication cycle of HBV where viral DNA is synthesized from an RNA template, an event mediated by the viral DNA polymerase, which therefore has reverse transcriptase activity. The advent of anti-HIV therapy has led to the development of a number of reverse transcriptase inhibitors, and both lamivudine and adefovir (Topics E4, E5) have both been shown to cause sharp reductions in HBV DNA synthesis.

Prevention

A **subunit** vaccine, consisting solely of HBsAg, is effective in inducing protective immunity against HBV infection. The standard regimen consists of three doses, at 0, 1 and 6 months. The degree of protection induced by the vaccine is directly proportional to the level of **anti-HBs** achieved in the recipient's serum.

The WHO recommends that HBV vaccination be included as a routine **universal vaccine of childhood**. Many countries have adopted this policy, but the UK currently has a selective policy – the vaccine is only recommended for those individuals at particular risk of acquiring HBV infection. These include babies of carrier mothers, sexual partners of chronic carriers, healthcare workers, and less easily reached subpopulations such as injecting drug users, sexually promiscuous individuals including sex workers, and male homosexuals.

In nonvaccinated individuals, or vaccine nonresponders, protection after an acute exposure incident can be given by **passive immunization** (Topic E10) with preformed anti-HBs. This is derived from blood donors, and is referred to as **hepatitis B immunoglobulin**, or HBIg. One particular use of HBIg is for babies of high risk (i.e. HBeAg positive) carrier mothers, where co-administration of both HBIg and vaccine offers the best chance of preventing mother-to-baby transmission.

An important additional measure to prevent the spread of HBV infection is the screening of all blood and organ donors for evidence of current HBV infection by testing their sera for the presence of HBsAg. For obvious reasons, blood or organs from an HBV-infected donor should NOT be used.

Hepatitis D virus   *Virology*

Hepatitis D virus (HDV) is an incomplete virus, consisting of a small piece of RNA, which encodes a single protein, its nucleocapsid, the **delta antigen**. The RNA plus delta antigen do not constitute a complete viral particle – an outer protein coat is required. HDV uses the hepatitis B virus surface antigen as its outer coat.

Epidemiology

Because HDV can only replicate in cells where HBV is also replicating (as otherwise it has no way of acquiring HBsAg), the risk factors for, and routes of transmission of, HDV are the same as those for HBV. Infection with HDV can occur either at the same time as HBV infection (**co-infection**), or it can occur in someone who is already an HBV carrier (**super-infection**).

Consequences of infection

Co-infection with HDV and HBV increases the risk that infection will become chronic. Patients who carry both HBV and HDV are at increased risk of serious liver disease and hepatocellular carcinoma.

Diagnosis

This is by detection of delta antigen, or antibodies to delta antigen, in peripheral blood.

Treatment

As for HBV infection – interruption of HBV replication will also have the effect of stopping HDV replication.

Prevention

As for HBV, anti-HBs will prevent both HBV and HDV infection.

B4 HEPATITIS C VIRUS

Key Notes

Hepatitis C virus	*Virology:* HCV is a flavivirus, with a positive ssRNA genome. Worldwide, there are six genotypes, and >100 subtypes, of HCV.
	Epidemiology: HCV is a blood-borne virus. Risk factors for infection are sharing of drug injection equipment, and receipt of blood or blood product transfusion. Prevalence varies around the world from <1% (e.g. UK) to 20% (e.g. Egypt). High prevalence areas have arisen through medical use of non-sterile needles.
	Consequences of infection: Most acute infections are asymptomatic. Around 75% of infections become chronic, which may lead to inflammatory liver disease, cirrhosis, and an increased risk of hepatocellular carcinoma.
	Diagnosis: Initial screening is for the presence of anti-HCV, a marker of past infection. Current infection is diagnosed by demonstration of HCV RNA in serum.
	Treatment: Combination therapy with pegylated interferon plus ribavirin leads to clearance of virus from 45% of patients with genotype 1 infection, and 80% of patients with genotypes 2 or 3 infection.
	Prevention: There is no vaccine or any form of passive immunization. Prevention strategies are aimed at interrupting transmission of infection (e.g. by screening of blood donors, needle-exchange schemes).
Related topics	Hepatitis A and E viruses (B2) Antiviral agents (E4) Hepatitis B and D viruses (B3)

By the mid-1970s, diagnostic assays for both hepatitis A and B viruses were widely available. However, not all patients with acute or chronic hepatitis that looked as though it was of viral origin had markers of HAV or HBV infection. Thus the search began for the so-called '**non-A non-B hepatitis**' virus or viruses. In 1989, the genome of **hepatitis C virus** (HCV) was published, followed in 1991 by that of **hepatitis E virus** (Topic B2).

Virology

HCV has a positive single-stranded RNA genome, and belongs within the Flaviviridae family (i.e. a distant relation of yellow fever virus). The genome, just over 9000 bases long, encodes **three structural proteins** (core and two envelope glycoproteins), and a handful of **nonstructural proteins**, including helicase and protease enzymes and an RNA-dependent RNA polymerase (NS5).

Not all HCV genome sequences are identical. HCV is now classified into six **genotypes** (denoted by Arabic numerals), each genotype differing from the others by ≥ 30% of genome sequence. In addition, there are **subtypes** within genotypes, denoted by alphabet letters, which differ by ~ 15% in genome sequence. More than 100 subtypes have been described.

As with HBV, it is easier to understand the epidemiology of HCV after discussing the consequences of infection.

Consequences of infection (Fig. 1)

Most acute (i.e. recent) infections are **asymptomatic**. Acute hepatitis may occur, but this is unusual. The majority (>75%) of HCV infections, whether in children or adults, are not cleared by the host, resulting in chronic infection.

Chronic infection is associated with the risk of inflammatory hepatitis, and, over a period of years, end-stage liver failure due to **cirrhosis**. It also predisposes to the development of **hepatocellular carcinoma**.

Factors which increase the chances of serious liver disease are older age at infection, male sex, and increased intake of alcohol.

Epidemiology

Chronic HCV infection is associated with continual excretion of mature virus particles by infected hepatocytes into the bloodstream. Thus, along with HBV and HIV, HCV is classified as a **blood-borne virus**, and this, to a large extent, explains the routes of transmission of the virus.

In the UK, over two-thirds of HCV infections are acquired by **injecting drug-users** through sharing of blood contaminated needles, syringes, or other paraphernalia. The increase in recreational drug use in the 'swinging 60s' and early 1970s has resulted in large numbers of individuals now presenting with symptoms and signs of chronic liver disease, following ~ 30 years or so of infection. Another important route is via **receipt of contaminated blood** or **blood products** (e.g. factor VIII for haemophiliacs) prior to the introduction of blood-donor screening for HCV infection. In contrast to HBV, there is very little evidence that HCV is spread sexually, and also the rate of transmission from carrier mother to baby is much less (of the order of 1 in 20 to 1 in 30) than for HBV-carrier mothers.

Prevalence surveys of HCV infection in different countries show considerable variation. Egypt has the highest reported rate (~ 20%). This is believed to have arisen through the widespread re-use of unsterilized needles in the 1950s and 1960s in large-scale campaigns for the eradication of bilharzia (schistosomiasis, see Topic D5). Similar unfortunate iatrogenic (i.e. caused by the medical profession) epidemics have arisen in many countries through the re-use of contaminated needles, or the use of HCV-contaminated intravenous gammaglobulin preparations. Areas of intermediate prevalence (e.g. 1–5%) include the Mediterranean and the USA. The UK and Northern Europe are low prevalence areas, with 0.5–1% of the population infected.

Diagnosis

The presence of IgG antibodies to HCV in serum indicates past infection (IgM tests are not routinely used). Current (as opposed to cleared) infection is diagnosed by demonstrating the presence of HCV RNA in serum using a genome amplification technique such as reverse transcriptase-polymerase chain reaction assay (Topic E1).

Treatment

Treatment of chronic HCV infection is with 6–12 months (depending on genotype) of interferon conjugated to polyethylene glycol (PEG-IFN), and ribavirin (Topic E4). The 'pegylation' ensures slow steady release of the IFN, such that it need be given by subcutaneous injection only once weekly. Response rates depend on viral genotype – only ~ 45% of patients infected with type 1 will clear virus, whereas cure rates with types 2 or 3 approach 80%. New approaches to therapy include the development of protease inhibitors.

Prevention   There is no vaccine yet which induces protective immunity against HCV infection. Unfortunately, unlike with HBV, there is also no form of passive immunization for HCV.

Prevention strategies are therefore centered on an understanding of the modes of spread of infection, and adoption of appropriate measures to minimize such spread. Blood-donor screening and heat treatment of blood products serves to protect the blood supply. Needle-exchange schemes have been shown to reduce the spread of infection amongst injecting drug-users.

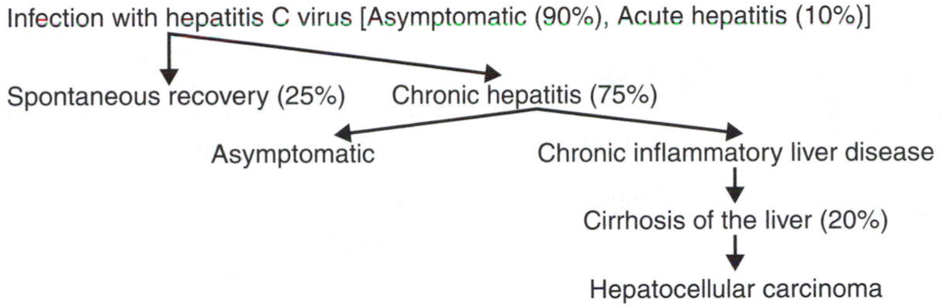

Infection with hepatitis C virus [Asymptomatic (90%), Acute hepatitis (10%)]

Spontaneous recovery (25%)   Chronic hepatitis (75%)

Asymptomatic   Chronic inflammatory liver disease

Cirrhosis of the liver (20%)

Hepatocellular carcinoma

Fig. 1. Schematic diagram of HCV outcome.

B5 HERPES SIMPLEX VIRUSES

Key Notes

Herpesviruses	*Virology* Herpesviruses are enveloped dsDNA viruses, with an icosahedral capsid. The genome encodes up to 200 proteins. All herpesviruses exhibit latency, so primary infection is followed by lifelong carriage and the possibility of secondary or reactivated infection. Eight herpesviruses have been identified as human pathogens.
Herpes simplex virus (HSV) -1 and -2	*Epidemiology* HSV infection is acquired through exposure to contaminated saliva or genital tract secretions. The site of latency is the nerve cell body.
	Consequences of infection Asymptomatic infection is the most likely outcome of primary infection; primary orolabial herpes presents with extensive ulceration in the oral cavity, and a systemic response, which may take up to 3 weeks to resolve. Recurrence at this site presents as a cold sore. Spread to the eye may cause conjunctivitis and keratitis. Recurrent keratitis eventually leads to blindness. Symptomatic primary genital tract infection is a debilitating disease with extensive ulceration and a systemic response. Recurrent genital lesions are much more trivial. Herpes encephalitis is a life-threatening infection arising from reactivation of virus in the brain substance. Neonatal herpes is usually acquired from passage through an infected birth canal, and has a poor prognosis due to internal dissemination of virus. A herpetic whitlow is infection of the nail-bed. Eczema herpeticum is a life-threatening complication arising in patients with chronic dermatitis, with potential internal organ infection as virus gains access to the bloodstream.
	Diagnosis HSV can be isolated in tissue culture from vesicle fluid or a swab of an ulcer base. Genome amplification of HSV DNA in cerebrospinal fluid is the method of choice for diagnosis of herpes encephalitis.
	Treatment Aciclovir and its derivatives.
	Prevention Vaccines are undergoing clinical trials.
Related topics:	Antiviral agents (E4) Infections in pregnancy and Other CNS infections (F4) neonates (F15)

Human herpesviruses (1): HSV

Virology

The Herpesviridae (i.e. family of herpesviruses) possess a dsDNA genome, surrounded by an icosahedral capsid and a loosely fitting lipid envelope.

Genome size is of the order of 150–200 kb, i.e. has the capacity to encode as many as 200 individual proteins (Topic A3).

Herpesviruses are widespread in nature. Thus far, eight herpesviruses have been identified as pathogens of humans (*Table 1*). On the basis of a number of biological properties, these are subgrouped into three subfamilies – the α-, β-, and γ-herpesviruses.

All herpesviruses exhibit **latency** (Topic A4). This means that infections may be primary (first exposure of an immunologically naïve host to the virus) or secondary (reactivation of latent virus), and the clinical consequences of these two types of infection may be very different from each other. The cellular site of latency is known for some, but not all the human herpesviruses.

Table 1. Human herpesviruses

α-Herpesviruses
    Herpes simplex virus type 1 (HSV-1)
    Herpes simplex virus type 2 (HSV-2)
    Varicella-zoster virus (VZV)
β-Herpesviruses
    Cytomegalovirus (CMV)
    Human herpesvirus type 6 (HHV-6)
    Human herpesvirus type 7 (HHV-7)
γ-Herpesviruses
    Epstein–Barr virus (EBV)
    Human herpesvirus type 8 (HHV-8)

Herpes simplex virus types 1 and 2

The biological properties of these two viruses are very similar, and at the genome level, they are > 80% identical. HSV infects at mucosal sites – intact skin is an effective barrier to infection. During primary infection, virus enters local nerve terminals, and ascends in their axons to reach the nerve cell bodies, the site of latency for these viruses.

Epidemiology

Infection with HSV-1 is common, occurring mostly in childhood. In the developed world, up to 90% of the adult population have antibodies to this virus. Spread is via exposure to saliva either directly (e.g. kissing) or indirectly (e.g. contact with objects that have been contaminated with saliva). HSV-2 is spread via exposure to genital tract secretions, and is therefore uncommon before puberty. Studies of antibody prevalence in the USA indicate that 10–15% of adults have been infected.

It used to be the case that the vast majority of HSV infections in the mouth were due to HSV-1, whereas HSV-2 caused the majority of genital tract infections. However, this distinction has become increasingly blurred with the practise of oral/safe sex, which has resulted in many genital tract infections being caused by HSV-1, and many oral infections by HSV-2.

Consequences of infection

(i)   *Asymptomatic infection.* The vast majority of primary HSV infections cause no illness at all (this is true of most virus infections). The individual, therefore, is not aware of having undergone infection – and yet virus will be latent, and he/she may suffer the consequences of reactivated infection, e.g. recurrent cold sores or genital herpes, without any history of a primary attack. Furthermore, reactivation itself is also often asymptomatic. In fact, most primary infections are acquired from a source undergoing an asymptomatic secondary infection, with transient excretion of virus into the saliva or genital tract, about which the individual is unaware.

(ii) *Orolabial herpes (gingivostomatitis)*. This arises from **primary infection** in the mouth, with extensive bilateral painful blistering and ulceration on the lips and inside the mouth. There may be some spread onto the facial skin. The local lymph nodes in the neck become swollen and painful (regional lymphadenopathy), and the patient may be systemically unwell with a fever. Untreated, the disease resolves in 2–3 weeks. **Secondary infection** arises from reactivation of latent virus in the trigeminal nerve cell body, following which virus tracks down the axons to reach the periphery, resulting in **recurrent herpes labialis**, more commonly known as a **cold sore**. Lesions are usually unilateral, occur in a cluster of no more than three or four blisters, there is no spread inside the mouth or onto the skin, there is no regional lymphadenopathy, and no systemic reaction, reflecting the fact that now the host immune system has some 'prior knowledge' of the virus. The lesions resolve in 5–6 days.

(iii) *Infection of the eye*. Patients with orolabial herpes may inadvertently spread virus from their mouth into their eye(s). Inoculation of the conjunctivae can result in a painful **conjunctivitis**. More seriously, the infection may involve the cornea – causing **herpetic keratitis**. Virus becomes latent in the nerve supplying the cornea, and may subsequently reactivate. Whilst initial lesions are superficial, subsequent reactivations, together with the host immune response to these, may result in severe inflammation of the cornea, which may become opaque. HSV keratitis is the commonest infectious cause of blindness in the UK.

(iv) *Genital herpes*. Symptomatic primary infection in the genital tract, can cause extensive areas of painful blistering and ulceration on the penis or the labia, vagina and cervix. Lesions near the urethral opening in women may be particularly distressing, and women are also at risk of herpes simplex **meningitis**, due to viral infection of the lumbosacral meninges. There will also be regional lymphadenopathy and systemic upset, with a fever, and recovery may take up to 3 weeks. Consequences of primary infection include recurrent genital tract lesions and asymptomatic shedding. Recurrences are usually much less severe, being unilateral and localized, without regional lymphadenopathy or systemic upset, and resolving in 5–6 days.

(v) *Herpes simplex encephalitis (HSE)*. HSV is the commonest viral cause of encephalitis (i.e. infection of the brain substance) in the UK. Virus most likely reaches the brain following reactivation from the trigeminal ganglion, traveling upwards to the brain rather than down the axons to the periphery. Replication of virus within brain cells results in cell lysis, spread to adjacent cells, and further cell lysis; untreated, herpes encephalitis is either fatal or leaves the patient with severe brain damage.

(vi) *Neonatal herpes*. Neonates may acquire HSV infection during passage through an infected birth canal. The immune system is not fully mature at birth, and neonatal herpes infection is potentially catastrophic. Virus may disseminate in the bloodstream to infect internal organs, causing herpes encephalitis, pneumonitis or hepatitis, all of which have a high mortality rate.

(vii) *Herpetic whitlow.* This refers to inoculation of the virus into the nailbed, e.g. when fingers are inserted into an infected oral cavity.

(viii) *Eczema herpeticum.* This complication of HSV infection arises in patients with eczema. Because the eczematous skin is not intact, virus spreads across the skin causing extensive areas of blistering. Virus may also gain access through the inflamed skin into the bloodstream, resulting in potentially life-threatening herpetic infection of internal organs.

Diagnosis

HSV grows well in tissue culture, therefore swabs of suspected ulcers, or vesicle fluid, should be sent to the laboratory for virus isolation. Diagnosis of HSE is difficult – the best approach is to demonstrate the presence of HSV DNA in cerebrospinal fluid by PCR assay (Topic E1).

Treatment

Aciclovir and its derivatives (Topic E4) are very effective anti-herpetic drugs and should be used in the management of patients with symptomatic primary orolabial or genital herpes. Their use in treating recurrent oral or genital herpes is more controversial, as these manifestations of HSV infection are relatively trivial, and the therapeutic benefit consequently is not so great. In life-threatening situations, such as herpes encephalitis, they should be given intravenously in high dosage.

Prevention

Vaccines designed to prevent genital herpes are currently undergoing clinical trials. Aciclovir prophylaxis is effective in immunosuppressed patients who would otherwise be at risk of severe manifestations or frequent recurrences of infection, e.g. bone marrow transplant recipients and HIV-infected individuals.

B6 VARICELLA-ZOSTER VIRUS

Key Notes

Virology	Varicella-zoster virus (VZV) is an α-herpesvirus
Epidemiology	More than 90% adults have evidence of VZV infection. Spread is via droplets excreted from the throat of patients with chickenpox, or through contact with vesicle fluid in chickenpox or shingles.
Consequences of infection	Primary infection may be asymptomatic, or may give rise to varicella (chickenpox). Complications, commoner in adults than children, include varicella pneumonia. Secondary infection presents as herpes-zoster or shingles, a blistering skin rash of dermatomal distribution. Complications depend on the site of reactivation, but include post-herpetic neuralgia, zoster keratitis (from ophthalmic shingles) or motor-nerve paralysis.
Diagnosis	Clinical presentation is usually sufficient. Laboratory confirmation is by virus isolation, electron microscopy, or antigen detection.
Treatment	Aciclovir and its derivatives.
Prevention	Active immunization is with a live attenuated vaccine, passive immunization is with zoster immunoglobulin (V2Ig).

Related topics	Antiviral agents (E4)	Infections in pregnancy and
	Vaccines and	neonates (F15)
	immunoprophylaxis (E10)	

Varicella-zoster virus

Virology
Varicella-zoster virus is a herpesvirus (Topic B5), and therefore exhibits latency (Topic A4).

Epidemiology
Infection usually occurs in **childhood**, and is extremely common: ~ 90% of adults have antibodies to this virus, indicating past infection. As with HSV, the **site of latency** of VZV is also the **nerve cell body**. Transmission of VZV is mostly via the **respiratory route**, but **vesicle fluid** in both chickenpox and shingles is also infectious. About 10 days after infection, virus appears in the bloodstream (**viremic spread**). Thus, the virus gains access to all parts of the body, and therefore goes latent in all the nerves of the spinal cord, as well as the cranial nerves.

Consquences of infection
(i) *Asymptomatic infection.* Primary infection with VZV does not always result in clinical disease.

(ii) *Chickenpox*. Primary infection, when symptomatic, causes the disease **varicella**, more commonly known as chickenpox. As a result of viremic spread of virus during the incubation period, the **vesicular rash** occurs all over the body, including within the mouth. There may be some generalized lymphadenopathy, and a febrile response. Complications are unusual in children, but in adults, chickenpox is a much more debilitating illness, and **varicella pneumonia**, where virus is present and replicating in the lungs, is a life-threatening possibility.

Chickenpox in a **pregnant woman** (Topic F15) may result in damage to the fetus/baby, such as **congenital varicella** (if the maternal infection is in the first 20 weeks of pregnancy) or **neonatal chickenpox** (if the maternal infection is in the last week of pregnancy).

As with many other virus infections, chickenpox may be more severe – and indeed life-threatening – in patients whose immune system is not intact (e.g. with HIV infection, or on immunosuppressive drugs post-transplantation).

(iii) *Shingles*. This acutely painful **blistering rash** is the manifestation of **secondary infection** with VZV, and is properly known as **herpes-zoster**. Virus, reactivated in the nerve cell body, travels down the nerve to reach the area of skin supplied by that nerve – the **dermatome**. This explains the unique dermatomal distribution of a shingles rash. Virus may be reactivated from any nerve, so the rash may appear on any part of the body, but it will always stop at the midline.

Risk factors for the development of herpes-zoster include:

(i) *Increasing age*. This reflects declining function of the host immune system. More than 20% of individuals aged > 80 years will give a history of having had shingles.

(ii) *Immunosuppression or -deficiency*. This also indicates that an intact immune system is necessary for prevention of reactivation of VZV. Possible causes include HIV infection, immunosuppressive therapy post-transplantation, chemotherapy, and underlying malignant disease, especially leukemia or lymphoma

(iii) *Stress*. For many patients, shingles arises at a stressful time of life, e.g. following a bereavement or loss of job. This hints at an as-yet-undefined link between the nervous system and the immune system.

The commonest and most distressing **complication** of herpes-zoster is **post-herpetic neuralgia**. This is defined as pain persisting in the area of a zoster rash after the rash itself has disappeared. Pathogenesis is poorly understood, but it can persist for weeks, months, or even years, and can have a significant deleterious impact on quality of life. Other possible complications depend on which particular nerve is affected and include:

(i) ophthalmic branch of the trigeminal nerve: virus may gain access to the cornea, causing **zoster keratitis**;

(ii) motor nerves: damage to **motor nerves** may result in paralysis, e.g. zoster of the facial nerve may cause facial nerve palsy.

Diagnosis

The vesicular rash of chickenpox or shingles is usually so characteristic that a **clinical diagnosis** can be made with reasonable certainty. If in doubt, swabs or vesicle fluid can be sent for **virus isolation** in tissue culture (takes several days), **electron microscopy**, or **antigen detection** by immunofluorescence (Topic E1).

Treatment
VZV is sensitive to the actions of **aciclovir** and its derivatives, although antiviral therapy is not usually indicated for uncomplicated chickenpox in children. Increased doses need to be given in comparison to the treatment of HSV infection. In immunocompromised or other at-risk individuals, treatment should be given intravenously.

Prevention
A live, attenuated varicella vaccine is licensed for general use in the USA. In the UK, this vaccine is reserved only for patients at risk of life-threatening complications of infection (e.g. children with leukemia), and for nonimmune health-care workers who are at occupational risk of acquiring infection, and also of passing on infection to their immunocompromised or pregnant patients. **Passive immunization** with **zoster immunoglobulin** (Topic E10) is also available for susceptible high-risk individuals exposed to the virus.

B7 CYTOMEGALOVIRUS, HUMAN HERPESVIRUS-6 AND -7

Key Notes

Cytomegalovirus	*Epidemiology* Infection is acquired via exposure to saliva or genital tract secretions, transplacentally, or via organ or bone marrow transplantation. Prevalence varies around the world, depending on the degree of over-crowding. About 40% of young adults in the UK have evidence of infection.
	Consequences of infection (i) Asymptomatic seroconversion (commonest). (ii) Acute infectious mononucleosis. (iii) Congenital infection: occurs in 1 in 300 pregnancies; 5–10% of infected babies have severe abnormalities evident at birth; 5–10% are normal at birth, but have suffered damage; the remainder are normal at birth and develop normally. (iv) In immunosuppressed patients, CMV is a multisystem pathogen causing for example retinitis (leading to blindness), pneumonitis, hepatitis, encephalitis.
	Diagnosis Serostatus is determined by testing for IgG anti-CMV antibodies. Infection in immunosuppressed patients is best sought in peripheral blood by genome amplification.
	Treatment Anti-CMV drugs include ganciclovir, cidofovir and foscarnet. These all have potentially serious toxicities.
	Prevention No vaccine is available at present. Serostatus of transplant donors and recipients should be determined, and appropriate prophylaxis offered to mismatched recipients.
Human herpesvirus-6 and -7	Infection is very common, acquired in childhood through exposure to saliva. There is usually no disease, but HHV-6 may cause roseola infantum.
Related topics	Antiviral agents (E4) Infections in pregnancy and Infection in immuno- neonates (F15) compromised patients (F14)

Cytomegalovirus   These are β-herpesviruses (Topic B5). Cytomegalovirus (CMV) acquires its name from the characteristic **cytopathic effect** (Topic A4) of swollen, rounded cells it induces in cell culture. The site of latency of CMV is not known with certainty, but the virus is found in peripheral blood white cells. It is a clinically important pathogen in two specific situations: in immunosuppressed patients, and in pregnancy.

Epidemiology
CMV is excreted in saliva, urine, and the genital tract, and is present in blood. Most infections occur through **close contact with young children** (aged 0–5 years) with exposure to a saliva/urine-contaminated environment, e.g. toys passed from one mouth to another. Post-puberty, virus is acquired through **kissing** or **sexual intercourse**. Recipients of **blood transfusions** are also at risk of CMV infection. A very important route of transmission is **transplacental**, from maternal blood to the fetus, resulting in congenital infection. Another rather specialized, though clinically important, route of transmission is via **organ transplantation** (e.g. kidney, liver).

In developed countries, ~ 40% of young adults (aged 20–25 years) have serological evidence of prior infection with CMV, and this increases by ~ 1% per year thereafter. In less developed countries, with overcrowding and poorer socioeconomic conditions, > 90% of the population may acquire CMV infection in childhood.

Consequences of infection
(i)   **Asymptomatic infection**. The vast majority of primary CMV infections in immunocompetent individuals do not cause clinically evident disease, and this is also the case for all secondary, or reactivated infections.
(ii) **Infectious mononucleosis (IM)** syndrome (Topic B8). A small minority of adults will present with a glandular fever/IM-like syndrome at the time of primary CMV infection.
(iii) **Congenital CMV infection** (Topic F15). This is the commonest **congenital infection** in the UK, affecting about **1 in 300 pregnancies**. Of all CMV-infected babies, 5–10% have serious congenital abnormalities evident at birth, e.g. microcephaly, hepatosplenomegaly. These babies have **cytomegalic inclusion disease**. Another 5–10% of babies are normal at birth, but as they develop it becomes apparent that they have suffered some adverse effects, e.g. nerve deafness. The remaining 80–85% of CMV-infected babies are normal at birth and develop normally (i.e. these babies are *infected*, but not *affected*).

Congenital infection can arise from mothers undergoing primary or secondary (reactivation) infections during pregnancy. Both types of maternal infection may give rise to an affected baby.
(iv) **CMV infection in the immunosuppressed.** CMV is a major **multisystem pathogen** in patients with a compromised immune system, especially where the cellular arm of the immune response is defective. There are increasing numbers of such patients. **HIV infection** results in an increased likelihood of reactivation of previously acquired latent CMV. The commonest manifestation of CMV infection in this setting is **retinitis** which, if unchecked, leads to blindness. However, CMV can cause disease in almost any organ of the body, resulting in for example esophagitis, colitis, hepatitis, adrenalitis or encephalitis.

The greatest risk for **solid organ transplant recipients** arises when the infection is transmitted by the organ itself to a previously uninfected recipient (i.e.

donor CMV positive, recipient CMV negative). The recipient will therefore undergo a primary infection precisely at the time when he/she is most heavily immunosuppressed. Conversely, in **bone marrow transplantation**, the most dangerous situation is a CMV-positive recipient (who will therefore have latent CMV on board, but whose own immune system is ablated prior to the transplantation) receiving bone marrow from a CMV-negative donor (whose T cells will therefore have no activity against CMV). The commonest life-threatening manifestation of CMV infection in transplant recipients is **pneumonitis**. Other manifestations include fever, neutropenia, and hepatitis.

The CMV status of donor and recipient should be determined prior to transplantation. If there is a mismatch, then recipients should be given anti-CMV prophylaxis (see below), and they must be monitored very closely in the post-transplantation period for evidence of CMV infection.

Diagnosis
Prior infection with CMV is determined by detecting **IgG anti-CMV antibodies**. **Genome detection** assays are the most sensitive methods for detection of CMV infection in transplant recipients. Alternatives include **isolation of virus** in tissue culture, or **CMV antigen detection** in peripheral blood white cells. As isolation of CMV may take several days, the **detection of early antigen fluorescent foci** (DEAFF) test was developed to speed up the process (Topic E1).

Treatment
Anti-CMV drugs include **ganciclovir** (and its derivative **valganciclovir**, better absorbed orally), **cidofovir** and **foscarnet** (Topic E4). None of these is ideal, as they all have toxic side-effects.

Prevention
There is no vaccine currently available. High-risk patients such as a CMV-positive bone-marrow recipient being given marrow from a CMV-negative donor should be given **prophylaxis** with anti-CMV drugs. Otherwise, transplant recipients may be carefully monitored, and given anti-CMV drugs at the first indication of CMV infection. This strategy is referred to as **pre-emptive therapy**, as the therapy is initiated before the appearance of symptoms and signs of disease.

As CMV may be transmitted by transfusion, blood and blood products may be screened to insure high-risk recipients only receive CMV-negative material.

Human herpesviruses type 6 and 7

These viruses were discovered relatively recently (1986 and 1991 respectively), and are genetically closely related to each other and to CMV. Infection is very common (e.g. >80%), and most likely occurs in early childhood. The viruses are present in saliva and other bodily fluids.

Infection with HHV-6 is usually asymptomatic, but can give rise to the syndrome *roseola infantum*. Roseola (also known as exanthem subitum) is a disease of 0–2-year-olds, the distinguishing feature of which is the sudden onset of a fever with no localizing signs, which resolves after 3–5 days, concurrent with the appearance of a nonspecific red rash. Of itself, this infection is of relatively little importance, but the fever may give rise to febrile convulsions.

HHV-6 may be a cause of encephalitis in children, but as yet this remains unproven. There are also speculative data concerning a possible link between HHV-6 infection and multiple sclerosis. HHV-7 is an orphan virus, i.e. no particular disease has yet been associated with it.

B8 EPSTEIN–BARR VIRUS AND HUMAN HERPESVIRUS 8

Key Notes

Epstein–Barr virus (EBV)

Epidemiology
More than 90% of the global population is infected. Most infections are acquired in the first 5 years of life. EBV causes a productive infection of epithelial cells. Infection is acquired through contact with saliva or genital tract secretions. The site of latency is the peripheral blood B lymphocyte.

Consequences of infection
(i)   Asymptomatic seroconversion.
(ii) Glandular fever (infectious mononucleosis). Features include fever, pharyngitis, generalized lymphadenopathy and atypical mononuclear cells (activated CD8-positive T cells) in peripheral blood. Complications include jaundice, respiratory obstruction, splenic rupture.
(iii) Association with malignancy. EBV is associated with the development of lymphomas in immunosuppressed patients, Burkitt's lymphoma, nasopharyngeal carcinoma and some types of Hodgkin's lymphoma.

Diagnosis
By serology, either through the detection of IgM anti-EBV antibodies, or in the Monospot/Paul Bunnell tests, the detection of red cell agglutinins.

Treatment
There are no antiviral drugs with specific activity against EBV.

Prevention
Vaccines are at an experimental stage only.

Human herpesvirus 8

Initially known as Kaposi's sarcoma (KS)-associated herpesvirus. Infection is acquired through sexual transmission, is usually asymptomatic, but can give rise to KS, especially in immunosuppressed patients, e.g. due to HIV infection. Diagnosis of KS is by histology of tissue biopsy. There is no specific antiviral therapy or vaccine available at present.

Epstein–Barr virus

Virology
This γ-herpesvirus (Topic B5) was discovered in 1964, in a biopsy specimen of Burkitt's lymphoma (see below).

Epidemiology
More than 90% of adults throughout the world have antibodies to the virus. The vast majority of the population acquire infection in the first 5 years of life. In more developed countries, with less overcrowding, infection may be delayed into young adulthood, or even until much later in life.

EBV infects the lining **epithelial cells** of the buccal and other mucosal sites, causing a productive, lytic infection (Topic A4). The **site of latency** is the peripheral blood **B lymphocyte**. Infection of B cells *in vitro* results in polyclonal B-cell activation, i.e. antibody production, and **transformation** (Topic A4).

Replication and release of virus at mucosal sites results in virus being present in saliva and genital tract secretions. Thus, infections are acquired through contact with **saliva** – either indirectly through contaminated objects, or directly through kissing – or through **sexual intercourse**.

Consequences of infection

(i)   *Asymptomatic seroconversion.* The vast majority of primary EBV infections, especially those acquired in childhood, do not cause clinical disease.

(ii) *Infectious mononucleosis.* If infection is delayed until adolescence, there is a 50% chance that it will be associated with the syndrome known as **glandular fever**, or infectious mononucleosis. Patients present with a **fever**, **sore throat** (**pharyngitis**, evidenced by grossly swollen tonsils covered in a white patchy membrane), and painful generalized lymph node swelling (**lymphadenopathy**), most prominently in the neck (cervical lymph nodes). Examination of a peripheral blood film reveals the presence of **atypical mononuclear cells**. These are **activated CD8-positive T lymphocytes** (Topic A2). Such cells may arise in small numbers in a variety of acute viral infections, but acute EBV infection is associated with particularly large numbers, e.g. 5–10% of all white cells present in the peripheral blood.

Acute glandular fever (also known as 'kissing disease') is a debilitating illness which may take some time to recover from. In a significant percentage of sufferers, it may be several months before full health is restored. Other possible complications include:

- **jaundice:** in ~ 10% of patients, as a result of an underlying EBV-induced hepatitis;
- **respiratory obstruction** due to excessive tonsillar swelling;
- **splenomegaly** and even **splenic rupture**, as part of the generalized lymphadenopathy.

(iii) *Malignant diseases.* EBV is a **transforming virus** – it is very easy to demonstrate *in vitro* immortalization of B cells by the virus. *In vivo,* uncontrolled cell proliferation is kept in check by a very potent **cytotoxic T-cell response** which recognizes and kills proliferating EBV-infected B cells. However, there is now a huge body of evidence indicating that EBV is associated with a variety of human malignant diseases:

(a)   **Lymphomas in immunosuppressed patients**, e.g. transplant recipients, AIDS patients. The immunosuppression, through whatever cause, removes the T-cell defense mechanism, thereby allowing uncontrolled proliferation of transformed B cells. Initially this is polyclonal, but eventually, a single clone becomes predominant.

(b)   **Burkitt's lymphoma**. This is a rare B-cell lymphoma except in certain parts of the world – sub-Saharan Africa and Papua New Guinea, where it is the commonest tumor of childhood.

(c)   **Nasopharyngeal carcinoma** (NPC). The geographical distribution of this undifferentiated tumor is also quite distinct, being rare in most countries, but common in China and South East Asia.

The pathogenesis in these latter two tumors is complex. The tumor cells, B cells in Burkitt's lymphoma, and epithelial cells in NPC, contain multiple copies of the viral genome, but other **co-factors** must act together with EBV infection resulting in malignant change, in order to explain their peculiar geographic distributions. One such co-factor in the development of Burkitt's lymphoma is the presence of **hyperendemic malaria** (i.e. malaria present all-year-round). Possible co-factors for NPC include a **genetic predisposition**, and chronic exposure of the nasopharynx to **chemical carcinogens** or **physical irritants**, e.g. salted fish (dietary exposure) or smoke or dust (occupational exposure).

(d) **Hodgkin's lymphoma**. EBV is associated with 40% of cases of this lymphoma – the role of EBV in its pathogenesis is the subject of much current research.

Diagnosis

Recent infection is diagnosed by the demonstration of **IgM class antibodies** against the virus, whilst IgG anti-EBV antibodies are a marker of past infection (Topic E1). In addition, the diagnosis of EB virus-induced glandular fever can be made using the **Paul Bunnell** or **Monospot** tests. These tests detect the presence of antibodies which agglutinate red blood cells from a different species (e.g. sheep). They rely on the fact that EBV infection of B cells causes polyclonal B cell activation, and therefore acute infection is associated with a variety of unusual antibodies, including red cell agglutinins. Only ~ 90% of patients acutely infected with EB virus will make these antibodies, so a negative Monospot test does not rule out acute EB virus infection.

Treatment

There are no specific antivirals for EBV infection. Steroids may be given in acute glandular fever to reduce the tonsillar swelling. Treatment of malignant disease associated with EB virus infection is with standard chemotherapy.

Prevention

There is a great deal of interest in developing a vaccine to protect against EBV infection, as this would potentially be an 'anti-cancer' vaccine. As yet, none has reached the stage of being licensed for clinical use.

Human herpesvirus type 8

The DNA of this herpesvirus was discovered in 1996 in biopsy tissue from a Kaposi's sarcoma (KS) lesion (see below) using extremely sophisticated molecular biological technology. Originally known as Kaposi's-sarcoma-associated human herpesvirus, or KSHV, it is now more properly referred to as HHV-8.

It is possible to be infected with this virus without having KS, and rates of infection vary in different countries, being highest in the Mediterranean region. Most people appear to acquire infection through sexual transmission.

HHV-8 is the causative agent of KS. This is a tumor of abnormal vascular structures, and was an unusual neoplasm until HIV infection and the AIDS epidemic emerged. It is the most common neoplasm affecting HIV-infected individuals. The commonest site is the skin, although it can occur elsewhere, e.g. in the hard and soft palate, and gums within the mouth. Diagnosis is by clinical and histological appearance.

There is no specific antiviral drug with activity against HHV-8, and no vaccine is currently available for prevention of HHV-8 infection.

B9 RHINOVIRUSES AND CORONAVIRUSES

Key Notes

Rhinoviruses	

Virology
Belong to the *Picornaviridae*. Genome is positive ssRNA. There are > 100 serotypes.

Epidemiology
Transmission is via droplets and aerosols. Infection is extremely common throughout the world and throughout life.

Consequences of infection
(i) Coryza – the common cold;
(ii) exacerbations of asthma and chronic bronchitis;
(iii) rarely, lower respiratory tract infection.

Diagnosis
Clinical features. Isolation of virus in tissue culture.

Treatment
Symptomatic treatment, e.g. with folk remedies. No licensed specific antiviral agents.

Prevention
No vaccine available.

Coronaviruses	

Virology
Large positive ssRNA genome. Morphology is characteristic, with envelope glycoprotein spikes forming a crown (corona). Two major serotypes have been known for years. These have been joined by severe acute respiratory syndrome coronavirus (SARS CoV), identified in 2003 as the causative agent of SARS.

Epidemiology
Infection is common. Most adults have serological evidence of infection.

Consequences of infection
(i) the common cold;
(ii) SARS. This is a severe life-threatening lower respiratory tract infection with a high mortality rate (5–15% depending on age).

Diagnosis
SARS was originally a diagnosis of exclusion, but laboratory tests are now available, e.g. virus isolation, serology, genome detection.

Treatment
Most SARS patients are treated with steroids.

Prevention
Infection control procedures to prevent nosocomial spread of SARS CoV, and quarantining of contacts of SARS patients are the mainstay of prevention. Vaccine development is being actively pursued.

Related topics   Upper respiratory tract   Lower respiratory tract
                              infections (F6)   infections (F7)

Respiratory tract   **Respiratory tract infections** are extremely common, at least in part because of the
infections   large number of pathogens, especially viruses, able to infect at this site. Clinically,
                        there is a major distinction to be made between infections of the **upper respiratory
                        tract** (URTI, see Topic F6) which are relatively trivial, and those of the **lower
                        respiratory tract** (LRTI, see Topic F7), which may be life-threatening.
                        Rhinoviruses and **coronaviruses**, although belonging to quite distinct virus
                        families, are common causes of URTI, and are considered together in this chapter
                        as they account for the majority of **coryzal illnesses**, i.e. the common cold. In
                        addition, a coronavirus is the causative agent of the newly described syndrome of
                        severe acute respiratory syndrome, or SARS.

Rhinoviruses   *Virology*
                        Rhinoviruses carry a positive ssRNA genome and belong to the Picornaviridae
                        family (pico = small, RNA). There are > 100 different serotypes, a serotype being
                        defined on the basis of neutralization by antibody.

                        Epidemiology
                        Transmission is via **aerosols** and **droplets**, most efficiently generated by
                        sneezing. Infection occurs worldwide and throughout life – different serotypes
                        circulate in different years, and infection with one serotype does not induce
                        protective immunity against other serotypes. On average, in the UK, common
                        colds occur at the rate of almost one per person per year, although this varies
                        with age, being higher in the first few years of life.

                        Consequences of infection
                        (i)   *The common cold.* Presenting symptoms include **rhinitis** (runny nose),
                              sneezing, **headache**, **sore throat** and **cough**. Systemic symptoms and signs
                              (fever, myalgia) are usually absent, which distinguishes this syndrome from
                              classical influenza. A clear and watery nasal discharge usually becomes
                              mucopurulent and tenacious after a few days. Illness and cough peak after
                              3–5 days, but complete recovery usually takes 7–10 days. Although a trivial
                              illness, the common cold has a considerable economic impact on the work-
                              force, through days lost at work.
                        (ii) *Exacerbations of asthma and chronic bronchitis.* These viruses are respon-
                              sible for a large majority of wheezy attacks in children.
                        (iii) *Bronchiolitis or pneumonia.* Rarely, rhinoviruses are isolated from the
                              lower respiratory tract, usually in samples from young babies.

                        Diagnosis
                        This is usually made clinically, but rhinoviruses can be isolated in tissue culture
                        from respiratory tract samples (e.g. a nasopharyngeal aspirate or throat swab).

Treatment
Despite intensive efforts on the part of the pharmaceutical industry, no antirhinovirus agents have yet made it to the clinic. Symptomatic relief is with appropriate analgesics, sympathomimetic nasal decongestants, antitussives and antihistamines. A range of over-the-counter medicines are available, although hot whiskey or other home-brewed cocktails may be just as effective. Antibiotics are definitely *not* indicated!

Prevention
There are no rhinovirus vaccines. One major barrier to the development of an effective vaccine is the diversity of serotypes.

Coronaviruses

Coronaviruses are the second commonest cause of the common cold (behind rhinoviruses). Until very recently, this merited only a cursory mention in most microbiology textbooks, but the SARS epidemic has given this virus family a huge publicity boost.

Virology
Coronaviruses have a large positive ssRNA genome, and a characteristic morphology on electron microscopy. Viral particles are surrounded by an outer envelope studded with spikes, resembling a crown (**corona**). They infect a wide range of animal hosts, including humans. Prior to the SARS epidemic, two major serotypes were known to infect humans, known as the 229E and OC43 coronaviruses. To this duo must now be added the SARS coronavirus (identified in 2003), and NL63 (identified in 2004).

Epidemiology
The majority of adults have antibodies to the two major serotypes of coronavirus, indicating that infection is common.

Consequences of infection
(i) **The common cold**. This is clinically indistinguishable from the same syndrome caused by rhinovirus infection.
(ii) **Severe acute respiratory syndrome (SARS)**. This new syndrome emerged in 2003. Patients presented with severe life-threatening lower respiratory tract infection for which no causative agent could be identified. Cases appeared first in China and Hong Kong, but very soon spread around the globe. Clinical features included a sudden onset of high fever (>38°C), cough or difficulty breathing, with chest X-ray findings of pneumonia and no response to standard antimicrobial therapy. Most gave a history of contact with a patient with SARS in the previous 10 days. Within 3 months of the WHO declaring a state of global alert, a collaborative effort by several virology laboratories around the world had identified the causal agent as a previously unknown coronavirus, now known as the SARS coronavirus (SARS CoV). The SARS epidemic appears to have arisen from cross-species transfer of SARS CoV into humans in China, in November 2002, although the natural host of SARS CoV remains a matter of conjecture (possibly the civet cat). SARS had an alarmingly high mortality rate: of the order of 5% in the < 60-year age-group, but 15% in those > 60 years.

Diagnosis

Coronavirus-induced coryzal illness is a clinical diagnosis – the serotypes responsible do not grow in routine tissue culture. The diagnosis of SARS relies on the prevailing World Health Organization case definition. A positive laboratory diagnosis can be made by isolation of virus in tissue culture (dangerous!), detection of viral RNA by genome amplification (e.g. RT–PCR), or by demonstration of the presence of antibodies to the virus in patients' sera. These tests are currently carried out in reference laboratories only, and the optimal approach to diagnosis has not yet been agreed upon.

Treatment

There are no specific anticoronavirus agents. A variety of treatment regimens were used on patients during the SARS epidemic. Steroids are useful as anti-inflammatory agents, but there is controversy as to whether ribavirin (Topic E4) is effective or not.

Prevention

Management of SARS requires intensive **infection control** procedures to prevent **nosocomial** spread (Topic E9), and quarantining of contacts of SARS cases. Implementation of these measures across the world has led to the elimination of human SARS CoV infection at the time of writing, but it may re-emerge in the future.

Intensive efforts are being made to develop a vaccine.

B10 RESPIRATORY SYNCYTIAL VIRUS, HUMAN METAPNEUMOVIRUS, AND PARAINFLUENZA VIRUSES

Key Notes

Respiratory syncytial virus (RSV)	*Virology* A member of the paramyxovirus family, has a negative ssRNA genome encoding a handful of proteins including two surface glycoproteins, F (fusion) and G.
	Epidemiology Infection is seasonal (winter) and extremely common worldwide. Spread is via inhalation of droplets or direct inoculation into the pharynx, and may arise nosocomially.
	Consequences of infection. (i) 0–12 months – bronchiolitis or pneumonia; mortality is low except in babies with congenital lung or heart disease, or immunodeficiency; (ii) children and adults – upper respiratory tract infection, with cough predominant; (iii) elderly or immunosuppressed – influenza-like illness and pneumonia.
	Diagnosis Antigen detection on nasopharyngeal aspirates.
	Treatment Supportive unless patient is in high-risk group; ribavirin, given by continuous inhalation of aerosolized drug, reduces mortality in the latter.
	Prevention No vaccine currently available. Passive immunization with monoclonal anti-RSV antibody is available for high-risk babies. Strict infection control procedures should prevent nosocomial transmission.
Human metapneumovirus (hMPV)	hMPV grows poorly in tissue culture, and there are no routine diagnostic tests available at present. Second only to RSV as a cause of lower respiratory tract infection in small babies.
Parainfluenza viruses	There are four serotypes. Parainfluenza viruses are the commonest cause of croup, or acute laryngotracheobronchitis. Occasionally cause life-threatening lower respiratory tract infections, particularly in immunosuppressed patients. Ribavirin may be therapeutically useful.
Related topics	Upper respiratory tract infections (F6) Lower respiratory tract infections (F7)

Respiratory syncytial virus, human metapneumovirus, and parainfluenza viruses. Respiratory syncytial virus (RSV) is a major cause of **lower respiratory tract infection** in humans. In addition, several other members of the same virus family (the *paramyxoviridae*) also infect the human respiratory tract.

Respiratory syncytial virus

The name RSV is derived from the characteristic virus-induced cytopathic effect (multinucleate giant cells, or **syncytia**) in cell culture.

Virology

As a member of the **paramyxovirus** family, RSV has a negative ssRNA genome encoding, amongst other proteins, two surface glycoproteins, F (for fusion) and G (for glycoprotein). The F protein explains the ability of RSV to induce syncytia formation, i.e. to fuse cells together into a multinucleate giant cell.

Epidemiology

Infection with RSV is **seasonal**, occurring in the winter in temperate climates. Infection is extremely common – by 2 years of age, virtually the entire population has evidence of infection with this virus. Thus, even if only a small proportion of babies suffer severe disease, there will nevertheless be a large number requiring hospitalization due to RSV infection each winter. RSV causes **re-infections** throughout life at 3–4-yearly intervals (due to antigenic drift, see Topic B11). Transmission is via release of virus from the respiratory tract, and subsequent **inhalation** of droplets, or by **direct inoculation** into the pharynx, e.g. from contaminated hands. **Nosocomial spread** is a well-recognized hazard, the risk of acquiring RSV infection in this way being directly proportional to the length of hospitalization.

Consequences of infection

These are dependent on the age of the host. As with most viruses, the majority of infections are **asymptomatic**, or cause only mild disease.

1st year of life – bronchiolitis and pneumonia

RSV is a particularly important pathogen in the first 12 months of life, when it can cause severe lower respiratory tract disease such as **bronchiolitis** or **pneumonia**. Symptoms and signs include fever, poor feeding, increased respiratory rate and use of accessory muscles of respiration. One important factor predisposing to severe disease at this age is the small size of the airways – blockage arising from any inflammatory response will be more likely to occur the narrower the airway.

The **prognosis** for hospitalized babies with RSV is generally **good**, even for those admitted to hospital (mortality < 0.5%). However, certain groups of babies have a much worse prognosis, with mortality rates of > 50%. These include babies with:

* **congenital lung disease** (including that arising from prematurity);
* **congenital heart disease**;
* **congenital immunodeficiency**.

There is some controversy as to whether babies who suffer severe RSV infections will be more likely to develop asthma in older life. Certainly there are data to suggest these babies may suffer bouts of wheezing in childhood.

Children and adults – upper respiratory tract infection
Re-infection with RSV after year 1 is much less likely to cause lower respiratory tract disease. Presentation is usually with a **common-cold-like illness**, perhaps with more emphasis on cough and sore throat than on rhinitis.

The elderly and the immunosuppressed – influenza-like illness and pneumonia
Outbreaks of severe influenza and **pneumonia** in nursing homes for the care of the elderly may be due to infection with RSV as well as influenza viruses. This increased pathogenicity in the elderly may be related to declining host immune function. Indeed, younger patients who are **immunosuppressed**, e.g. following bone marrow or organ transplantation, can also succumb to severe lower respiratory tract RSV infection.

Diagnosis
This is usually made by **antigen detection** in respiratory samples using **indirect immunofluorescence** (Topic E1). The best sample is a **nasopharyngeal aspirate**, obtained by passing a fine tube up the nose into the nasopharynx, and aspirating the secretions using a syringe. This is a straightforward procedure in small babies, but is not usually performed in older children or adults, where a throat swab is the norm.

Antigen detection is a rapid diagnostic technique. In the RSV season, many laboratories will run an on-call diagnostic service to allow accurate diagnosis of babies who need admission to hospital, so that they can be properly isolated and nursed.

Treatment
Ribavirin (Topic E4) is a broad-spectrum antiviral agent that has activity against RSV. Treatment may reduce mortality considerably in the 'at-risk' groups of babies listed above. However, in otherwise immunocompetent and healthy babies, the benefit is minimal. Ribavirin must be inhaled in order to reach the site of virus replication, and is therefore administered via a small-particle aerosol generator, with the baby placed in a head-box into which the aerosol is released.

Prevention
There is currently no vaccine available for the prevention of RSV infection. At-risk babies can be offered passive protection with a monoclonal anti-RSV antibody (although this is prohibitively expensive). Current preventive strategies are focussed on minimizing spread of virus in a hospital setting where nosocomial transmission is known to be a risk. All RSV-infected babies should be nursed on the same ward, and looked after by the same staff, well away from other babies. Carers should be educated as to the importance of handwashing after handling patients.

Human metapneumovirus

This virus was discovered in 2001. It is also a paramyxovirus. It is difficult to culture in the laboratory (which is probably why it took so long to discover), and currently there are no widely available laboratory reagents to allow identification of infection by antigen detection. Preliminary genome detection data suggest that hMPV infection is common, and gives rise to clinical manifestations similar to those of RSV in small babies.

Parainfluenza viruses

These are also paramyxoviruses. There are four serotypes of parainfluenza virus (1–4). These are the commonest cause of the syndrome known as **croup**, more properly referred to as **acute laryngotracheobronchitis**. Croup presents with nonspecific symptoms such as a runny nose, followed by the development of an **inspiratory stridor** (as air passes through an inflamed and partially obstructed larynx), and a barking or **'croupy' cough**, classically worse at night. The degree of stridor may be sufficient to merit hospitalization, although with conservative management, the vast majority of cases make an uneventful recovery. Occasionally, the parainfluenza viruses may infect lower down in the respiratory tract, causing bronchiolitis or pneumonia. This risk is increased in immunodeficient or -suppressed patients, e.g. bone marrow transplant recipients. Diagnosis is by antigen detection, usually on nasopharyngeal aspirates. There are anecdotal reports that ribavirin may be useful in this setting.

B11 INFLUENZA VIRUSES

Key Notes

Virology	Influenza virus genomes have eight negative ssRNA segments, encoding 10 or 11 proteins. **Types** (A, B or C) of influenza viruses are based on the nature of the internal proteins. **Subtypes** of type A viruses are based on the nature of the surface glycoproteins, hemagglutinin (H, of which 15 have been described) and neuraminidase (N, nine).
Epidemiology	**Epidemics** occur each winter, due to **antigenic drift**. This arises from spontaneous mutations in key regions of the H and N proteins which allow escape from immunity in the population induced by the previous year's strain. **Pandemics** occur every 30 years or so, due to **antigenic shift**. Shift refers to the emergence of a new influenza A subtype, arising from **genetic reassortment** between human and avian strains.
Consequences of infection	The illness influenza comprises both **respiratory tract** and **systemic** symptoms and signs. The latter arise from virus-induced **interferon** induction. Complications include **primary viral pneumonia**, **secondary bacterial pneumonia**, myocarditis, and rare neurological manifestations. Mortality is increased in patients with pre-exisiting pulmonary, cardiac, endocrine, or renal disease, the immunosuppressed, and the elderly.
Diagnosis	**Isolation** of influenza viruses in tissue culture; **antigen detection** by immunofluorescence; **genome detection** by PCR-based assays; **serological** diagnosis by demonstration of an antibody rise in paired acute and convalescent sera.
Treatment	**Amantadine** blocks **uncoating** of influenza A (but not B) viruses, but is poorly tolerated due to its stimulatory side effects. **Neuraminidase inhibitors** are effective if given early enough, but are currently reserved for the treatment of severe and complicated influenza virus infections.
Prevention	By means of a subunit vaccine containing the H and N proteins from currently circulating types and subtypes. The vaccine is changed annually to take account of antigenic drift. UK policy is for selective vaccination of patients with underlying lung, heart, renal, endocrine or immunodeficiency disease, and all individuals aged > 65 years.
Related topics	Antiviral agents (E4) Upper respiratory tract Vaccines and infections (F6) immunoprophylaxis (E10) Lower respiratory tract infections (F7)

Respiratory viruses (3): Influenza viruses

Virology

Influenza viruses carry their genome as eight **segments** of negative single-stranded RNA, which encode 10 or possibly 11 proteins.

Influenza viruses are classified into **types:**

- A (widespread in nature, infecting a number of species including humans, swine, horses and especially birds), subdivided into subtypes;

- B (an exclusively human pathogen);
- C (not a serious human pathogen).

Typing is on the basis of the nature of the **internal** viral proteins (e.g. nucleo-capsid protein, see *Fig. 1*).

Subtyping of type A viruses is on the basis of the nature of the two surface glycoproteins, the **hemagglutinin** (H or HA) and **neuraminidase** (N or NA). There are 15 distinct H and nine N molecules which have been identified in influenza A viruses. Each H or N molecule differs by ≥ 20% in amino acid sequence from all other H or N molecules. Thus, reference to an influenza A virus must indicate which subtype it is, e.g. influenza A H1N1, influenza A H3N2 etc.

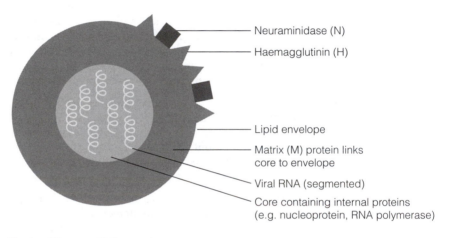

Neuraminidase (N)

Haemagglutinin (H)

Lipid envelope

Matrix (M) protein links core to envelope

Viral RNA (segmented)

Core containing internal proteins (e.g. nucleoprotein, RNA polymerase)

Fig. 1. Diagram of influenza virus

Epidemiology

Influenza has an unusual epidemic pattern. Each winter there is an outbreak (or **epidemic**) of infection, but superimposed on this annual cycle are unpredictable **pandemics** – huge outbreaks which spread around the world with very high rates of infection and associated mortality. Three pandemics occurred in the 20th century (*Table 1*) – in the 1918–19 pandemic, there were more deaths from influenza (estimated 20–40 million) than in World War I.

The explanation for this unusual pattern of infection lies in the phenomena of **antigenic drift** and **antigenic shift**

Antigenic <u>drift</u> is responsible for the generation of new epidemic strains of both influenza A and B viruses each winter, and arises from **random sponta-neous mutation** in the genes encoding the **H** and **N** molecules, resulting in ~ 1% differences in their amino acid sequences each year. These minor genetic

Table 1. Pandemic influenza of the 20th century

Year	Virus	Name
1918–1919	H1N1	Spanish flu
1956–1957	H2N2	Asian flu
1968	H3N2	Hong Kong flu

changes allow the virus to avoid some of the immune response induced by circulation of the previous year's virus, and explains why each year a large percentage of the population can become infected by influenza viruses.

Antigenic shift gives rise to the pandemic strains every 30 years or so. This occurs only in influenza A viruses. The mechanism involves co-infection of a cell with two different subtypes of virus, e.g. a human and an avian strain, with subsequent shuffling between the eight gene segments, a process known as **genetic reassortment**. This leads to the production of a new viral subtype in which the genes encoding the internal proteins are derived from the human 'parent' virus – and therefore the 'new' virus is well adapted to growth within human cells – but with one or both of the external glycoprotein genes derived from the avian 'parent' strain. The new surface protein(s) are so different from those previously circulating amongst humans that the population has no effective immunity against them, hence infection is very widespread, and a new pandemic emerges.

Influenza B viruses undergo antigenic drift, but not antigenic shift – there is no animal reservoir of B viruses with which human strains can undergo genetic reassortment.

Three human influenza viruses currently co-circulate in each winter: influenza A viruses H1N1 and H3N2, and influenza B. In any one season, one of these three viruses tends to dominate, but the dominant virus may change from season to season. The emergence of a new pandemic subtype of influenza A is unpredictable. There have been major scares recently, with several cases of H5N1 infection of humans reported in Hong Kong (1997), and Vietnam and Thailand (2004). However, the human cases of H5 infection appear to have arisen from direct transfer of viruses from infected chickens to humans, and as these viruses had not undergone genetic reassortment, they were not well suited for human-to-human spread. The outbreaks have been controlled by mass slaughter of chickens.

Consequences of infection

The disease 'influenza' classically comprises both **respiratory tract** symptoms and signs (runny nose, sore throat, irritating cough) and a **systemic response** [fever, headache, muscle aches and pains (myalgia), severe malaise]. The latter arise through viral induction of cytokines and interferons which circulate in the bloodstream.

Life-threatening complications of influenza virus infection include:

(i) **Primary influenzal pneumonia**. This arises from spread of the virus itself into the lower respiratory tract, with a potent inflammatory response filling the alveoli with cells and fluid. This can occur in previously healthy individuals of any age.

(ii) **Secondary bacterial pneumonia**. This is more common than (i). Virus-induced lysis of respiratory epithelial cells effectively strips off the mucus-secreting cells, and also cells bearing cilia (small, hairlike structures which 'beat' particulate matter upwards), which are two important defense mechanisms that help to prevent inhalation of particulate matter into the lower respiratory tract. This damage therefore enhances access of bacteria to the lower respiratory tree, resulting in pneumonia, and an intense polymorphonuclear infiltrate into the air spaces. This complication tends to occur in the elderly, and in patients with pre-existing lung disease, e.g. chronic bronchitis.

(iii) Other, less common, complications include **myocarditis**, **encephalitis**, and **Guillain-Barré syndrome**.

Diagnosis

Influenza viruses can be isolated in tissue culture, but may take several days to induce a cytopathic effect. Antigen detection using immunofluorescence is a more rapid approach (Topic E1). RT–PCR assays for the amplification and detection of influenza virus RNA are increasing in popularity, being both sensitive and fast. Retrospective diagnosis can be made serologically by demonstrating an antibody rise in sera taken several days apart.

Treatment

Amantadine (Topic E4) inhibits **uncoating** of influenza A (but not B) viruses. Amantadine is effective both as a form of therapy, and for prophylaxis of influenza A in unvaccinated individuals. However, the CNS stimulatory side-effects of this drug make it poorly tolerated, especially in the elderly.

Zanamavir and **oseltamivir** are **neuraminidase (NA) inhibitors**, a relatively new class of antiviral drugs. These prevent release of newly formed virus particles from the cell surface. They have activity against all known subtypes of influenza NA.

Current recommendations in the UK are to use NA inhibitors for the prevention and treatment of influenza-associated complications in at-risk individuals (i.e. those targeted for vaccination, see below), rather than for uncomplicated influenza virus infection alone in otherwise healthy individuals.

Prevention

Currently licensed influenza vaccines are **subunit vaccines** containing the surface proteins from each of the co-circulating viruses. They are ~ 70% effective in protecting against infection, but protection lasts for one winter only, as the circulating viruses exhibit antigenic drift from year to year. Vaccines are revised each year, on the basis of recommendations made by the World Health Organization, and **at-risk individuals** (see below) need a dose of vaccine each autumn.

The UK policy is for selective rather than universal vaccination. Immunization is therefore targeted at patients at increased risk of serious infection:

(i) all individuals aged > 65 years;
(ii) all aged > 6 months in the following risk groups:
  • chronic respiratory disease, including asthma;
  • chronic heart disease;
  • chronic renal disease;
  • diabetes mellitus and other endocrine disorders;
  • immunosuppression due to disease or treatment.

Immunization is also recommended for individuals within residential homes or other institutions where rapid spread is likely to follow introduction of infection, and for healthcare workers in acute care hospitals.

B12 ADENOVIRUSES

Key Notes

Virology	Nonenveloped dsDNA viruses with icosahedral capsids. More than 50 serotypes are subdivided into six subgenera, A–F.
Epidemiology	Infection with adenoviruses (Ad) 1–7 is extremely common, but with other serotypes less so. Most Ad are excreted from the respiratory tract, and also in feces. Nosocomial transmission may arise if instruments contaminated with virus are not properly decontaminated.
Consequences of infection	(i) Upper respiratory tract infection, e.g. pharyngitis, tonsillitis; (ii) lower respiratory tract infection, e.g. pneumonia; (iii) infections of the eye – conjunctivitis, keratitis, particularly subgenus B viruses, e.g. Ad 8; (iv) gastroenteritis – associated specifically with subgenus F viruses, Ad 40 and Ad 41; (v) multisystem infections in the immunosuppressed.
Diagnosis	Most (but not all) Ad can be isolated in tissue culture, but this is slow. Antigen or genome detection are more rapid assays. Serological diagnosis is by demonstration of an antibody rise in acute and convalescent sera. Diagnosis of infection with Ad 40 and 41 is by electron microscopy or antigen detection using stool samples.
Treatment	No specific anti-Ad antiviral drugs exist, but ribavirin has had limited success in the treatment of serious infections in immunosuppressed patients.
Prevention	No vaccine widely available. Infection control procedures are important in preventing nosocomial spread.
Related topics	Upper respiratory tract infections (F6) Lower respiratory tract infections (F7) Gastroenteritis and food-poisoning (F8)

Virology

The Adenoviridae are nonenveloped, dsDNA viruses with icosahedral capsids. There are > 50 serotypes grouped into six subgenera (labelled A–F) based on a variety of viral properties. Viruses in each subgenus therefore behave slightly differently from those in the other subgenera.

Epidemiology

Infection with serotypes 1–7 is extremely common and occurs in childhood. Infection is usually acquired from virus originating in the respiratory tract, e.g. in droplets, although adenoviruses (Ad) are also excreted from the gastrointestinal tract. Ad particles are not inactivated by many standard disinfection protocols. Thus, contamination of instruments (e.g. tonometers, used to measure intraocular pressure) is a well-recognized cause of outbreaks of adenoviral conjunctivitis, as the virus is inadvertently spread nosocomially within a hospital setting.

Consequences of infection

In immunocompetent hosts, most adenovirus infections are mild and transient. Infection in immunosuppressed patients may, however, be life-threatening. Syndromes arising from infection include:

(i) **Upper respiratory tract infection** – often in young children, Ad 1–7 cause pharyngitis, tonsillitis, adenoidal enlargement, conjunctivitis, nasal congestion, cough.

(ii) **Lower respiratory tract infection**, such as laryngotracheobronchitis, bronchiolitis, pneumonia. This is relatively unusual in children, but may occur in outbreaks in young adults crowded together, e.g. military recruits.

(iii) **Infections of the eye**. These may present as conjunctivitis (inflammation of the conjunctivae), which usually occurs with other features such as pharyngitis, or, more seriously, as keratitis (infection of the cornea). Particular subtypes (e.g. Ad 8 from subgenus B) are associated with epidemic keratitis. Resolution of infection may be accompanied by the development of corneal opacities and scarring.

(iv) **Gastroenteritis**. Ad 40 and 41, the only members of subgenus F, are the second commonest cause (behind rotaviruses) of diarrhea and vomiting in young children. These particular subtypes are thus referred to as the **enteric adenoviruses**.

(v) **Multisystem disease in immunosuppressed patients**, e.g. bone marrow recipients. Adenovirus infections in this patient group can give rise to hepatitis or pneumonitis, which carry a high mortality.

Diagnosis

Most Ad will grow in standard cell cultures, so swabs taken from the site of infection can be sent to the laboratory for virus isolation. However, growth is fairly slow (e.g. 10–14 days), so antigen detection by immunofluorescence, or genome detection by DNA amplification are preferred. Serology can also be useful, by demonstration of a rise in antibody titer in samples taken a few days apart. Ad 40 and 41 do not grow in culture, but can be identified in feces by electron microscopy or antigen detection.

Treatment

Only a very small minority of Ad infections are serious enough to warrant therapeutic intervention. No antiviral agents have undergone proper clinical trials in these settings, but there are anecdotal reports of the successful use of ribavirin, or, more recently, cidofovir (Topic E4).

Prevention

Vaccines have been tried in the setting of military recruits, but these are not generally available. Adherence to disinfection protocols is essential in eye clinics to prevent nosocomial spread of adenoviral conjunctivitis.

B13 MEASLES AND MUMPS VIRUSES

Key Notes

| Measles virus |

Virology
A paramyxovirus, therefore has negative ssRNA.

Epidemiology
Humans are the only host. Transmission is via aerosols released from the respiratory tract. The World Health Organization estimates there are 10^6 deaths annually due to measles, mostly in the developing world.

Consequences of infection
Measles is not a trivial illness. Patients present with a prodromal febrile illness with respiratory tract symptoms, followed by a red rash appearing first on the face and spreading down the trunk. Serious complications include secondary bacterial infections (e.g. otitis media, purulent conjunctivitis, bronchopneumonia), acute post-infectious encephalitis (1 in 5000 cases, has a poor prognosis), subacute sclerosing panencephalitis (rare, occurs years later), and giant cell pneumonia (in immunosuppressed patients, almost always fatal). In undernourished children in the developing world, significant mortality arises from secondary infections such as diarrhea, tuberculosis and pneumonia.

Diagnosis
Demonstration of a specific IgM antibody response; antigen detection in respiratory secretions.

Treatment and prevention
There is no specific form of therapy. Prevention is by use of a live attenuated vaccine, usually combined with mumps and rubella (MMR) and administered at 12–18 months of age, with a preschool booster.

| Mumps virus |

Virology
Mumps is a paramyxovirus.

Epidemiology
Infection is spread via respiratory secretions. Most infections occur in children, but infection at an older age is associated with significantly more complications.

Consequences of infection
Mumps usually presents as an infection of the parotid glands (parotitis). Serious complications include meningitis, meningoencephalitis (better prognosis than measles encephalitis), orchitis and oophoritis, and pancreatitis, the latter complications arising only in post-pubertal hosts.

Diagnosis
Demonstration of an IgM antibody response, or by virus isolation from buccal swab or urine.

Treatment and prevention
Management is supportive only. Mumps is preventable by means of a
live attenuated vaccine (part of MMR).

Related topics	Vaccines and	Other CNS infections (F4)
	immunoprophylaxis (E10)	

Measles virus

Virology
Measles virus has a genome of ssRNA of negative polarity. It belongs to the
Paramyxoviridae family (as does mumps virus).

Epidemiology
Humans are the only natural host. Prior to the introduction of vaccination, widespread epidemics occurred globally every 2–3 years. Measles is extremely conta-
gious. Transmission is via aerosols released from the respiratory tract.

Infection rates vary around the world depending on local vaccination policies.
In the developing world, most infections occur in the first 5 years of life. The
World Health Organization estimates there may be $> 10^6$ measles-related deaths
per year, the vast majority arising in children in the poorer countries of the
world.

Consequences of infection
Typical measles presents with a **prodromal illness** of **upper respiratory tract
symptoms** (runny nose, cough, conjunctivitis) and a **fever**. Koplik's spots,
bluish-white spots on the buccal mucosa (the inside of the cheeks), appear
during this phase of the illness, and fade by the time the rash appears. Two to
three days later, the characteristic **red maculopapular blotchy rash** appears,
firstly on the face and neck, then spreading down the trunk, and finally
involving the whole body surface. The rash fades over the following 4 or 5
days.

Children with measles are usually quite unwell, but it is the legion of **complications** that sets measles out as a feared infection. These include:

(i) **Secondary bacterial infection** of the necrotic epithelium of the respiratory
tract causing purulent conjunctivitis, laryngotracheitis, otitis media, bronchitis and, most serious, bronchopneumonia.

(ii) **Encephalitis**. There are several clinical forms of this, the most common (1
per 1000–5000 cases of measles) being **acute encephalitis**. This presents
~ 1 week after the onset of the initial illness with a recrudescence of fever,
headache, and irritability, followed by **declining levels of consciousness**.
Mortality is high, and survivors are left with severe brain damage. Measles
virus is not found in the brain tissue, so the damage is thought to arise from
an aberrant immune response to the virus which cross-reacts with brain
substance. This is therefore referred to as a **'post-infectious' encephalitis**.
Subacute sclerosing panencephalitis (SSPE) is another, but much rarer (1
in 10^6 cases of acute measles), form. This presents many years after the
initial infection with generalized intellectual deterioration and motor
dysfunction. In SSPE, measles virus is present in the brain, and the disease
arises through viral-induced death of brain cells.

(iii) **Giant cell pneumonia**. This is a viral (as opposed to a secondary bacterial) pneumonia arising in immunosuppressed patients (e.g. children with leukemia), which is almost always fatal.

(iv) In the developing world, mortality associated with measles is related to the immunosuppressive effects of the infection, resulting in a variety of life-threatening secondary infections such as bacterial pneumonia, tuberculosis, and diarrhea.

Diagnosis
Clinical suspicion of measles should be confirmed by demonstration of a specific **IgM antibody** response. Measles antigen can be detected in respiratory secretions by immunofluorescence. Virus can be isolated in routine tissue culture.

Treatment
There is no specific antiviral agent with activity against measles virus.

Prevention
A live attenuated vaccine is available for prevention of measles virus infection. This is usually given as part of the triple **measles-mumps-rubella** (MMR) vaccine, recommended for all children at age 12–18 months. A single dose may not be sufficient to provide long-lasting protection in all vaccinees, and therefore a preschool booster dose is also advisable. Owing to the highly infectious nature of measles virus, high levels of vaccination coverage (>85%) are needed to prevent circulation of infection. The recent decline in MMR vaccination rates in the UK, due to a misplaced anxiety over possible complications of vaccination, has resulted in the re-emergence of outbreaks of measles, with consequent damage to children suffering severe complications of infection.

Mumps virus

Virology
Mumps virus is also a paramyxovirus with a negative ssRNA genome.

Epidemiology
Transmission is via respiratory secretions. Although usually a disease of childhood, infection can occur at any age. Many of the complications of mumps (see below) are age-related, being more common in post-pubertal children and adults.

Consequences of infection
Subclinical infection occurs in ~ 30% of cases. In symptomatic cases, there is often a nonspecific prodrome of malaise and fever lasting up to a week, most commonly followed by **parotitis**, i.e. inflammation of the parotid salivary glands (uni- or bilaterally), which become swollen and tender. The submandibular and sublingual salivary glands are also occasionally involved. The most important complications of mumps are:

(i) **Central nervous system (CNS)** infection. This usually manifests as **acute meningitis**. This may occur in the absence of parotitis. Virus is present in the meninges, and may spread into the brain substance to cause a **meningo-encephalitis**. The prognosis of this is much better than with measles acute encephalitis, but nevertheless there are occasional fatal cases, and some survivors are left with cerebral damage, most commonly nerve deafness.

(ii) **Orchitis** and **oophoritis** (infection of the testes or ovaries). These occur only in post-pubertal children and adults. Resultant testicular atrophy may cause sterility, but the risk of this is low.

(iii) **Pancreatitis**. Again, this is more common in adults than children

Diagnosis

Clinically suspected cases should be confirmed by demonstration of a mumps **IgM** antibody response. Virus can be isolated from swabs of the buccal mucosa, cerebrospinal fluid or urine.

Treatment

There is no specific antiviral agent with activity against mumps virus.

Prevention

A live attenuated vaccine is included in the MMR vaccine. The rationale for universal vaccination against mumps virus is in order to prevent the central nervous system complications of infection.

B14 ROTAVIRUSES AND CALICIVIRUSES

Key Notes

Virology	Rotaviruses belong to the Reoviridae, contain dsRNA, and have a characteristic wheel-shaped morphology. The Caliciviridae include noro- and caliciviruses, and have positive ssRNA.
Epidemiology	Rotaviruses are the commonest cause of gastroenteritis in children < 5 years old, resulting in > 1×10^6 deaths per year worldwide. Transmission is via the fecal–oral route, and in temperate climates, is more common in winter. Noroviruses are the commonest cause of outbreaks of nonbacterial gastroenteritis in older children and adults. Transmission is via contaminated food or water, and person-to-person spread. Nosocomial outbreaks in hospital wards and nursing homes are a particular problem.
Consequences of infection	Gastroenteritis (defined as acute onset vomiting and diarrhea) is the predominant clinical manifestation of infection, leading to potentially life-threatening fluid lens and electrolyte imbalance. Rotavirus-infected children may also be febrile, and take up to 7 days to recover. Norovirus gastroenteritis usually resolves in 24–48 hours.
Diagnosis	Diagnosis is by demonstration of virus in feces, by electron microscopy, antigen or genome detection techniques.
Treatment	Management requires fluid and electrolyte replacement.
Prevention	Prevention is by appropriate infection control practices. Rotavirus vaccines are in development – a new one has been licensed for use in Mexico in 2005.
Related topics	Gastroenteritis and food-poisoning (F8)

Rotaviruses and caliciviruses

The term **'gastroenteritis'** describes a syndrome of acute onset vomiting and diarrhea, for which there is a large number of possible causes, including several different virus infections This chapter will describe the salient features of two of the most important viral causes, namely **rotaviruses** and **caliciviruses**. Enteric adenoviruses are considered in Topic B12.

Virology

Rotaviruses contain dsRNA genomes, present in 11 discrete segments, and belong to the Reoviridae family. They have a characteristic 'wheel' morphology (hence the name rotavirus), with an inner core connected to an outer shell via a number of protein 'spokes'.

The Caliciviridae are an extensive viral family all of which carry a positive ssRNA genome. The name is derived from the cup-like hollows (Greek, calyx) on the surface of viral particles. The family includes two important genera – the **noroviruses** (previously known under a number of different names, including Norwalk viruses or small round-structured viruses) and the **caliciviruses.**

Epidemiology

Diarrheal illness is second only to respiratory tract infection as the major cause of death worldwide, resulting in millions of infections annually, and an annual estimated $>10^6$ deaths, predominantly in the developing world, although there is also significant mortality in the developed world. Most deaths occur in children, but the elderly are also vulnerable to severe illness.

Rotaviruses are the most commonly identified pathogen associated with gastroenteritis in children aged < 5 years (enteric adenoviruses are the second commonest). About 125×10^6 cases and 440,000 deaths occur annually. Severe illness is unusual after the age of 3 years. Transmission is via the fecal–oral route. Infection is most common in the winter months.

Caliciviruses may cause gastroenteritis in children, but can also infect older individuals. The noroviruses in particular appear to affect older age groups, and cause significant outbreaks in many settings including nursing homes, on cruise ships, and in hospital wards. Transmission is usually by contaminated food (especially shellfish) and water, but aerosol spread has also been suggested. Outbreaks occur throughout the year, but with a preponderance in winter.

Consequences of infection

Rotavirus infection in children presents with frequent vomiting and fever, which may precede watery diarrhea by 48 h. Symptoms may last for 5–7 days. In severe infections, excessive fluid loss may result in dehydration and electrolyte imbalance.

Calici- and norovirus infections also result in vomiting and diarrhea, but illness usually resolves in 24–48 h.

Diagnosis

Electron microscopy of diarrheal feces (or vomitus) was the mainstay of diagnosis of viral gastroenteritis, as the viral culprits are excreted in huge numbers in feces, and can be identified by their characteristic morphologies. However, there are now antigen detection kits available for rota-, noro- and adenoviruses, which obviate the need for electron microscopy, and PCR-based genome detection assays are also being increasingly used.

Treatment

There is no specific therapy for viral gastroenteritis. Management is dependent largely on oral (or intravenous, if necessary) rehydration with solutions containing glucose and electrolytes.

Prevention

Gastroenteritis viruses are easily transmissible from person to person, and may cause nosocomial outbreaks unless infection control policies are correctly followed. The mainstay of these is adherence to strict hand-washing protocols for hospital staff and visitors. Patients should ideally be nursed in single rooms. Admissions to, and transfers from, affected wards should be restricted during outbreaks.

Considerable effort has been expended in the development of an effective rotavirus vaccine. One particular vaccine, containing a rhesus-human reassortment virus, showed very promising results in large-scale clinical trials, and was licensed in the USA in 1998. Unfortunately, it had to be withdrawn because of an unacceptably high rate of serious side-effects. A new vaccine has been licensed for use in Mexico in 2005. It is to be hoped that this proves to be an effective and safe vaccine, and its use extended across the world soon.

B15 ENTEROVIRUSES

Key Notes

Virology	Enteroviruses have a positive ssRNA genome. They include polioviruses, coxsackie A and B viruses, and echoviruses.
Epidemiology	Enteroviruses enter and exit the host through the enteric tract, and thus transmission is fecal–oral.
Consequences of infection	Infection is common, and usually asymptomatic, but can lead to a number of clinical manifestations, including skin rashes with fever, ulceration in the mouth (herpangina), hand, foot and mouth disease, poliomyelitis, meningitis, encephalitis, myo- and pericarditis, and upper and lower respiratory tract infection.
Diagnosis	Diagnosis is by virus isolation from throat swab, feces, or CSF, although not all enteroviruses grow in routine tissue culture. RNA detection by RT–PCR is an alternative.
Treatment	There is no specific treatment.
Prevention	Poliovirus infection is preventable by means of a live attenuated (Sabin) or an inactivated (Salk) vaccine. A global poliomyelitis eradication campaign, run by the World Health Organization, is nearing completion.

Related topics

Vaccines and immunoprophylaxis (E10)

Meningitis (F3)
Other CNS infections (F4)

Virology

The enteroviruses are a large group of viruses (>70 currently described), all with positive ssRNA genomes. The group contains the **polioviruses**, **coxsackie A** and **B viruses**, **echoviruses**, and one or two others which are given the prefix EV, and a number. These different viruses are serologically distinct, i.e. infection with one enterovirus does not induce protective immunity against any of the other enteroviruses. The group belongs within the family Picornaviridae ('pico-RNA' = small RNA), along with the rhinoviruses (Topic B9).

Epidemiology

These viruses are termed enteroviruses because of their transmission via the enteric (gastrointestinal) tract – they enter through the mouth, and are excreted in the feces. This route of transmission is termed 'fecal–oral', as infection is acquired through ingestion of contaminated food or water.

Enterovirus infection occurs commonly worldwide. However, a World Health Organization (WHO)-led polio vaccination campaign is edging towards the global elimination of poliovirus infection.

Consequences of infection

Enteroviruses can infect a number of different body systems, giving rise to a wide range of clinical syndromes, although despite their name, they have not been proven to cause disease of the gastrointestinal tract itself. Most commonly, infection is asymptomatic. Clinical manifestations include:

(i) *General.* **Fever** and nonspecific **rash** – often in children.

(ii) *Skin and mucous membranes.* Painful **ulceration** in the mouth is known as **herpangina**, usually due to coxsackie A viruses. If the ulcers are also present on the hands and feet, the disease is known as **hand, foot and mouth disease**. (Note: this is *not* the same infection as foot and mouth disease of cattle). Some enteroviruses (coxsackie A24, EV70) cause **conjunctivitis**.

(iii) *Central nervous system.* The disease **poliomyelitis** is a syndrome of flaccid paralysis arising from virus-induced death of lower motor neurones. It occurs in ~ 1–2% of poliovirus infections. Coxsackie A and B and echoviruses are now the commonest viral causes of **meningitis** (since the advent of MMR vaccination has virtually eliminated mumps meningitis). Spread into the brain substance may result in **encephalitis**.

(iv) *Others.* Coxsackie B viruses are the commonest viral cause of **pericarditis**, and may also infect the heart muscle itself (**myocarditis**). Infection of the chest wall musculature may give rise to severe chest pain (**Bornholm's disease**). Upper respiratory tract infection may present as a **common-cold like illness**, and in small babies, enteroviruses may also cause severe lower respiratory tract infections such as **bronchiolitis** and **pneumonia**. Infection in the **newborn** can be life-threatening owing to the multisystem nature of the infection in this patient group (e.g. myocarditis, hepatitis, encephalitis).

Diagnosis

Many, but not all, enteroviruses will grow in routine **tissue culture**, so viral culture of throat swabs, feces, or cerebrospinal fluid (in cases of meningitis) should be instituted in suspected cases. A more sensitive (but also more expensive) approach is **enteroviral RNA detection** by genome amplification. Demonstration of coxsackie **virus-specific IgM** in an acute serum sample, or a 4-fold rise in enteroviral-specific IgG in paired acute and convalescent serum samples, may also be possible.

Treatment

There is no specific antiviral agent for the treatment of enterovirus infection.

Prevention

Live attenuated (Sabin) or **killed** (Salk) **poliovirus vaccines** are effective in preventing infection with the three types of poliovirus, but do not provide cross-protection against infection with other enteroviruses. The WHO poliovirus elimination campaign is now in its end stages – in 2003, poliovirus infections were described in only six countries.

B16 RUBELLA VIRUS AND PARVOVIRUS B19

Key Notes

Rubella virus

Virology
Rubella virus is a positive ssRNA virus, belonging to the Togaviridae.

Epidemiology
Infection usually occurs in childhood, with transmission via respiratory secretions.

Consequences of infection
In children, rubella is a mild illness, with a nonspecific rash, fever, and cervical lymphadenopathy. In adults, especially females, there may also be arthralgia and arthritis. In pregnant women, the virus crosses the placenta and infects the fetus. In early pregnancy, this may result in major developmental abnormalities involving the CNS, eye, and heart (the congenital rubella syndrome).

Diagnosis
Diagnosis is by demonstration of an IgM antibody response.

Treatment and prevention
There is no specific treatment – women with rubella in early pregnancy may opt for therapeutic abortion. Rubella is preventable by means of a live attenuated vaccine.

Parvovirus B19

Virology
This is a ssDNA virus, belonging to the family Parvoviridae.

Epidemiology
Transmission is usually via the respiratory route, but parvovirus can be transmitted by blood transfusion.

Consequences of infection
In children, infection may give rise to erythema infectiosum, with a prominent malar rash. Disease in adults may be clinically indistinguishable from rubella, with a rash and joint pains, the latter more common in females. Parvovirus arthritis may be multiarticular, prolonged, and severe. Parvovirus infects rapidly dividing cells in the bone marrow, and in patients with chronic hemolytic anemia can cause a life-threatening bone marrow arrest (aplastic crisis). Infection in pregnancy increases the risk of spontaneous miscarriage, and may give rise to fetal hydrops, but does not cause congenital malformations.

Diagnosis
Diagnosis is by demonstration of an IgM antibody response or by detection of parvovirus DNA.

> *Treatment and prevention*
> There is no specific treatment. There is no vaccine available for prevention.
>
> **Related topics**   Infections in pregnancy and neonates (F15)

Rubella virus and parvovirus B19

Virologically speaking, these two viruses are very different from each other – one is an RNA virus, the other contains DNA – but they are grouped together because of similarities in their clinical manifestations of infection.

Rubella virus

Virology
Rubella virus has a positive ssRNA genome and belongs to the Togavirus family.

Epidemiology
Rubella virus infection usually affects children aged 6–12 years. However, high vaccine coverage rates (see below) in many countries have resulted in prevention of circulation of virus amongst children. Exposure to rubella in the community is therefore rare, unless the virus is introduced (e.g. from abroad), and the average age at infection is much higher, as most young children are immune. Transmission is via respiratory secretions.

Consequences of infection
(i)   Infection may be asymptomatic.
(ii) In children, the disease 'rubella' (also known as German measles) presents with fever, a nonspecific red **rash**, and **swollen lymph glands**, particularly in the neck. Illness is not usually severe, and recovery is complete after a few days.
(iii) In adults, the disease is similar, but, especially in females, an additional feature is painful joints (**arthralgia**), or even visibly swollen joints (**arthritis**).
(iv) The most important clinical consequence of rubella infection is in **pregnant women**, where virus may cross the placenta and give rise to severe damage to the developing fetus. The **congenital rubella syndrome** encompasses a wide range of abnormalities including growth retardation, cardiac defects, cataracts, deafness and mental impairment. The likelihood of damage to the fetus is highly dependent on the timing of the maternal infection – if in the first 12 weeks of the pregnancy, then multiple defects are almost certain, but if after 18 weeks, then the chance of any damage is almost nil.

Diagnosis
The cornerstone of rubella diagnosis is serology (the detection of **IgM anti-rubella antibodies** in a serum sample taken soon after the onset of illness). It is imperative that in any pregnant woman with a rash, a diagnosis of rubella is either confirmed or refuted by appropriate testing. In children suspected of having rubella, the importance of confirming the diagnosis lies in monitoring

the effectiveness of vaccination. It is possible to detect rubella IgM in saliva, which is a much easier specimen to take from a small child than peripheral blood.

Treatment
There is no effective antiviral against rubella virus, and, for a pregnant woman who contracts rubella early in pregnancy, the choice is between therapeutic abortion or continuing with the pregnancy with a high risk of fetal damage.

Prevention
Rubella vaccine is a **live attenuated vaccine**, recommended in the UK for all 12–18-month-old children as part of the measles-mumps-rubella (MMR) vaccine. The aim of rubella vaccination is to prevent the tragedy of babies born with congenital rubella. This is achieved by interrupting the circulation of rubella in the community (hence the need to vaccinate both boys and girls) such that any nonimmune woman has very little risk of coming into contact with the virus whilst pregnant, and also by achieving high rates of immunity such that very few women of childbearing age are not immune.

Parvovirus B19   *Virology*
This has a ssDNA genome and belongs to the family Parvoviridae.

Epidemiology
Acquisition of infection is usually in childhood. Transmission is via the respiratory route. As there is a viremic stage during infection, parvovirus can be transmitted by blood transfusion.

Consequences of infection
The majority of parvovirus infections are **subclinical**. However, infection may result in a variety of clinical manifestations:

(i) **Slapped cheek syndrome**, also known as **fifth disease** or **erythema infectiosum**. This syndrome of fever and a red rash most prominent on the cheeks (i.e. a **malar** rash) arises in children. Rash is less prominent in adults – where parvovirus infection may be clinically indistinguishable from rubella virus infection.

(ii) **Arthralgia/arthritis**. As with rubella, this is much commoner in adults, especially females. Parvovirus arthritis may be quite severe and prolonged, mimicking rheumatoid arthritis, although it is usually self-limiting.

(iii) **Aplastic crisis**. Parvovirus preferentially infects rapidly dividing cells such as those in the bone marrow, causing an arrest of the production of blood cells of all types. In otherwise healthy individuals, this is not clinically significant, but in patients with chronic hemolytic anemia, the resultant aplastic crisis with loss of red cell formation is life-threatening.

(iv) **In pregnancy**. The consequences of acute parvovirus infection in pregnancy are very different from those of rubella. Parvovirus can cross the placenta, and infected fetuses may become anemic, sufficient to develop heart failure. This presents as an accumulation of fluid (hydrops fetalis). Intrauterine blood transfusion can be life-saving. In contrast to rubella, however, fetal infection does not cause developmental abnormalities, but there is an increased risk of spontaneous miscarriage.

Diagnosis

Acute infection should be confirmed by demonstration of parvovirus-specific **IgM antibodies**, or by detection of parvovirus DNA by genome amplification. It is particularly important to make an accurate diagnosis in a pregnant woman with a rash, where the differential diagnosis includes rubella infection.

Treatment

There is no specific treatment for acute parvovirus infection.

Prevention

There is no vaccine available.

B17 ARENA-, FILO-, FLAVI-, AND RHABDOVIRUSES

Key Notes

Exotic infections

A number of viruses, each with distinct geographical distributions, may give rise to hemorrhagic fevers (Lassa, Ebola, Marburg, Yellow fever, Dengue HF) or encephalitis (West Nile, Rabies), both life-threatening diseases. Returned travellers with suspected viral hemorrhagic fever should be transferred to high security infectious diseaseunits for management.

Arenaviruses

These contain ambisense ssRNA. Rodents are the natural host for Lassa fever virus, which is excreted in urine. Human infection presents with fever and pharyngitis, and can progress to severe hemorrhagic manifestations, with mortality rates of 2–25%. Treatment is with ribavirin.

Filoviruses

These have negative ssRNA genomes. The natural hosts for Ebola and Marburg viruses are not known. Human infection arises from contact with monkeys, or person-to-person spread. Mortality rates are > 50%, with no specific antiviral therapy. Ebola vaccines are undergoing clinical trials.

Flaviviruses

These have a positive ssRNA genome. Yellow fever virus is transmitted to humans via mosquitoes from a monkey reservoir, and can give rise to severe multisystem disease, including liver failure. Overall, mortality rates are 1–5%. Infection is preventable by a live attenuated vaccine.

There are four dengue virus serotypes. Dengue hemorrhagic fever occurs in children, with a mortality of 1–5%.

West Nile virus naturally infects birds, and is transmitted to humans via mosquitoes. Recent spread into the USA has occurred. Disease is usually mild, but a fatal encephalitis may occur in the elderly or immunosuppressed.

Rhabdoviruses

These have negative ssRNA genomes. Rabies virus occurs in many warm-blooded animals worldwide. Humans acquire infection after being bitten by rabid animals, resulting in a fatal encephalitis. Inactivated rabies vaccine is effective in preventing disease, given either pre-exposure to those at risk, or post-exposure following a bite, when human rabies immunoglobulin should also be administered.

Related topics

Other CNS infections (F4)

Exotic infections

This chapter deals with a number of relatively rare but potentially serious virus infections which occur naturally in certain specified areas of the world, but for the most part are not endemic in the UK, Europe or the USA. There is, of course, always the possibility that a traveller may inadvertently bring one of these infections back into his/her country of origin. The viruses concerned belong to a diverse group of virus families including the **Arena-, Filo-, Flavi-,** and **Rhabdoviridae**. The diseases they give rise to can be broadly categorized into two groups – **hemorrhagic fevers** (Lassa, Ebola, Marburg, Yellow fever, Dengue) and **encephalitis** (Rabies, West Nile). Occurrence of these infections in a returned traveller poses an infection risk to healthcare staff, especially those who handle blood and other patient samples. Patients with suspected viral hemorrhagic fever should be transferred to specialized high security infectious diseases units for investigation and management.

Arenaviruses

These viruses contain segments of ambisense (i.e. both positive and negative) ssRNA. The most widely known is **Lassa fever virus**, endemic in West Africa. **Rodents** are the natural reservoir, with virus being excreted in urine. Human infection presents with an insidious onset of fever and malaise – an influenza-like illness. Severe illness is heralded by high fever, pharyngitis, hypotension and hemorrhage. Mortality varies from 2 to 25% in different outbreaks. Person-to-person spread is associated with overcrowding and poor infection control. Treatment is with ribavirin (Topic E4).

Filoviruses

Ebola and **Marburg** viruses belong to the family *Filoviridae*. They have negative ssRNA genomes. Geographic distribution is limited to sub-Saharan, central and eastern Africa, although the natural host for these viruses is not known. Human infection has been acquired from contact with monkeys, or via person-to-person spread. Infection has a dramatic presentation with high fever, prostration, headache, abdominal pain and hemorrhage, and most outbreaks have had very high mortality rates: > 50% for the Zaire and Sudan strains of Ebola virus. Treatment is supportive only. Vaccines against Ebola virus are currently undergoing clinical trials.

Flaviviruses

The **Flaviviridae** are a large group of viruses with positive ssRNA genomes. Exotic members of this family include the hemorrhagic fever viruses yellow fever and dengue, and also West Nile virus, which causes encephalitis. All three are transmitted to humans from an animal reservoir via mosquitoes.

Yellow fever is endemic in Central and South America and Africa. The principal hosts are **monkeys**, and the virus is transferred to humans by the *Aedes aegypti* **mosquito**. An initial nonspecific febrile prodrome is followed in a minority of cases by severe **multisystem** disease with jaundice (and liver failure) and hemorrhagic manifestations. Overall mortality is of the order of 1–5%. The disease is preventable by means of a **live attenuated vaccine** – indeed, travel to yellow-fever-endemic countries is not allowed without certification that vaccine has been administered.

Dengue viruses are also transmitted to humans by *Aedes* **mosquitoes**. Dengue infection is now widespread in South East Asia, the Caribbean, and Africa. Clinical disease presents in two forms. **Classical dengue fever** is an acute febrile illness with rash, severe headache, muscle and bone/joint pains (breakbone fever). Full recovery is the usual outcome. **Dengue hemorrhagic fever** (DHF), thought to arise following re-infection with a different serotype of virus, occurs in young children, with fever, profound prostration, hypotension and hemorrhage, and has a mortality of 1–5%. Treatment is supportive, and there is no vaccine available.

The reservoir of **West Nile** virus is **wild birds**, which also act as the mode of geographical spread through migration. Until recently, this infection was confined to Africa, the Middle East and parts of Asia, but in 1999 the virus was introduced into the USA, where it has spread to involve virtually all states, and is now endemic. Humans are infected via the bite of *Culex* **mosquitoes**. Most infections (80%) are asymptomatic. More severe disease presents with fever, rash, headache and muscle weakness. In the most severe cases (<1 in 100, and usually the elderly or immunosuppressed) encephalitis may develop, which can be fatal. There is no specific treatment. Prevention relies on appropriate mosquito control programs. Experimental vaccines are being developed.

Rhabdoviruses   **Rabies** virus belongs to the **Rhabdoviridae** family, and carries a negative ssRNA genome. It has a characteristic bullet shape on electron microscopy. It can infect many warm-blooded animals, and infection is found worldwide. Transmission to humans is via inoculation through the skin following a bite, with **dogs**, **cats**, **foxes** and **bats** being the commonest source. Virus replicates locally in the muscle, and then crosses the neuromuscular junction to enter nerves. This is a relatively slow process, so the incubation period between the bite and clinical presentation with disease may be several weeks. Once in the peripheral nerves, virus is transported upwards into the central nervous system. Spread around the brain results in an invariably fatal encephalitis. Pre-mortem, virus travels back down the nerves to the periphery, and hence is present in saliva, explaining its transmission through bites. Once symptoms appear, there is no specific treatment.

A number of **rabies vaccines** have been used to prevent this horrifying disease over the years. The most effective and safest is based on virus grown in human diploid cells and then killed. **Pre-exposure prophylaxis** should be offered to those at risk of acquisition of infection (e.g. vets, staff at animal quarantine centres). **Post-exposure prophylaxis** (i.e. vaccine given after a bite), is also effective, because of the prolonged incubation period. In addition to active vaccination with rabies vaccine, passive protection can be given with human rabies immunoglobulin. This should be infiltrated around the site of the bite, and acts to neutralize virus in the first few days, before the appearance of protective antibodies induced by the vaccine.

B18 POXVIRUSES

Key Notes

Virology	Poxviruses are large viruses with dsDNA genomes.
Epidemiology	Variola virus is the causative agent of smallpox. This was eradicated in 1977, but may reappear as a result of bioterrorism. Infection is spread via the respiratory route. Vaccinia virus is used as a vaccine against smallpox. Monkeypox virus was restricted to parts of Africa, but has recently appeared in the USA. Rodents are probably the natural host. Orf and pseudocowpox viruses are occupationally acquired infections from sheep and cattle respectively. Molluscum contagiosum virus is spread by direct contact.
Consequences of infection	Smallpox presents with a prodromal fever followed by vesicular and pustular skin lesions. Viremic spread results in internal organ infection and mortality rates may approach 50%. Monkeypox has clinical similarities to smallpox but a much lower capacity for person-to-person spread, and lower mortality rates. Orf and pseudocowpox cause vesicular lesions on the exposed skin of farm workers. Molluscum lesions are small waxy nodules, of cosmetic importance only.
Diagnosis	This is by electron microscopy of vesicle fluid or skin scrapings, or by genome detection techniques.
Treatment	Most poxvirus lesions resolve spontaneously. Cidofovir may be of use in smallpox, but there are no clinical data to support this.
Prevention	Smallpox can be prevented by vaccination with live vaccinia virus. This has a number of serious adverse effects, with an estimated five deaths per million vaccinees.

Virology

The **Poxviridae** (poxviruses) are the largest viruses that infect humans. They carry a dsDNA genome.

Epidemiology

Variola virus, the causative agent of **smallpox**, was the most important poxvirus infection of humans, but, as a result of the first ever successful global eradication campaign, natural infection was last recorded in 1977. However, laboratory stocks of the virus remain, and this is now high on the list of possible bioterrorist agents. **Vaccinia virus** is of importance as it is used as a vaccine to protect against smallpox.

Monkeypox virus is the causative agent of **monkeypox**, a disease which, until 2003, had only ever been described in the tropical rainforests of West and Central Africa. Despite the name, monkeys are probably not the primary host. Importation of infected exotic rodent species led to an outbreak of monkeypox in the USA in 2003. It is possible, but not yet known, that the virus has therefore become endemic in North American rodents.

Orf and **pseudocowpox** are very similar viruses, principally distinguished from each other according to whether infection was acquired from sheep and goats (orf) or cattle (pseudocowpox).

Molluscum contagiosum infects humans, being spread by direct contact including sexual transmission.

Consequences of infection

Smallpox is a potentially life-threatening disease. Infected individuals shed virus from the upper respiratory tract, and susceptible contacts acquire infection via the same route. During the 12-day incubation period, viremia results in seeding of internal organs and then the skin. A **prodromal fever** heralds the appearance of the vesicular and pustular **skin lesions**. Prognosis is dependent on the particular strain of virus – mortality may be as high as 20–50% in previously unvaccinated individuals.

Vaccinia virus originated as a laboratory strain of cowpox. In recent years, this was only administered to laboratory workers handling pox viruses, but vaccination of the general population now has to be considered as protection against a bioterrorist release of smallpox.

Monkeypox may be clinically indistinguishable from smallpox, but has a lower mortality (0–10%). It also exhibits a much lower capacity for case-to-case spread. The outbreak of monkeypox in the USA involved 82 human cases. No deaths were reported.

Orf and **pseudocowpox** are occupationally acquired infections, presenting with isolated, relatively painless, papules on exposed skin around the hands. Enlargement to nodules may occur with a weeping surface. Healing occurs spontaneously over a few weeks, but re-infection may occur.

Molluscum contagiosum infection results in the formation of waxy, smooth, dome-shaped nodules, often with an umbilicated appearance. The infection is of little more than cosmetic importance, and many lesions will resolve spontaneously. Particularly florid lesions may occur in HIV-infected individuals.

Diagnosis

The various papules, vesicles and scabs arising from poxvirus infections are best diagnosed by electron microscopic examination of fluid or skin scrapings, although it may not be possible to ascertain with certainty which pox virus is responsible on the basis of morphology alone. Specific genome amplification tests are being developed, principally for the rapid and accurate diagnosis of smallpox, should this contagion reappear.

Treatment

Most poxvirus infections resolve spontaneously without the need for intervention. There is some evidence from animal models that cidofovir can inhibit smallpox replication, but whether this agent will prove useful in the management of smallpox is not known.

Prevention

There is no doubt that vaccination with vaccinia virus can prevent smallpox, although the duration of the immune response induced by vaccination may only be short term. However, there are serious adverse events arising from such vaccination (*Table 1*).

Decisions about if and when to introduce mass vaccination campaigns have to weigh the risks of re-emergence of smallpox against the certainty that mass vaccination will result in significant morbidity and mortality, e.g. five deaths per million vaccinees. Clearly, the availability of a safer and equally effective vaccine against smallpox would be most welcome.

Table 1. Adverse effects of vaccinia vaccination

Effect	Comment
Vaccinia keratitis	Arises from inadvertent transfer of virus from the inoculation site to the eye
Generalized vaccinia	Arising from bloodstream spread of the virus
Eczema vaccinatum	Arising from spread of virus through nonintact skin
Post-vaccinial encephalitis	Aberrant immune response cross-reacts with neural antigens
Progressive vaccinia	In immunocompromised individuals, with a high fatality rate
Myocarditis	Recognized after recent large-scale vaccination in the USA

B19 PAPILLOMA- AND POLYOMA-VIRUSES

Key Notes

Papillomaviruses

Virology
The Papovaviridae, which include papillomaviruses and polyomaviruses, have dsDNA genomes. Human papillomaviruses (HPV) are classified into > 80 types on the basis of differences in genome sequence.

Epidemiology
All HPV are tropic for epithelial surfaces, but different types infect at different sites. Spread is by direct contact or sexual.

Consequences of infection
Infection results in epithelial cell overgrowth, leading to a papilloma, or wart. Cutaneous warts occur commonly on the hands and feet (verrucae). Mucosal warts may occur in the mouth, larynx or genital areas. Benign genital warts are due to HPV types 6 or 11. Malignant changes are associated with HPV types 16 and 18, leading to cancer of the cervix or other genital structures.

Diagnosis
Common warts are diagnosed by their clinical appearance. Diagnosis of genital tract infection is by genome detection techniques, which can also determine which HPV type is responsible.

Treatment
Warts resolve spontaneously in many cases. Physical treatments such as cryotherapy or topical podophyllotoxin may speed resolution.

Prevention
Vaccines designed to protect against genital tract infection (and therefore cancer of the cervix) are under development.

Polyomaviruses

JC and BK viruses are the only known human polyomaviruses. Infection is common but usually of no clinical consequence. JC virus may reactivate in the brain of immunosuppressed individuals leading to demyelination, whilst BK virus may reactivate in the kidneys of renal transplant recipients and cause BK nephropathy.

Papillomaviruses

Virology
The **papilloma** and **polyoma** viruses are both dsDNA viruses, belonging to the same family (the Papovaviridae). They are, however, biologically and genetically distinct and do not share antigenic structure. Papillomaviruses (PV) are by

far the most important in terms of human infection. They are classified on the basis of differences in DNA sequence into > 80 types, referred to as H (for human) PV followed by a number.

Epidemiology

All HPV are tropic for epithelial cells but different types are associated with infections of different tissues (see below). HPV infections occur worldwide and at all ages. Transmission is by **direct contact**, or, in the case of the genital HPV, by **sexual contact**.

Consequences of infection

HPVs infect and replicate in **squamous epithelial cells** on both keratinized and mucosal surfaces, giving rise to a papilloma, more commonly known as **a wart**. Clinical presentation may be with:

(i) **Cutaneous warts** – due to infection with HPV types 1–4. These usually occur on the hands and feet (a **verruca**), but may occur elsewhere.

(ii) **Mucosal papillomas** – may occur in the mouth, or as benign epithelial tumors of the larynx. Benign genital warts on the penis, vulva, cervix and perianal areas are due to infection with HPV types 6 and 11.

(iii) **Malignant change** – certain types of HPV are associated with malignant change. HPV 16 or 18 infection may result in cancer of the uterine cervix, a common and important tumor, or the much rarer cancers of the vulva, vagina and penis. In immunosuppressed patients, HPV 5 or 8 may result in squamous cell carcinoma of the skin.

Diagnosis

Diagnosis of common warts is usually clinically obvious. It is not possible to grow HPV in the laboratory. HPV-infected cells may be identified cytologically (koilocytes) in a Papanicolaou smear, but infection in the genital tract is best demonstrated by a **genome amplification technique**, which can also determine which particular type is involved.

Treatment

The natural history of most warts is for spontaneous regression, although this may take several months. Most warts do not require therapy. Small genital warts can be removed by **cryotherapy**, or treatment with **topical podophyllin**, a cytotoxic agent. More extensive warts may respond to **interferon therapy** or require **surgery**.

Prevention

Genital infection with certain HPV types has now been shown unequivocally to be the cause of cancer of the uterine cervix. Considerable effort is therefore being expended in trying to develop an HPV vaccine. Experimental vaccines are in advanced stages of clinical trials and the hope is that genital HPV infection, and therefore cancer of the cervix, may soon be a preventable disease.

Polyomaviruses   JC and BK viruses, both named after the initials of the patients from whom they were first isolated, are the two most important polyomaviruses in humans. In immunocompetent individuals, infection with these viruses is common, but of no clinical consequence. In immunosuppressed hosts, however, the viruses can reactivate and cause disease. JC virus is the etiological agent of a rare degenerative brain disease, particularly in HIV-infected patients (progressive multifocal leukoencephalopathy, PMLE, see *Table 2* in Topic B1). BK virus replicates in the urogenital tract, and can cause damage to transplanted kidneys (so-called BK nephropathy), or hemorrhagic cystitis in bone marrow transplant recipients.

C1 STAPHYLOCOCCI

Key Notes

Microbiology	Staphylococci are Gram-positive cocci that form clusters and produce the enzyme catalase. The genus consists of several species. *Staphylococcus aureus (Staph. aureus)*, the most virulent species, produces the enzyme coagulase, whereas all the other, less virulent species, are collectively named coagulase-negative staphylococci (CoNS).
Pathogenesis	Staphylococci possess carbohydrate capsules and produce numerous cell-wall-bound and secreted proteins that are designed to help the organism survive in various environments and invade the host. *Staph. aureus* produces several potent toxins which are responsible for the wide range of clinical syndromes caused.
Epidemiology	Most people carry staphylococci on their skin and in their anterior nasal nares. The bacteria also colonize most animals and are transmitted by direct or indirect contact. Methicillin-resistant *Staph. aureus* (MRSA) strains colonize hospital inpatients and may cause epidemic outbreaks of disease among high-risk patients, such as those with post-operative wounds.
Clinical infections	*Staph. aureus* causes serious infections of the skin, soft tissues, bone, lung, heart, brain or blood. CoNS species are mainly associated with infected medical implants, although they can cause invasive disease in the immunocompromised host.
Laboratory diagnosis	Appropriate specimens are Gram-stained and cultured. Catalase-positive, Gram-positive cocci are labelled *Staph. aureus* if they are positive for coagulase and DNAase enzymes. CoNS are negative in the latter two tests. The antibiotic susceptibilities of all isolates are determined. Strains associated with serious illnesses or outbreaks of infection (e.g. MRSA) are further characterized by typing and DNA fingerprinting.
Treatment	The majority of staphylococci are resistant to penicillin, but sensitive to some of its derivatives, including methicillin and flucloxacillin. Alternative antibiotics for resistant organisms (e.g. MRSA) include vancomycin, erythromycin and gentamicin. Some strains become resistant to multiple antibiotics.
Prevention and control	Difficult to prevent or control. In hospitals, strict rules of handwashing are applied to minimize the spread of multiresistant strains. Patients infected with such strains are isolated from other patients during treatment.

Microbiology   Staphylococci are Gram-positive cocci that divide three-dimensionally to form clumps or **clusters** of cells. They are nonmotile, nonsporulating and facultative anaerobes which grow well on most media. All staphylococci produce the enzyme **catalase** which is used in the laboratory for rapid identification.

Staphylococcus aureus (*Staph. aureus*) is the most virulent member of this genus and its defining characteristic is its ability to produce the enzyme **coagulase**. All the other species, collectively named coagulase-negative staphylococci (CoNS), are weaker pathogens. The best known and most prevalent CoNS species is *Staph. epidermidis*. Another medically important CoNS is *Staph. saprophyticus* which causes urinary tract infection in sexually active women.

Pathogenesis   All the staphylococci are common colonizers of healthy humans, but given the opportunity they can invade the host and cause disease. The genome sequences of *Staph. aureus* and *Staph. epidermidis* reveal the impressive genetic make-up of the organisms and their wide range of virulence and antibiotic resistance genes which can spread between strains. Some 'hypervirulent' staphylococcal strains may accumulate a large number of these genes and regulate their expression more efficiently than others.

The staphylococci produce **carbohydrate capsules** and **slime**. The latter may be produced in sufficient quantities to create a **'biofilm'** which protects the organism from host immune or antibiotic attacks (Topic A6). The CoNS strains are particularly efficient in creating biofilms around medical devices (implants).

Staph. aureus produces a range of potent protein-based enzymes (toxins) that may cleave host molecules or damage host cells. Examples are given in *Table 1*.

Table 1.   Some of the well-characterized secreted toxins of Staph. aureus

Secreted toxin	Effect or function
Coagulase	Clots plasma
DNAase	Hydrolyses DNA
Enterotoxins	Cause vomiting and diarrhea
Epidermolytic toxins A and B	Exfoliation of the skin
Fibrinolysin	Lyses plasma clots
Hemolysins α, β, γ	Lyse red blood cells
Hyaluronidase	Degrades hyaluronic acid in connective tissues
Leukocidin	Kills white blood cells
Lipase	Lyses lipid (lipolytic)
Toxic shock syndrome toxin 1	Shock, skin rash, desequamation

Epidemiology   Staphylococci are highly successful colonizers of humans and animals. They reside mainly on the **skin,** particularly in moist areas such as the anterior nares (nose), axilla and groin. Between one-third and three-quarters of individuals carry these organisms at any one time. Staphylococcal infections occur worldwide, and newly emerging hypervirulent or multiresistant strains spread rapidly over wide geographical areas. Transmission is by direct or indirect contact. The bacteria survive in the air, on objects or in dust for days, therefore they can contaminate environments (such as hospitals) and continue to be transmitted over long periods of time. Some individuals may shed the organism more heavily than others. Staphylococcal infections are acquired from either self (**endogenous**) or external (**exogenous**) sources.

Clinical infections   Staphylococci can infect immunocompetent and immunocompromised individuals. *Staph. aureus* can produce a variety of clinical manifestations. Depending on the site of infection, the clinical syndromes can vary (*Table 2*).

CoNS species are less virulent than *Staph. aureus*; however, they are ubiquitous and readily colonize medical implants such as intravascular or peritoneal dialysis catheters, pacemakers, artificial heart valves, joint replacements and neurosurgical devices.

Table 2.   Common Staph. aureus *infections*

Infection site	Clinical syndrome
Skin and soft tissue	Boils, carbuncles, abscesses, impetigo, pyomyositis, scalded skin syndromes, wound infection
Bone	Osteomyelitis
Joint	Septic arthritis
The internal lining of the heart	Endocarditis
Blood	Septicemia, toxic shock syndrome
Brain	Brain abscess
Lungs	Pneumonia
Gut	Enterocolitis, food-poisoning

Laboratory diagnosis   Appropriate specimens should be taken from the site of infection. Pus samples or tissue debris are usually Gram-stained on receipt and examined under the microscope for a rapid assessment. Cultured bacteria are Gram-stained and tested for the production of catalase. *Staph. aureus* strains are coagulase and DNAase positive and CoNS species are negative . The bacteria are then tested for their susceptibility to several antibiotics. Isolates from severe infections, particularly those that are associated with outbreaks or multiple antibiotic resistance, will be further characterized, i.e. by serotyping, phage typing (old), molecular typing (increasingly used) or molecular fingerprinting.

Treatment   Most staphylococci produce the β-lactamase enzyme which inactivates penicillin. The bacteria are also capable of developing resistance to almost any antibiotic. Until recently, most staphylococcal strains were sensitive to methicillin and related antibiotics. However, **methicillin-resistant *Staph. aureus* (MRSA)** strains are now very common in hospitals and the community, therefore alternative antibiotics are used. These include vancomycin, teicoplanin, rifampicin, gentamicin, clindamycin, fusidic acid or erythromycin. In severe life-threatening infections combinations of antibiotics are often used. Vancomycin is the most effective and widely used antibiotic against severe cases of MRSA infections. However, **vancomycin-resistant *Staph. aureus* (VRSA)** strains have also been isolated from patients.

Prevention and control   Handwashing is key for the prevention of the spread of multiresistant strains in hospital environments. Isolating patients with hypervirulent and multiresistant strains such as MRSA is essential, as is rapid diagnosis and treatment.

C2 STREPTOCOCCI AND ENTEROCOCCI

Key Notes

Microbiology
Streptococci and enterococci are catalase-negative Gram-positive cocci, which are seen in pairs or chains. Partially (α-) hemolytic streptococci include *Streptococcus pneumoniae* and the 'viridans' streptococci. Fully (β−) hemolytic species include *Strep. pyogenes, Strep. agalactiae* and others. Enterococci can be α-, β- or nonhemolytic. Streptococci are also divided serologically into 20 different 'Lancefield' groups A-H and K-V.

Pathogenesis
Most but not all streptococci produce polysaccharide capsules. They secrete toxins which are considered vital for bacterial survival and host invasion. Some, e.g. *Strep. pneumoniae* and *Strep. pyogenes,* are more virulent than others.

Epidemiology
All streptococci are normal human commensals colonizing the mouth and upper respiratory tracts. All enterococci are normal commensals of the human and animal gut.

Clinical infections
Strep. pneumoniae is the commonest cause of pneumonia and causes a wide range of other serious diseases, including upper respiratory tract infection, meningitis and septicemia. Viridans streptococci cause dental caries and infective endocarditis. *Strep. pyogenes* causes a wide range of clinical syndromes, including upper respiratory tract infections, skin and soft tissue infections and septicemia. It can also induce post-infection (reactive) complications, such as rheumatic fever, arthritis and glomerulonephritis. Enterococci can cause urinary tract infection, endocarditis and abdominal wound infections.

Laboratory diagnosis
Gram stain, typical colonial morphology, hemolysis, biochemical reactions, Lancefield grouping and antibiotic sensitivity tests are all used for the identification of the various *Streptococcus* species.

Treatment
Almost all streptococci are sensitive to penicillin and a range of other antibiotics. Enterococci are slightly more tolerant of penicillin but are usually sensitive to amoxicillin. Vancomycin-resistant *Enterococcus* (VRE) is a major hospital-acquired infection that is often difficult to treat.

Microbiology
The streptococci and enterococci are Gram-positive cocci that divide two-dimensionally, and form short or long chains. They are nonmotile and most strains grow under both aerobic and anaerobic conditions. Some species are strictly anaerobic (*Peptostreptococcus)* and others require CO_2 for good growth.

Unlike staphylococci, streptococci are catalase negative. Streptococci are separated into different species by their ability to **hemolyse** blood, by serology and/or by biochemical tests. Most streptococci do not tolerate bile (except *Strep. agalactiae*), whereas all enterococci are bile tolerant and live in the human gut.

Classification on the basis of hemolysis is crude, yet extremely useful clinically. On a blood agar, streptococcal colonies are surrounded by either **partial (α), complete (β)** or **no zone of hemolysis** (also known as **γ-hemolysis**).

Strains that produce a greenish zone of α-hemolysis include *Strep. pneumoniae* and a large number of other species collectively named 'viridans' streptococci (see below).

β-Hemolytic strains producing clear (transparent) zones of total hemolysis include *Strep. pyogenes, Strep. agalactiae* (some times α– or γ-hemolytic) and *Strep. dysgalactiae*.

Enterococcus species are largely nonhemolytic although isolates can be α- or β-hemolytic. The medically important species are *Enterococcus faecalis, E. faecium, E. durans* and *E. avium*.

Streptococci and enterococci are also differentiated by **Lancefield grouping**. This technique involves the use of specific antibodies which react with bacterial polysaccharide or glycerol-teichoic acid antigens present in the cell wall. Some 20 groups, A-H and K-V, are recognized. Often streptococci are named by their Lancefield grouping instead of their long names. For example *Strep. pyogenes* and *Strep. agalactiae* are also called Group A and B streptococci, respectively. *Strep. bovis* and all enterococcal species are labelled as Lancefield group D. Viridans streptococcal strains may possess none or any one of the Lancefield group antigens.

The streptococci are resistant to nalidixic acid (and aminoglycosides), which is incorporated into culture media to suppress contaminating Gram-negative bacteria.

α-Hemolytic streptococci

Strep. pneumoniae Pneumococci

Strep. pneumoniae is a very common **colonizer of the upper respiratory tract**, carried in the mouth and pharynx of at least one-third of people at any one time. It is, however, an opportunistic pathogen, capable of causing a wide range of infections.

Pneumococci are seen mostly in pairs (**diplococci**) and produce large quantities of a carbohydrate capsule, which is one of its most important survival/virulence factors. Probing the organism with specific antiserum causes the capsule to swell and become visible under the microscope (Quellung reaction). Different strains express antigenically different capsular types. At least **84 capsular types** have been recognized so far, but these are not equally common. Pneumococcal capsules can generate protective immunity and there are vaccine preparations available which consist of some (usually 6 or 23) of the most prevalent capsular types. Other virulence factors include enzymes (toxins) such as pneumolysin.

Unlike other α-hemolytic streptococci, all pneumococci are **sensitive to optochin** (an antibiotic), readily lysed by bile salts and able to ferment inulin. These properties have been exploited in the laboratory for a rapid differentiation of pneumococci from viridans streptococci.

Pneumococci are by far the most common cause of community-acquired **lobar pneumonia**. They also cause acute exacerbation of chronic bronchitis, otitis media, sinusitis, conjunctivitis, peritonitis, severe (often fatal) meningitis, septicemia and other clinical syndromes.

Pneumococci are sensitive to several antibiotics, including penicillin and cephalosporins. However, penicillin-resistant strains are now spreading worldwide.

'Viridans' streptococci

'Viridans' streptococci consist of a heterogeneous group of α-hemolytic strepto-cocci which show greening on blood agar, or occasionally no hemolysis (non-hemolytic). They are normal residents of the mouth and oropharynx. Examples of species within this group include: *Strep. mutans, Strep. mitior (mitis), Strep. sanguis, Strep. milleri, Strep. salivarius.* To identify clinical isolates to species levels, commercial kits are available which distinguish different species on the basis of their biochemical activities.

Of particular note is *Strep. milleri,* which may also be Lancefield group A, C, F, or G. It is more widely distributed (found also in gut and vagina) and able to cause suppurative conditions, such as **deep abscesses** of abdomen, liver and brain.

Viridans streptococci (particularly *Strep. mutans*) are of low virulence but implicated in the two most prevalent diseases in man, dental caries and peri-odontal disease. They are also considered the commonest causes of **infective endocarditis** with *Strep. mitior, Strep sanguis* and *Strep. mutans* implicated in 65% of cases. The organisms enter the bloodstream and may lodge in damaged or abnormal heart valves arising from rheumatic or congenital heart disease. Vegetations form and valve function becomes further impaired. Infective endo-carditis has a high mortality rate.

Viridans streptococci are sensitive to many antibiotics, including penicillin. In cases of endocarditis, the combination of penicillin and gentamycin is used for better, **synergistic**, effect. Penicillin (or suitable alternatives) is also used for prophylaxis during dental procedures in patients with a suspected heart valve abnormality.

β-Hemolytic streptococci

Strep. pyogenes **(group A streptococci)**

This is a major human pathogen present in the nasopharynx of healthy adults and children. *Strep. pyogenes* can be further subdivided into Griffiths' types, using antibodies raised against three surface protein antigens M, R and T. Virulent strains contain the antigenically variable M protein which inhibits phagocytosis; > 100 different M proteins have been described.

This organism is the only β-hemolytic streptococcus sensitive to bacitracin (an antibiotic). This property has been used in the laboratory for rapid diagnosis.

Strep. pyogenes produces a number of toxins and enzymes (Table 1), which are key virulence factors essential for the colonization and invasion of the host.

Table 1. Some of the well-characterized toxins produced by Strep. pyogenes

Toxin	Function/effect
Hemolysin (streptolysin)	Kills red blood cells
Leukocidin	Kills white cells
Nicotinamide adenine dinucleotidase (NADase)	Degrades NAD
Streptokinase	Lyses fibrin
Hyaluronidase	Degrades hyaluronic acid of connective tissues
Deoxyribonucleases A, B, C, D (DNAases)	Degrade foreign DNA
Erythrogenic toxin	Causes erythematous rash in scarlet fever

The bacterium is capable of causing a wide range of infections in humans, including **skin and soft tissue infections**, upper respiratory tract infections, septicemia, post-natal sepsis and many others. Most infections come under one of two main categories:

(i) **Suppurative** infections where the organism causes damage directly. These include sore throats, tonsillitis, peritonsillar abscess, pharyngitis, cellulitis, erysipelas, otitis media, mastoiditis, sinusitis, scarlet fever, impetigo, erysipelas, puerperal sepsis, necrotizing fasciitis (a medical emergency) and wound infection.

(ii) **Nonsuppurative** (post-streptococcal) complications, where host antibodies (generated in response to the bacterial infection) may damage uninfected parts of the body, for example the heart valves (rheumatic fever), joints (reactive arthritis) and the kidneys (glomerulonephritis).

Fortunately, *Strep. pyogenes* remains fully sensitive to penicillin, cephalosporins and many other antibiotics.

Strep. agalactiae (Group B streptococci)

Strep. agalactiae is another widespread organism that inhabits mainly the female genital tract. It is usually harmless, but, it is capable of causing serious invasive disease, particularly in neonates and pregnant women. *Strep. agalactiae* is a relatively common cause of **neonatal meningitis and/or septicemia**. It also causes septic abortions, puerperal sepsis and infection of female pelvic organs. Prepartum screening programmes for *Strep. agalactiae* genital tract carriage has greatly reduced the incidence of neonatal meningitis caused by this organism.

Enterococci (Lancefield group D streptococci and enterococci)

The Lancefield **group D** organisms include all enterococcal species and *Strep. bovis*. Their normal habitat is the human and animal gut, but they are occasionally found in the mouth or external genitalia. They can grow in the presence of high concentrations (e.g. 40%) of bile salts or sodium chloride (6.5%). They are normal commensals of humans, but are also opportunistic pathogens causing diseases such as urinary tract infection (most commonly), biliary tract infection, abdominal wound infection and infective endocarditis (less commonly).

Enterococci are sensitive to amoxicillin but moderately resistant to penicillin, whereas *Strep. bovis* is penicillin sensitive. Some enterococci have become highly resistant to several antibiotics, including vancomycin, hence the name **vancomycin-resistant enterococci (VRE)**. These cause serious problems in hospitalized patients.

C3 BACILLUS SPECIES

Key Notes

Microbiology

The *Bacillus* genus consists of numerous species, only two of which are commonly associated with disease in humans. *B. anthracis* causes anthrax and *B. cereus* causes food-poisoning. Under harsh environmental conditions, the organism forms tough and durable spores which can vegetate (i.e. revert to growth) once environmental conditions become favorable.

B. anthracis

Pathogenesis
The key virulence factors in pathogenic *Bacillus* species are the powerful toxins released from vegetative bacteria. *B. anthracis* spores vegetate in human tissues, especially lymph nodes, and secrete a group of toxins that together induce edema, tissue damage and hemorrhage.

Epidemiology
Anthrax is a zoonotic disease transmitted from animals to humans via occupational routes, affecting butchers and workers in the textile industry. *B. anthracis* spores have been exploited for production of biological weapons for military or bioterrorist use.

Clinical manifestations
When anthrax spores come in contact with human skin they may vegetate, giving rise to cutaneous anthrax. If anthrax spores are inhaled, they cause 'pulmonary anthrax' or 'wool-sorter's disease', which is usually fatal. 'Intestinal anthrax' follows ingestion of anthrax spores.

Laboratory diagnosis
Tissue specimens or body fluids from infected sites can be Gram-stained and cultured. Bacterial DNA or antigens can be detected in the laboratory.

Treatment
B. anthracis is sensitive to many antibiotics, including penicillin. However, these must be given early in the disease otherwise they may not be effective.

Prevention and control
Anthrax can be prevented by vaccination, which is usually offered only to people at risk. The vaccine consists of cell free components including the 'protective antigen'.

B. cereus

B. cereus spores vegetate in left-over food and release toxins which accumulate in sufficient quantities to cause gastrointestinal upset after ingestion.

Microbiology   *Bacillus* species are large (4-8 × 1.5μm) cylindrical Gram-positive bacteria. They grow best under aerobic conditions. There are a large number of *Bacillus* species, but not many are capable of causing disease in humans. These include *B. anthracis*, the causative agent of anthrax, and *B. cereus*, a cause of food-poisoning. The genus also includes some potentially useful species such as *B. stearothermophilus* and *B. megaterium* which are used for testing of the efficiency of autoclaves and ethylene oxide (respectively) for sterilizing medical equipment.

Bacillus species form tough and durable **spores** that help the organisms survive harsh environmental and nutritional conditions. The spore remains viable but its biological activity comes to an almost complete standstill. This status can last for months, years or decades, until a favorable environment becomes available, when the organism **vegetates** and begins to grow.

B. anthracis is a nonmotile (unlike other species of this genus), facultative anaerobe that can grow on ordinary laboratory media. It forms long chains that inter-tangle and look like a 'medusa head' under the microscope.

Anthrax (*Bacillus*   *Pathogenesis*
anthracis   The key virulence factors in pathogenic *Bacillus* species are the powerful toxins
infection)   released from vegetating (i.e. growing) bacteria. *B. anthracis* spores vegetate in human tissues, especially lymph nodes, and secrete a group of **toxins** which together induce edema, tissue damage and hemorrhage.

The toxin components include the **edema factor, lethal factor** and **protective antigen**. The latter protects the former two from body proteases before entering host cells, and it also induces protective antibodies when used as a vaccine. All three toxins are encoded by genes that are present on a **plasmid**. Another essential element in pathogenesis is the presence of a capsule (also plasmid-encoded), without which the organism is harmless.

Epidemiology
Anthrax is largely a zoonotic disease, i.e. acquired from animals. Infected animals die and generate billions of anthrax spores which contaminate other carcasses, or the wool or hair of live ones. These are occasionally transmitted to humans. In industrialized countries, anthrax is a rare imported disease that affects mainly workers in the textile industry who handle animal wool or hair.

The disease is more common in developing countries. Major animal and human **outbreaks** of anthrax have occurred in many parts of the world, including Africa and the Middle East. The robust nature of anthrax spores has been exploited by the military. The organism is now top of the list of potential biological warfare and bioterrorism agents.

Clinical manifestations
When anthrax spores come in contact with human skin, they vegetate and cause a lesion (**cutaneous anthrax**) consisting of a swollen, hemorrhagic and necrotic (but painless) area called a '**malignant pustule**'. The local hemorrhage gives the black color effect, hence the name '*anthracis*' (i.e. 'black' in Greek). This is not normally fatal, unless totally neglected.

In contrast, inhalation of anthrax spores may result in severe and often fatal disease. The inhaled spores are taken via lymph vessels to the **mediastinal lymph nodes** in the middle of the chest. They may remain dormant for days or weeks before they vegetate and release their potent toxins. The lungs remain clear on X-ray but the mediastinum is expanded, a classical diagnostic feature.

This is 'pulmonary anthrax' or 'wool-sorter's disease'. Hemorrhagic fluid may accumulate in the lung and the organisms and their toxins will then disseminate in the blood to the rest of the body, leading to an overwhelming and almost invariably fatal septicemic shock.

When anthrax spores are ingested, 'oropharyngeal' or 'intestinal anthrax' may follow. There may be local tissue damage and in half of the cases, patients develop septicemia and die.

Laboratory diagnosis

Tissue specimens or body fluids from infected sites can be Gram-stained and cultured. Specimens should be handled inside safety cabinets. Gram stain shows the characteristic chains of Gram-positive bacilli. Other serological or molecular techniques are also available to detect antigens or DNA traces, respectively.

Treatment

B. anthracis remains sensitive to many antibiotics, including penicillin, erythromycin, tetracycline, chloramphenicol and ciprofloxacin. However, these must be given early in the disease otherwise they may not be of benefit. Importantly, some of the weaponized B. anthracis strains (particularly in Russia), have been engineered to be resistant to penicillin and tetracycline. Ciprofloxacin is the drug of first choice following bioterrorist use of anthrax.

Prevention and control

Several vaccines are available against anthrax. Those manufactured in the UK and USA consist of cell-free components that contain the protective antigen. These vaccines are usually offered only to people at risk of infection.

Bacillus cereus food-poisoning

B. cereus is found in soil and contaminates rice and other seeds. When rice is boiled in bulk and subsequently pan-fried rapidly (usually in less than a minute), spores can remain viable and later vegetate after the rice is stored at room temperature for a lengthy period. Toxins are then released, which can accumulate in sufficient quantities to cause disease after ingestion.

A notable example is food-poisoning following consumption of contaminated (fried) rice in Chinese 'take-aways'. The main features include vomiting and gastrointestinal upset. B. cereus food-poisoning is usually self-limiting and can be treated with simple rest and fluid replacement. It can be prevented by paying attention to good kitchen practises and safe cooking and storing of rice.

C4 CORYNEBACTERIA

Key Notes

Microbiology	The *Corynebacterium* species are Gram-positive bacilli. Groups of bacterial cells form angles with each other, looking like Chinese characters under the microscope. There are many *Corynebacterium* species. Most are harmless, causing disease only in the immunocompromised host, or are detected when they contaminate laboratory cultures. However, *C. diphtheriae* is a most important human pathogen. Three subspecies of *C. diphtheriae* are known, *gravis, intermedius* and *mitis*.
Pathogenesis	Pathogenic strains of *C. diphtheriae* produce 'diphtheria toxin' that causes cell damage or death. Nontoxigenic strains are not pathogenic and are commonly found in the normal flora.
Epidemiology	Diphtheria has now virtually disappeared in industrialized countries as a result of mass vaccination. The organism, however, is still widespread and can cause disease in the absence of vaccination. Transmission from person to person is via respiratory droplets. The disease has a relatively high mortality rate, especially among children and the elderly.
Clinical infections	Patients present initially with generalized nonspecific symptoms before developing a thick, adherent 'pseudomembrane' around the tonsils or pharyngeal area, associated with neck swelling. Systemic intoxication may affect the heart, motor nerves and the adrenal glands.
Laboratory diagnosis	Throat swabs are cultured. Identity of the organism is confirmed with carbohydrate and enzymatic tests. The Elek test is carried out to confirm toxigenicity of the isolate.
Treatment	Neutralization of the toxin, using antitoxins, and administration of antibiotics, such as erythromycin.
Prevention and control	A safe and highly effective vaccine is available

Microbiology

Corynebacterium species are Gram-positive non-spore-forming **pleomorphic** (i.e. many forms) bacilli. They snap and bend on division. Adjacent cells lie at angles and in small groups, arranged in shapes that mimic Chinese characters. There are many species within the *Corynebacterium* genus, but only few cause serious infections in humans. These include *C. diphtheriae, C. ulcerans* and *C. pseudotuberculosis. C. diphtheriae*, the causative agent of diphtheria, is the most important human pathogen.

C. diphtheriae is a facultative anaerobe and grows best at 37°C on blood- and serum-containing media, e.g. blood tellurite agar. The latter contains potassium tellurite which is reduced by the organism to tellurium, hence the gray/black discoloration of the colonies. The organism forms polymerized phosphate deposits (called metachromatic granules) that are best demonstrated with special staining methods (Neisser's methylene blue).

Three subspecies of C. diphtheriae are known, including *gravis, intermedius* and *mitis*. All of these cause diphtheria.

Pathogenesis

The key virulence factor in pathogenic C. *diphtheriae* is the production of a powerful toxin, diphtheria toxin, the gene for which is carried on a bacterio-phage (a virus which infects bacteria). C. *diphtheriae* isolates that lack the toxin are harmless. Toxigenicity of isolates can be demonstrated by a gel-precipitation method called the 'Elek' test. Here, toxin-antitoxin interaction is visualized on a serum agar.

Diphtheria toxin prevents protein biosynthesis in human cells leading to cell death due to its ADP-ribosylation of eukaryotic elongating factor. Locally secreted toxin can damage or destroy cells at the site of infection in the throat and upper respiratory tract resulting in production of a gray adherent 'pseudomembrane', consisting of bacteria, fibrin, epithelial cells and dead white blood cells. The toxin is also released into, and disseminated via, the blood-stream. Certain body tissue cells are more vulnerable than others, particularly heart muscle, motor nerves and the adrenal glands.

Epidemiology

Diphtheria has now virtually disappeared in developed countries as a result of mass immunization. Elsewhere, the disease is becoming increasingly uncommon. The toxigenic and nontoxigenic forms of the organism, however, are still prevalent worldwide. In the former Soviet countries, > 50 000 cases were reported after the collapse of the Union, due to breakdown in the healthcare system and poor vaccine uptake.

Transmission from person-to-person is via respiratory droplets (e.g. cough, sneeze) or close contact. Asymptomatic carriage is common and is considered the major factor in the spread of infection, as actual cases of diphtheria are usually rapidly identified and treated. The disease has a relatively high mortality rate, especially among children and the elderly.

Clinical infections

A few days after exposure to the organism, patients present with generalized nonspecific symptoms, including sore throat, high temperature and general weakness. A thick, adherent pseudomembrane appears around the tonsils or pharyngeal area. Neck swelling due to edema involving cervical lymph nodes may give the appearance of a 'bull neck'. Laryngeal involvement may lead to serious obstruction of the airways. Systemic intoxication may take the form of **myocarditis** (causing irregular heartbeats and heart failure), **peripheral neuritis** (causing temporary paralysis of limbs) and **thrombocytopenia**. Death is usually from heart-related complications.

Laboratory diagnosis

Throat swabs should be obtained and cultured. Blood agar and a selective medium containing tellurite are normally inoculated. Gram stain of typical colonies will reveal the Chinese characters. Identity of the organism is confirmed with tests of carbohydrate utilization and enzymatic activities. The Elek test is carried out to confirm toxigenicity of the isolate. **PCR** may be used to confirm the presence of the gene encoding for the toxin. Antibiotic sensitivity tests are also carried out.

Treatment

The most important part of treatment is neutralization of the toxin, using horse or human antitoxin sera. The organism should be eradicated with penicillin or erythromycin. The latter is preferred because it eliminates throat carriage; however, some C. *diphtheriae* isolates may be resistant to erythromycin, therefore laboratory isolation and antibiotic sensitivity tests are vital.

Prevention and control

A safe and highly effective vaccine is available and has been incorporated into the World Health Organization's program of childhood immunization since the 1940s. The vaccine consists of diphtheria toxin that is rendered harmless (toxoid). With high rates of vaccine uptake, the disease can be eliminated from any given community. Patients should be isolated and promptly treated. Suspected carriers should be vaccinated and offered prophylactic antibiotics.

Coryneform bacteria

Numerous other corynebacteria colonize humans, mainly the skin and mucous membranes. They commonly contaminate laboratory cultures and are collectively described as 'coryneforms' or 'diphtheroids'. These are harmless in immunocompetent hosts. However, they may infect surgical implants causing implant failure, or occasionally cause more serious illnesses (e.g. endocarditis, bone infection) in immunocompromised hosts. They may cause peritonitis in patients receiving chronic ambulatory peritoneal dialysis.

C5 LISTERIA

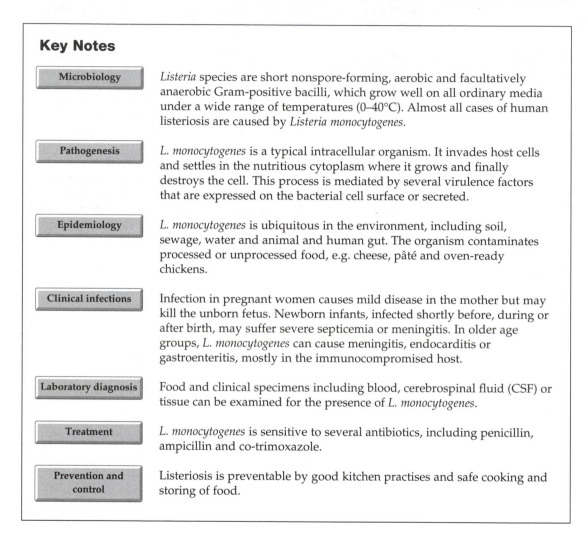

Key Notes

Microbiology	*Listeria* species are short nonspore-forming, aerobic and facultatively anaerobic Gram-positive bacilli, which grow well on all ordinary media under a wide range of temperatures (0–40°C). Almost all cases of human listeriosis are caused by *Listeria monocytogenes*.
Pathogenesis	*L. monocytogenes* is a typical intracellular organism. It invades host cells and settles in the nutritious cytoplasm where it grows and finally destroys the cell. This process is mediated by several virulence factors that are expressed on the bacterial cell surface or secreted.
Epidemiology	*L. monocytogenes* is ubiquitous in the environment, including soil, sewage, water and animal and human gut. The organism contaminates processed or unprocessed food, e.g. cheese, pâté and oven-ready chickens.
Clinical infections	Infection in pregnant women causes mild disease in the mother but may kill the unborn fetus. Newborn infants, infected shortly before, during or after birth, may suffer severe septicemia or meningitis. In older age groups, *L. monocytogenes* can cause meningitis, endocarditis or gastroenteritis, mostly in the immunocompromised host.
Laboratory diagnosis	Food and clinical specimens including blood, cerebrospinal fluid (CSF) or tissue can be examined for the presence of *L. monocytogenes*.
Treatment	*L. monocytogenes* is sensitive to several antibiotics, including penicillin, ampicillin and co-trimoxazole.
Prevention and control	Listeriosis is preventable by good kitchen practises and safe cooking and storing of food.

Microbiology

Listeria species are short, nonspore-forming Gram-positive bacilli (0.5–2 μm) which grow well on all ordinary media. They are aerobic and facultatively anaerobic. The genus contains six species, but almost all cases of human listeriosis are caused by *L. monocytogenes*. There are several **serovars** of this species; serovars 4b, 1/2a and 1/2b are the most frequent causes of human disease.

   L. monocytogenes can **grow in cold or warm water**, ranging from 0 to 40°C. The bacterium is motile in colder temperatures and moves in a jerky fashion (at 25°C), creating what is described as '**tumbling motility**'. It is nonmotile at 37°C. Some strains produce toxins that lyse red blood cells (β-hemolytic).

Pathogenesis

L. monocytogenes is an intracellular organism and has been studied as a model bacterium that **gains entry to human cells**, settling in the cytoplasm where it evades the humoral immune defense system. It uses one of its surface proteins (internalin) to enter host cells, and releases listeriolysin O and phospholipase C to escape from the encapsulating host membrane into the cytoplasm. Here the bacterium obtains nutrition and replicates rapidly. It then exploits the host cell actin filament apparatus to eject itself into neighboring host cells.

Epidemiology

L. monocytogenes is **ubiquitous** in the environment, including the soil, sewage, water and the gastrointestinal tract of animals and humans. The organism can **contaminate food** (processed or unprocessed), including soft cheese, pâté, sandwich meats and oven-ready chickens.

The bacterium may continue to **grow slowly in domestic refrigerators**. It survives dry or moist surfaces and tolerates salt. Therefore, food is the main route of infection, except in unborn and newborn children. The peak incidence of listeriosis occurs in the summer months.

L. monocytogenes causes disease mainly in the immunoimmature and immunocompromised hosts, including **neonates** (>70% of cases), elderly and pregnant women. In **pregnant women** it can infect the fetus (10–20% of cases), leading to abortion or stillbirth.

Clinical infections

Intrauterine infection is rare before the 20th week of gestation. The mother may suffer symptoms such as high temperature, back pain, sore throat, headache or diarrhea. The unborn child, however, may die *in utero* (leading to **abortion** or **stillbirth**).

Newborn infants may suffer severe listeriosis soon after birth. This may be in the form of **septicemia** or **meningitis**. 'Early-onset' disease occurs when the child is infected *in utero* and develops symptoms within 2 days of birth. The clinical features are largely of septicemia. If the infant becomes infected during or after birth, he/she will suffer 'late-onset' disease, a largely meningitic picture which appears ≥ 5 days after birth. Early-onset disease is more common and more serious than late-onset disease, with a mortality rate of 30–60%, compared to ~ 10%.

In older age groups, including children and adults, *L. monocytogenes* can rarely cause serious (often fatal) meningitis, endocarditis and gastroenteritis.

Laboratory diagnosis

Food and clinical specimens are examined for the presence of *L. monocytogenes*. Gram-positive bacilli producing a narrow zone of hemolysis and demonstrating tumbling motility (seen under the microscope) at room temperature are highly suggestive of *L. monocytogenes*.

In early-onset disease of neonates, the bacterium can be isolated from amniotic fluid and the child's blood and superficial and deep tissues. In late-onset disease and cases of meningitis in older age groups, the organism may be seen in Gram stains of cerebrospinal fluid (CSF) deposits (Topic E2). To increase the chance of detection, the CSF is centrifuged and the deposit is Gram-stained and examined. The CSF will also have an increased number of cells (mainly polymorphonuclear cells), increased protein and reduced sugar content. In all cases of listeriosis, including endocarditis, blood cultures must be taken.

Treatment

L. monocytogenes is sensitive to several antibiotics, including penicillin, ampicillin, tetracycline, co-trimoxazole, aminoglycosides (e.g. gentamicin) and chloramphenicol. Penicillin (or ampicillin) is the drug of choice which is often combined with gentamicin for synergistic effect. Chloramphenicol is effective, but in meningitis it may be associated with relapse and is known to cause serious side-effects, such as aplastic anemia.

Prevention and control

There are no vaccines against *L. monocytogenes*. Listeriosis is preventable by good kitchen practises and safe cooking and storing of food.

C6 MYCOBACTERIA

Key Notes

Microbiology	Mycobacteria are strictly aerobic, slow-growing, intracellular bacilli. They are Gram positive, but do not stain easily, therefore, a special staining method (Ziel–Neelson) is used which includes a strong acid for decolorization (hence named acid-fast bacilli). They require special media for growth, e.g. Löwenstein–Jensen (LJ) medium which contains egg, asparagine and glycerol. Several mycobacterial species cause disease in humans, including *Mycobacterium tuberculosis, M. bovis, M. leprae* and a large number of weakly pathogenic species, collectively named 'atypical mycobacteria'.
Pathogenesis	Mycobacteria are intracellular pathogens, difficult to kill. They possess a tough lipid-containing cell wall.
Epidemiology	All mycobacterial infections are of increasing importance worldwide and are most common among the poor and the immunocompromised.
Clinical manifestations	*M. tuberculosis* primarily infects the lungs, leading to pulmonary tuberculosis. It may infect other body organs, including bone, kidneys, meninges. *M. leprae* targets the peripheral nerves and causes leprosy. Atypical mycobacteria are opportunistic organisms and infect mainly the immunocompromised.
Diagnosis	The acid-fast bacilli can be detected in sputum smears or tissue biopsies. The organism is often isolated and grown in broth or on agar. Detection of mycobacterial DNA, e.g. by PCR, is also carried out.
Treatment	Requires multiple antimicrobial drugs which are given over several months. Tuberculosis is treated with isoniazid, rifampicin, pyrazinamide and ethambutol. Leprosy is treated with dapsone and clofazime.
Prevention	Involves health promotion, isolation of patients, effective treatment, contact tracing, antibiotic prophylaxis and administration of the BCG vaccine.

Microbiology

Mycobacteria are Gram-positive, strictly aerobic, nonsporing intracellular bacilli. They are resistant to decolorization by mineral acid or alcohol after staining with carbol fuchsin, hence called 'acid-fast bacilli'. The latter forms a strong complex with the mycolic acid content of the cell wall, therefore the organism is only decolorised with 20% sulfuric acid and alcohol. This is the basis of the **Ziel–Neelson (ZN) stain**.

Medically important mycobacterial species grow slowly. Their generation (doubling) time ranges from a few hours to > 2 days, compared to 40–60 mins for most other bacteria. Colonies may take 2–3 weeks to develop from the time of inoculation of culture media.

Mycobacteria do not grow on ordinary laboratory media but grow well on **Löwenstein–Jensen (LJ) medium**, containing egg, asparagine, glycerol and malachite green. The latter inhibits contaminants. Colonial morphologies on LJ agar slopes vary widely among mycobacteria and are used in identification of species.

The most common and important mycobacterial species affecting humans is *Mycobacterium tuberculosis* (Mtb), which causes the disease 'tuberculosis'. There are several uncommon species which cause opportunistic infection in the immunocompromised, collectively named 'atypical' mycobacteria. *M. bovis*, which causes tuberculosis in cattle, can infect humans via ingestion of milk. *M. leprae* is the causative agent of leprosy.

Tuberculosis

Tuberculosis (TB) is a slow-progressing chronic disease of human and animals caused mainly by *M. tuberculosis*, less commonly by *M. bovis* and rarely by *M. microti* and *M. africanum*. These bacteria enter the body via the respiratory or alimentary tracts, or by contact, and form 'tubercles' in the lung or other parts of the body. The basic lesion, consisting of mycobacteria inside host defense cells, surrounded by numerous other white cells, is known as a **granuloma or 'the tubercle'**. With progression of infection, the centre of granulomas may liquefy (**caseate**) into pus.

Epidemiology
Tuberculosis is one of most common diseases **worldwide**, with high morbidity and mortality rates, particularly in developing countries. It is now of increasing importance in the Western world, especially among the immunocompromised (e.g. in HIV-infected patients) and poorer communities.

Clinical manifestations
After the first encounter with the organism (**primary infection**), the patient may remain asymptomatic or develop primary pneumonia, meningitis, osteomylitis, disseminated (miliary) tuberculosis or a range of other manifestations.

In immunocompetent hosts, the primary infection remains asymptomatic for a long time (years) until the organism finds the opportunity (e.g. immunocompromised status) to reactivate. **Reactivation** of old infection or acquisition of new infection and development of disease is called '**post-primary' tuberculosis**. Pulmonary tuberculosis is the most common manifestation and lesions usually occur in the upper lobes of the lung.

In a minority of cases, tuberculosis may resolve spontaneously; however, in most cases it progresses and requires treatment. Patients often produce blood-containing sputum (hemoptysis) and suffer persistent high temperature, night sweats, loss of appetite and loss of weight. If untreated, patients may die.

Diagnosis
Clinical suspicion and typical **chest X-ray** findings are often sufficiently convincing for initiation of treatment before laboratory confirmation. X-ray may show lesions in the upper parts of the lung with evidence of destruction of lung tissue (cavitating lesions). Exposed or infected patients produce immune responses against the organism, which can be detected by skin tests

(**Mantoux or Heaf tests**) that use purified protein derivative from the bacterial cell wall to demonstrate cell-mediated immunity.

In cases of pulmonary tuberculosis, **early morning sputum** is most appropriate for microbiological examination. Microscopy can be performed using conventional **ZN** or **auramine staining**. The latter, examined under a fluorescent microscope, is a sensitive but not very specific method of detection. Suspected positive sputum is then stained with ZN (examined under light microscope) and cultured on LJ slopes.

Broth cultures may speed up isolation and molecular techniques (e.g. PCR) may accelerate diagnosis. **Biopsies** of other infected tissues may reveal the organism and/or caseating granulomata when stained with ZN or other histological stains, and may yield the organism when cultured. PCR may also be used for the detection of rifampicin resistance.

Treatment

Treatment requires the use of **multiple drugs** for long periods of time (6–9 months). Commonly used regimens include 2 months of isoniazid, rifampicin, pyrazinamide and ethambutol followed by 4–6 months of the first two drugs alone. Some of the drugs have severe side-effects and many mycobacterial species develop resistance to the common antimicrobials. Multi-drug resistant *M. tuberculosis* is a major public health problem worldwide.

Prevention

General measures of prevention include: improvement of social conditions and nutrition; isolation of patients with active TB; notification of cases and the use of **BCG** (Bacille Calmette–Guérin) vaccine which is **70% protective in the UK**. BCG is not universally considered effective, especially in children. Other measures include effective treatment of all cases, tracing all close contacts and offering chemoprophylaxis to all who are skin test positive, HIV positive or neonatal contacts.

The low incidence of primary infection in some industrialized countries (e.g. USA) has led to the decision not to use widespread immunization. This strategy allows for a positive skin test to directly indicate exposure.

Leprosy

Leprosy is caused by *M. leprae* which cannot be grown on artificial medium; instead, it is grown in armadillo tissue. This is predominantly a disease of poorer countries. The organism attacks **Schwann cells** of the peripheral nerves in cold parts of the body, such as the extremities, tip of the nose or ears. This results in anesthesia and paralysis. Due to loss of sensation, infected parts become subjected to repeated trauma, infection and deformities.

Depending on the quality of the host immune response, **two extreme forms** of leprosy are described. **Tuberculoid leprosy** reflects a strong host cell-mediated immune response and lesions have small numbers of bacilli. **Lepromatous leprosy** reflects a weak host response and lesions contain large numbers of leprosy bacilli. There is a wide spectrum of clinical presentation between these two extremes.

Diagnosis is based primarily on clinical suspicion and histological features. The organism can be demonstrated in tissues.

Treatment consists mainly of surgical correction of deformities and the use of antimicrobial drugs, e.g. dapsone and clofazime.

Atypical mycobacteria

Atypical mycobacteria are largely opportunistic pathogens causing infection in immunocompromised patients, e.g. those suffering from HIV infection, chronic pulmonary disease or on long-term immunosuppressive drugs. They cause pulmonary infection or become disseminated throughout the body.

A common finding is a generalized or localized lymphadenopathy. Person-to-person spread is rare and laboratory diagnosis is by ZN stain and culture of appropriate specimens. These bacteria can be environmental contaminants, therefore interpretation of microbiology results has to be carried out with care, e.g. more than one specimen should usually be positive. The common atypical mycobacterial species and clinical conditions include:

- *M. marinum* causes a warty lesion which may resolve spontaneously. This usually occurs on the fingers of those who handle fish (**fish-fanciers finger,** or **fish-tank granuloma**).
- *M. avium* and *M. intracellulare* represent a group of similar species (also described as the **avium–intracellulare complex**). Infection is associated with profound cell-mediated immune depression and bacteremia, and is a **terminal event in AIDS**.
- *M. kansasii* causes respiratory infection in middle-aged or elderly patients with underlying chest disease, e.g. pneumoconiosis. Clinically it **mimics TB**.

C7 ENTEROBACTERIACEAE

Key Notes

Microbiology	Enterobacteriaceae (also known as enterobacteria or coliforms) include numerous genera and species with several properties in common. They are Gram-negative bacilli, tolerate bile salts, grow under aerobic or anaerobic conditions, ferment glucose and produce catalase but not oxidase enzymes. *Escherichia coli* is the most common, and best known, species of this family. Other medically important enterobacterial genera include *Citrobacter, Enterobacter, Hafnia, Klebsiella, Morganella, Proteus, Providencia, Salmonella, Serratia, Shigella* and *Yersinia*.
Pathogenesis	Most Enterobacteriaceae members are gut commensals, but they are also opportunistic pathogens. Some are more pathogenic than others. Virulence factors include capsules, endotoxins, motility elements, pili, numerous exotoxins and interactive surface molecules. They are antigenically variable and able to exchange DNA.
Epidemiology	Enterobacteria exist in very large numbers in the small and large intestine of humans and animals. Hospital environments usually have resident Enterobacteriaceae that are resistant to multiple antibiotics.
Clinical infections	Enterobacteriaceae cause a wide range of clinical conditions. Common ones include urinary tract infection, septicemia, gastroenteritis and hospital-acquired infections.
Laboratory diagnosis	Appropriate specimens are processed in the laboratory where Gram stain, colonial morphology, serology and biochemical reactions will all be used to identify the organism to species level, dependent on the seriousness of the condition and the clinical value of the information. Bacteria will be tested for their susceptibility to a range of antibiotics. Antibiotic choices depend on sensitivity of the isolate. Commonly used agents include amoxicillin, cephalosporins, gentamicin and ciprofloxacin. Hospital-resident Enterobacteriaceae can accumulate resistance against many of these antibiotics.
Prevention and control	Generally, keeping good personal and food hygiene. Vaccines are available against some members, such as *Salmonella typhi* and *Y. pestis*.
Related topics	Lower respiratory tract infections (F7) Intra-abdominal infections (F9) Gastroenteritis and food- poisoning (F8) Genital infections (F11)

Microbiology   The Enterobacteriaceae (commonly referred to as enterobacteria, see *Table 1*) include numerous species which have several properties in common. They are aerobic or **facultatively anaerobic Gram-negative bacilli**. They **tolerate bile salts** (and hence survive in the human gut), ferment glucose and produce catalase but not oxidase enzymes. *Escherichia* is the most common genus of this family, of which *E. coli* is the dominant and best studied species. Members of the Enterobacteriaceae are often described as '**coliforms**' denoting *E. coli*-like bacteria.

Most, if not all, members of the Enterobacteriaceae are opportunistic pathogens. The Enterobacteriaceae can grow **on simple culture media**, e.g. nutrient agar, without special nutritional or environmental requirements. They have different colonial morphologies on agar plates. For example, *Proteus* species swarm culture plates in waves of growth; *Klebsiella* species produce highly mucoid colonies (due to capsule overproduction); *Shigella*, *Salmonella* and *Proteus* species do not ferment lactose, therefore colonies look pale on MacConkey agar. They grow in a wide range of temperature and pH conditions. Most are motile, but some (e.g. *Shigella* and *Klebsiella*) are nonmotile. They are differentiated easily in the laboratory by their biochemical, antigenic and genetic differences. Also, they vary in terms of susceptibility to antibiotics.

The ability of Enterobacteriaceae to ferment (i.e. utilize for energy production) various carbohydrates or their alcohols has been exploited in the laboratory for differentiation between members of the genus. *Klebsiella* species ferment a wide range of carbohydrates, whereas *Shigella* and *Salmonella typhi* ferment a much narrower range of these molecules.

All Enterobacteriaceae members have antigenically diverse capsules (K antigen), lipopolysaccharides (also called endotoxin, or O antigen) and flagella (H antigen). These have been exploited for differentiation between various species and subspecies of the genus. For example, >2000 serotypes of *Salmonella* and > 800 of *Klebsiella* have been described.

Pathogenesis

Most Enterobacteriaceae members are gut commensals present in very large numbers (billions per gram of feces). They may play useful roles in excluding (by competition) other invading pathogens. However, given the opportunity (e.g. in immunocompromised individuals), they can invade and cause disease. Some are more pathogenic than others, e.g. *Salmonella typhi*, *Shigella dysenteriae* and *Yersinia pestis*. They all possess **virulence** factors that help organisms survive and overcome host immune defenses. These factors include the presence of a capsule, **endotoxin**, motility elements, numerous **exotoxins** and surface molecules that directly interact with host molecules.

The **genomes** of many of the medically important Enterobacteriaceae have been recently sequenced. Detailed analysis reveals a wide range of previously unknown virulence capabilities for many of these organisms. Many of the virulence and antibiotic resistance genes can move horizontally between different members of Enterobacteriaceae. Genes move individually, in small groups or in very large numbers (genomic pathogenicity islands). Bacterial viruses (**phages**) and **plasmids** may act as **vectors** between strains of the same species or sometimes between species or even genera. These, coupled with the ability of bacteria to mutate frequently and regulate expression of virulence genes, add to the pathogenicity of Enterobacteriaceae.

Epidemiology

Enterobacteriaceae colonize **the human and animal gut and the female genital tract**. *Salmonella typhi* (typhoid) is now rare in industrialized countries but common in Asia. *Shigella sonnei* is the most common *Shigella* species in Europe, whereas *Shigella dysenteriae* is rare. *Yersinia pestis* (plague), which is mostly carried by animals but causes disease in humans, is relatively common in Southern Africa and some parts of the Far East, but not in Europe. *E. coli* O157, carried by animals and passed on to humans, is endemic in many parts of the world, including the UK.

Hospital environments usually have resident Enterobacteriaceae which are resistant to multiple antibiotics and many have strong virulence capabilities. These strains colonize patients readily and are transmitted between patients in the same hospital or between hospitals locally, nationally and internationally.

Clinical infections *Table 1* lists a number of common clinical conditions caused by various members of the Enterobacteriaceae.

It is important to remember that, given the opportunity, any of the Enterobacteriaceae can cause infection at any site in the body (i.e. opportunistic pathogens). Breaks in physical infection barriers (e.g. skin, mucus membranes, urine flow) or general immunosuppression will allow the organism to reach normally sterile sites of the body and establish infection. Thus, almost all Enterobacteriaceae are able to cause urinary tract infection, wound infection and septicemia under appropriate circumstances. Many of these bacteria (e.g. *E. coli*, *Klebsiella*, *Enterobacter* and *Proteus* species) can develop resistance to multiple antibiotics. These bacteria are often described as 'hospital residents' where they not only colonize patients and staff but also contaminate the environment such as sinks, taps, toilets, intravenous fluids, medicines or surgical instruments.

Each genus within Enterobacteriaceae consists of diverse species which cause different sets of clinical syndromes. Even within individual species, strains may cause different diseases. For example, different strains of *E. coli* are associated with distinct forms of diarrhea. Of particular note is the so-called *E. coli* O157 (verocytotoxin-producing strain) which causes a serious colitis associated with severe (potentially fatal) anemia and renal failure.

Salmonella enterica Typhi (known as *Salmonella typhi*) causes typhoid, a severe and potentially fatal disease where the bacterium invades intestinal lymphoid tissue and may disseminate throughout the body. *Salmonella enterica* Paratyphi strains cause a milder form of typhoid illness. Most of the other salmonellae cause gastroenteritis (food-poisoning) characterized by diarrhea and vomiting.

Shigella species invade the large intestine causing a form of colitis described as 'bacterial dysentery' (loose stool mixed with blood and mucus).

Yersinia pestis causes plague, an invasive (often fatal) disease where the organism invades the lymphoid tissues, lung and/or blood.

Laboratory diagnosis Appropriate specimens, e.g. stool for gastroenteritis, urine for urinary tract infection (UTI), blood for septicemia, cerebrospinal fluid for meningitis and pus or wound swab for wound infections should be sent to the laboratory for culture. **Gram stain**, **colonial morphology**, **serology** and **biochemical reactions** will all be used to identify the organism to species level. The amount of effort and resources spent on detailed identification of organisms to species and subspecies level will depend on the seriousness of the condition and the clinical value of the information. It is important to remember that Enterobacteriaceae are **common colonizers** of moist sites, and may not be responsible for invasive disease despite their isolation from specimens. Therefore, **careful interpretation** of culture results in the light of clinical information is necessary before appropriate antibiotics are chosen.

Table 1.   Alphabetically listed enterobacteria genera and their most common species

Genus	Species	Subspecies/types	Clinical syndrome
Escherichia	coli	Entero-pathogenic	Infantile enteritis
		Entero-toxigenic	Diarrhea
		Entero-invasive	Dysentery (similar to Shigella)
		Entero-aggregative	Chronic diarrhea
		Verocytotoxin-producing (O157)	Diarrhea, hemorrhagic colitis, hemolytic-uremic syndrome.
		Most other E. coli strains	UTI, GE, HAI, septicemia, meningitis (in neonates), wound infection
Citrobacter	amalonaticus, freundii, koseri		HAI
Enterobacter	aerogenes, cloacae agglomerans		UTI, HAI
Klebsiella	oxytoca, pneumoniae	pneumoniae, aerogenes, ozaenae, rhinoscleromatis	Most infections are due to K. pneumonia ssp. UTI, HAI, pneumonia, septicemia, meningitis (in neonates), chronic URTI
Morganella	morganii		UTI, HAI
Proteus	mirabilis, vulgaris		
Providencia	rettgeri, stuartii		HAI
Salmonella	enterica	Typhi	Typhoid (enteric fever)
		Paratyphi (A, B, C)	Paratyphoid
		Others:>2000 serotypes, e.g Enteritidis, Typhimurium	Food-poisoning (GI). Rarely cause invasive diseases
		Rare ones, e.g. Dublin, Virchow	GI, invasive disease, e.g. septicemia
Serratia	liquefacien, marcescens		HAI
Shigella	dysenteriae, flexneri, boydii, sonnei		Dysentery (invasive diarrhea)
Yersinia	pestis, pseudotuberculosis, enterocolitica		Plague Ileitis, mesenteric lymphadenitis Enteritis

UTI, urinary tract infection; GE, gastroenteritis; HAI, hospital acquired infection, or nosocomial infection (this includes a variety of different clinical syndromes, such as UTI, septicemia, ventilation-associated pneumonia, wound infection and many others); URTI, upper respiratory tract infections.

Where necessary, genetic methods, e.g. **PCR**, will be employed to identify and fingerprint the organism. In cases of uncomplicated UTI or simple wound infections, identification may be restricted to genus level. In more serious cases, e.g. septicemia, all organisms will be identified to species and subspecies levels. Commercially available, miniaturized biochemical and serological tests are available to expedite the process.

In cases of outbreaks or severe cases with public health implications, (e.g. food-poisoning or the spread of multiresistant bacteria), all bacteria will be fully identified and sent to specialized laboratories for **typing or genetic finger-printing**. This will help determine the link between isolates from different patients from the same or different geographical locations.

Cultured bacteria are tested for their susceptibility to a range of antibiotics. Given the ability of Enterobacteriaceae to develop and exchange antibiotic resistance rapidly, constant **surveillance** of all multiresistant strains is essential.

Treatment

Not all enterobacterial infections require treatment. For example, food-poisoning due to *Salmonella* or *Shigella* is usually self-limiting and antibiotics are not recommended. In all other invasive clinical conditions, antibiotics are required. Some enterobacteria are more resistant than others, and this may change with time and place. For example, the majority of *E. coli* isolates in the community were sensitive to amoxicillin until a decade ago. Now, more than half the strains causing UTI in the community are resistant to this antibiotic.

Klebsiella spp. are all resistant to amoxicillin. Generally, most Enterobacteriaceae remain sensitive to second generation cephalosporins, gentamicin, ciprofloxacin and carbapenems. Exceptions are the hospital-resident Enterobacteriaceae which can accumulate resistance against many of these antibiotics.

Prevention and control

Generally, keeping good personal and food hygiene protects against infections such as UTI and food-poisoning. **Good hospital hygiene** is important and patients with multiresistant bacteria are usually separated from high-risk patients (e.g. those with open wounds). Precautions are taken to prevent the spread of infection between patients. Strict measures may be required to **control infection outbreaks** within the hospital (Topic E9). **Vaccines** are available against some members of Enterobacteriaceae, such as *Salmonella typhi* and *Yersinia pestis*.

Additional information on selected pathogens

Escherichia coli

Escherichia coli (*E. coli*) is the best known and most important species of the genus *Escherichia*, one of the most prevalent members of Enterobacteriaceae, and the most common opportunistic pathogen that lives in human and animal gut. Strains of *E. coli* are diverse, as confirmed by genome sequencing, and cause diverse clinical syndromes, including UTI, hospital-acquired infections (can be multiresistant), wound infection, septicemia and, in neonates, severe meningitis. Five additional and distinct types of strain are associated with gastroenteritis. These include entero-pathogenic *E. coli* (EPEC), entero-toxigenic *E. coli* (ETEC), entero-invasive *E. coli* (EIEC), entero-aggregative *E. coli* (EAEC) and verocytotoxin-producing *E. coli* O157 (VTEC). They demonstrate different phenotypic characteristics and virulence capabilities. VTEC strains live in animal gut and can cause major outbreaks of hemorrhagic colitis and hemolytic-uremic syndrome which can be fatal.

Salmonella

Almost all pathogenic *Salmonella* strains are now taxonomically included within *Salmonella enterica* species. There are > 2000 different subspecies; however, the vast majority of these are associated with food-poisoning and gastroenteritis.

Salmonellosis ranges clinically from the common self-limiting gastroenteritis (diarrhea, abdominal cramps, and fever) to enteric fevers (including typhoid and paratyphoid fevers) which can be life-threatening febrile systemic illnesses.

Contamination with human feces is the major mode of spread of *Salmonella*, and the usual vehicle is contaminated water or contaminated food (handled by asymptomatic carriers).

Salmonella typhi, which causes the more severe forms of enteric fever (typhoid), has no animal reservoir, therefore the epidemiology of enteric fevers primarily involves person-to-person spread. The symptoms begin after an incubation period of 10–14 days. The symptoms of enteric fevers are nonspecific and include fever, anorexia, headache, myalgias, and constipation. Focal infections and an asymptomatic carrier state occur.

Nontyphoidal salmonellosis is a worldwide disease of humans and animals. Animals are the main reservoir, and the disease is usually food-borne, although it can be spread from person to person.

General salmonellosis treatment measures include replacing fluid loss by oral and intravenous routes, and controlling pain, nausea, and vomiting. Typhoid fever and enteric fevers should be treated with antibiotics.

Salmonellae are difficult to eradicate from the environment. However, reducing the number of salmonellae harbored in animals (during processing, packaging and cooking) would significantly reduce human exposure. Vaccines are available for typhoid fever and are partially effective, especially in children. No vaccines are available for nontyphoidal salmonellosis.

Shigella

Shigellae are Gram-negative, nonmotile, facultatively anaerobic, nonspore-forming bacilli. They are differentiated from other Enterobacteriaceae (e.g. *E. coli*) on the basis of pathogenicity, physiology and serology. The genus is divided into four serogroups, including *S. dysenteriae*, *S. flexneri*, *S. boydii* and *S. sonnei*.

Infection is initiated by ingestion of shigellae (usually via fecal–oral contamination). The hallmarks of shigellosis are bacterial invasion of the colonic epithelium and inflammatory colitis. These manifest themselves as colitis in the rectosigmoid mucosa, with concomitant malabsorption, and classical signs of bacillary dysentery, namely scanty, unformed stools tinged with blood and mucus. These may be associated with abdominal pain, tenesmus, fever, vomiting and dehydration.

Shigellosis is endemic in developing countries where sanitation is poor. In developed countries, single-source, food- or water-borne outbreaks occur sporadically, although pockets of endemic shigellosis can be found.

Shigellosis is suspected in patients with fresh blood in the stool or neutrophils in stool fecal smears. Diagnosis is confirmed by isolation of the organism from stools.

Most cases are self-limiting but severe dysentery is treated with antibiotics. Prevention of fecal–oral transmission is the most effective control strategy.

Yersinia

Yersinia are small, Gram-negative coccobacilli showing bipolar staining. *Y. pestis*, *Y. enterocolitica* and *Y. pseudotuberculosis* are the medically important members of the genus. The latter two pathogens cause gastrointestinal diseases, whereas *Y. pestis* causes plague.

Y. pestis is primarily a rat pathogen and human infections are initially transmitted by rat fleas, leading to bubonic plague (swollen, blackened lymph nodes) which may develop into pneumonic plague (septicemia and hemorrhagic pneumonia and death) which spreads directly from human to human via respiratory droplets. Outbreaks are explosive in nature, and invariably lethal.

Early clinical diagnosis is essential, and the organism can be isolated from blood or sputum. Treatment is by antibiotics, such as sulfadiazine, streptomycin, tetracycline, and chloramphenicol.

Y. pestis is an extremely infectious hazard for nursing and laboratory personnel. Vaccines are available for laboratory personnel. Prevention by controlling rats and rat fleas is crucial.

C8 PSEUDOMONAS AND RELATED ORGANISMS

Key Notes

Microbiology	*Pseudomonas, Stenotrophomonas* and *Burkholderia* are motile, strictly aerobic, Gram-negative bacilli which are oxidase positive. They grow on minimal carbon and nitrogen sources, including simple media and moist surfaces. The most important species of these genera include *P. aeruginosa*, *S. maltophilia*, *B. capacia*, *B. mallei* and *B. pseudomallei*. They resist antiseptic solutions and many antibiotics.
Pathogenesis	These organisms are opportunistic pathogens and some are more pathogenic than others. They express motility elements, pili, endotoxins, outer membrane proteins and secreted exotoxins. They are antigenically variable and able to exchange virulence and antibiotic resistance genes between species. They are capable of forming biofilms *in vivo* and *in vitro*.
Epidemiology	They exist in very large numbers in the human environment. *Pseudomonas* colonizes human and animal gut. They are capable of inhabiting/contaminating water, moist surfaces and sewage. Hospital environments usually have resident *P. aeruginosa* that are resistant to multiple antibiotics. *S. maltophilia* and *B. cepacia* are environmental contaminants. *B. mallei* infects horses and is occasionally transmitted to humans. *B. pseudomallei* is endemic in South East Asia.
Clinical infections	All three genera are largely opportunistic organisms. *Pseudomonas* and *Stenotrophomonas* cause a wide range of clinical conditions, including wound infection, urinary tract infection, septicemia, and hospital-acquired infections in debilitated or immunocompromised hosts. *B. cepacia* causes chest infection in cystic fibrosis patients. *B. mallei* causes mild infection acquired from horses, and *B. pseudomallei* causes melioidosis.
Laboratory diagnosis	Appropriate specimens are processed in the laboratory where Gram stain, colonial morphology, serology and biochemical reactions will all be used to identify the organism to species level.
Treatment	Antibiotic choices depend on sensitivity of the isolate. *Pseudomonas* and *Burkholderia* are usually sensitive to ceftazidime, carbapenems, fluoroquinolones and aminoglycosides. However, they can develop resistance very rapidly. *S. maltophilia* is resistant to carbapenems but sensitive to co-trimoxazole.
Prevention and control	Protecting the immunocompromised and good hospital hygiene are important. Surgical equipment should be sterilized.

Microbiology   *Pseudomonas*, *Stenotrophomonas* and *Burkholderia* were all previously grouped together in the *Pseudomonas* genus, and hence are referred to as *Pseudomonas* and related organisms (PRO). They are motile, **oxidase-positive**, strictly aerobic and long, cylindrical, Gram-negative bacilli. They do not ferment lactose but can grow on minimal carbon and nitrogen sources, including simple media and moist surfaces.

The medically important species of these genera include *P. aeruginosa*, *P. fluorescens*, *P. putida*, *S. maltophilia*, *B. capacia*, *B. mallei* and *B. pseudomallei*.

Pathogenesis   These organisms are opportunistic pathogens and some have greater pathogenic potential than others. Like Enterobacteriaceae, the PRO express antigenically diverse capsules (K antigen), lipopolysaccharides (O antigen) and flagella (H antigen). They also express pili which are essential for adhesion and DNA uptake, outer membrane proteins, some of which act as receptors for human molecules, and secrete a range of exotoxins which contribute significantly to the pathogenesis of disease.

They **resist antiseptic** solutions and many antibiotics. Some species, including *P. fluorescens and P. putida*, produce blue/green and yellow fluorescent colors, respectively, and are able to grow at 4°C (in refrigerators), thereby contaminating and discoloring stored fluids (including blood).

The PRO are antigenically variable and have large genomes (chromosomes and plasmids) which contain numerous mobile elements able to move virulence and antibiotic resistance genes between species and strains.

Epidemiology   *P. aeruginosa* is the most common species of this group. It is ubiquitous worldwide and inhabits the human and animal gut, water, soil, sewage and moist surfaces. It **contaminates hospital environments** and most hospitals have resident, **multiresistant**, strains which are transmitted between patients in the same hospital or between hospitals (locally, nationally and internationally).

The fluorescent pseudomonads (*P. fluorescens*, *P. putida*), *S. maltophilia* and *B. cepacia* may also become hospital residents and can accumulate multiresistance genes. *B. cepacia* is endemic among cystic fibrosis patients. Constant surveillance of all multiresistant strains is essential.

B. mallei infects horses and is rarely transmitted to humans. *B. pseudomallei* is endemic in South East Asia, lives as a saprophyte in the soil and infects animals and humans.

Clinical infections   *P. aeruginosa* is a major opportunistic pathogen of the immunocompromised, causing a wide range of hospital-acquired infections. These include infections of burns, post-operative wounds, urinary tract (especially in patients with catheters), lower respiratory tract (particularly those with cystic fibrosis), ears, eyes (including contact lense-associated corneal ulceration) and other skin infections (hot-tub folliculitis). Infection with the organism frequently leads to severe sepsis and death can occur due to septicemic shock.

S. maltophilia and *B. cepacia* also cause hospital-acquired infection in the immunocompromised. *B. cepacia* is particularly common among patients with cystic fibrosis. *B. mallei* causes glanders in horses and in humans. *B. pseudomallei* causes melioidosis in animals and humans. Most cases are asymptomatic; however, the bacterium may cause a severe pneumonic picture with skin lesions, and may kill as a result of septicemia.

Laboratory diagnosis

Appropriate specimens, depending on the site of infection, are obtained and cultured in the laboratory. The Gram stain and colonial morphologies are highly suggestive. Strictly aerobic, nonfermenting and oxidase-positive, Gram-negative bacilli are presumptively labelled *Pseudomonas*.

Isolates are identified and speciated on the basis of their biochemical and serological reactions. Miniaturized kits are available for identification. Molecular techniques are available for identification, typing and fingerprinting, which are particularly useful during outbreak investigations. Once bacteria are grown, they are tested for their susceptibility to a range of antibiotics.

Treatment

Most strains of PRO are resistant to several antibiotics. *Pseudomonas* is usually sensitive to piperacillin, ticarcillin, aztreonam, ceftazidime (a third generation cephalosporin), aminoglycosides, fluoroquinolones, carbapenems and colistin. Exceptions are the hospital-resident pseudomonads which can accumulate resistance against many of these agents.

S. maltophilia strains are sensitive to co-trimoxazole but resistant to carbapenems, which are often used in selective media for rapid isolation and identification. *B. cepacia* is sensitive to the anti-*Pseudomonas* antibiotics and co-trimoxazole, chloramphenicol and minocycline. *B. mallei* and *pseudomallei* are sensitive to anti-*Pseudomonas* drugs.

Prevention and control

Protecting the immunocompromised and good hospital hygiene are important. Surgical equipment should be sterilized.

C9 PARVOBACTERIA

Key Notes

Microbiology

Parvobacteria are small and fastidious Gram-negative coccobacilli which require enriched media for growth. They consist of a diverse group including *Haemophilus*, *Brucella*, *Bordetella*, *Pasteurella*, *Francisella*, *Actinobacillus*, *Cardiobacterium*, *Eikenella*, *Gardnerella* and *Streptobacillus*.

Pathogenesis

Many of these species are capsulated. They all express antigenically variable outer membrane proteins and secrete enzymatically active proteins. Some are more virulent than others.

Epidemiology

Most of the medically important haemophili, actinobacilli, *Cardiobacteria* and *Eikenella* species are normal commensals of the human upper respiratory tract. *Gardnerella vaginalis* is a human female genital commensal. All the others are zoonotic infections. *Brucella* infection is acquired from cattle, goats, sheep and pigs. *Pasteurella* infection is acquired from dogs and cats. *Francisella* and *Streptobacillus* infections are acquired from rodents.

Clinical features

H. influenzae cause upper and lower respiratory tract infections, meningitis and epiglottitis. *H. aegyptius* causes conjunctivitis. *H. ducreyi* causes the sexually transmitted 'chancroid'. *Bordetella pertussis* causes pertussis (whooping cough). *Brucella* species cause brucellosis. *Pasteurella multocida* causes wound infection and endocarditis. *Francisella tularensis* causes tularemia. *Cardiobacterium hominis*, *A. actinomycetemcomitans*, *Eikenella corrodens* cause endocarditis. *Gardnerella vaginalis* causes bacterial vaginosis.

Diagnosis

Isolation of the organism, detection of antibodies in serum or detection of DNA in specimens (by PCR).

Treatment

The *Haemophilus*, *Pasteurella*, *Eikinella* and *Cardiobacterium* species are sensitive to β-lactams, cephalosporins, aminoglycosides and fluoroquinalones. *Bordetella* species are sensitive to erythromycin. *Francisella tularencsis* is sensitive to aminoglycosides and tetracycline. *Gardnerella vaginalis* is sensitive to metronidazole.

Prevention

It is difficult to prevent infection with human commensals. Vaccines are available against *H. influenzae* type b (Hib) and against pertussis. Avoiding patients infected with pertussis and animals infected with *Francisella tularensis* is advised. Avoiding dog bites prevents *Pasteurella multocida* infection.

Related topics

Eye infections (F5)   Bacteremia and septicemia (F12)
Upper respiratory tract infections (F6)

Parvobacteria

Parvobacteria (parvus = small) are **small, fastidious**, nonsporing Gram-negative coccobacilli which require enriched media for growth. They include a diverse group of bacteria which are not necessarily related, the important genera being: *Haemophilus*, *Brucella*, *Bordetella*, *Pasteurella*, *Francisella*, *Actinobacillus*, *Cardiobacterium* and *Eikenella*, *Gardnerella* and *Streptobacillus*.

Haemophilus

The *Haemophilus* genus consists of *H. influenzae* (most common) and a number of other species listed in *Table 1*. For optimal growth, these organisms require humidity, CO_2 and temperatures close to 37°C.

Haemophilus species only grow in enriched media, including chocolate or blood agar containing **X-factor** (hemin or another iron-containing source) and/or **V-factor** (di- or triphosphopyridine nucleotidase). In the laboratory, disks pre-impregnated with X- and V-factors are added to pre-seeded nutrient agars. Growth detected around either or both of these disks indicate dependency on the respective factors and assist in bacterial identification (Fig. 1). V-factor may also be provided by other co-cultured bacteria, such as *Staphylococcus aureus*. Molecular methods are also available for diagnosis of various *Haemophilus* species.

H. influenzae is the main pathogenic species of the genus. The bacteria may or may not express an antigenically diverse capsule which is used in classifying strains to serotypes (a–f Pittman types). **It is a normal commensal** of the upper respiratory tract of humans. *H. influenzae* strains which are capsulated with the serotype b (*H. influenzae* **type b**, Hib) are considered most pathogenic and are responsible for serious, often fatal, diseases (in children <5 years), including **epiglottitis, meningitis and pneumonia**. They also cause other upper respiratory tract infections (*Table 1*). The majority of the strains are, however, non-capsulated and cause disease only in the immunocompromised, e.g. those with pre-existing chest problems such as chronic bronchitis or bronchiectasis.

The bacterium is sensitive to ampicillin, cephalosporins, fluroquinolones and macrolides. β-Lactamase-producing strains are widespread, therefore co-amoxyclav may be used. **Hib vaccine** is now part of the triple vaccine given to all children in the UK.

Table 1.   Haemophilus *species, their X/V dependence and clinical syndromes*

Species	X/V dependence	Clinical syndromes
H. influenzae	X and V	Pneumonia, acute exacerbation of chronic bronchitis, sinusitis, otitis media, meningitis, epiglottitis, arthritis
H. parainfluenzae	V	Rarely causes upper respiratory tract infection
H. haemolyticus	X and V	Rarely causes upper respiratory tract infection
H. parahaemolyticus	V	Upper respiratory tract infections
H. aegyptius	X and V	Conjunctivitis
H. ducreyi	X	Chancroid (STD)

X, X-factor (hemin or another iron-containing source); V, V-factor (di- or triphosphopyridine nucleotidase); STD = sexually transmitted disease.

Brucella

Brucella are short, slender, **pleomorphic**, nonmotile and **noncapsulated** Gram-negative bacteria. They chronically infect domestic animals and are transmitted

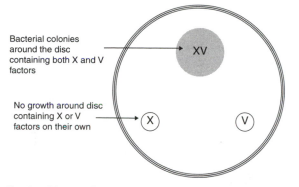

Fig. 1.   Diagram illustrating a simple test on a nutrient agar place to confirm the identity of Haemophilus Influenzae which requires the presence or both X and V factors for growth.

to humans via **unpasteurized milk** or its products. The organism causes a protracted illness associated with fever (**undulant fever**).

Three main species are known, including *Brucella melitensis* (infecting goats and sheep, in Mediterranean countries), *Brucella abortus* (infecting cattle world-wide) and *Brucella suis* (infecting pigs, mainly in USA and Denmark). Each species is divided into several biotypes. The species are differentiated by biochemical tests or their sensitivity to certain dyes. Antisera raised against the common 'M' (for *Brucella melitensis)* or 'A' (for *Brucella abortus)* antigens are also used to differentiate between these species.

Bacterial isolation or detection of specific antibodies are used in diagnosis of disease. Treatment is by combining tetracycline and rifampicin.

Bordetella

Bordetella species (*Bordetella pertussis, Bordetella parapertussis*) cause pertussis, commonly known as whooping cough. The bacteria are **fastidious** Gram-negative coccobacilli which produce 'split pearls'-like colonies and are capsulated. They require charcoal or a high concentration of blood (e.g. Bordet–Gengou medium) for growth.

Bordetella pertussis is more common and important because it causes more severe disease than *Bordetella parapertussis*. Illness is characterized by **paroxysmal coughing and breathlessness**. The highest mortality occurs in the first 2 years of life with epidemics every 4 years. **Complications** include pneumonia, bronchiectasis (usually localized), hemorrhage (e.g. epistaxis, subconjunctival bleeding) and cerebral anoxia leading to brain damage.

The disease is now rare in the UK due to the highly effective **vaccines available**. In the 1970s, vaccine compliance dropped due to bad publicity, and cases and fatalities increased dramatically. The currently available whole cell killed or subunit vaccines are safe and effective. Three main serotypes of *Bordetella pertussis* are known, including types 1,2; 1,3 and 1,2,3 (denoting the co-expression of antigens 1, 2 and 3).

The disease is diagnosed clinically and by bacterial isolation. Post-nasal (not throat or nasopharyngeal) swabs are appropriate for cultures. Patients should be isolated and treated with erythromycin.

Other parvobacteria

There are several other fastidious small Gram-negative organisms, which vary in terms of their prevalence and the severity of the clinical syndromes to which they give rise. These are listed in *Table 2*.

Table 2: Some members of parvobacteria which are either uncommon or cause mild illnesses

Genus	Species	Reservoir	Microbiological characteristics	Clinical syndromes	Antibiotic treatment
Pasteurella	*multocida*	Dogs	Oxidase positive, capsulated, grows on nutrient agar but not on MacConkey	Wound infection (e.g. dog bite), endocarditis	Penicillin
Francisella	*tularensis*	Rodents	Grows in presence of cystine and glucose.	Tularemia (similar to plague)	Aminoglycosides, tetracycline.
Actinobacillus	*actinomycetemcomitans*	Human mouth	Facultatively anaerobic, grow on nutrient agar	Periodontal infection, endocarditis	Penicillin
Cardiobacterium	*hominis*	Human mouth	Requires CO_2 (capnophilic), grows slowly on blood or chocolate agar	Endocarditis	Penicillin
Eikenella	*corrodens*	Human mouth	Slow-growing, facultatively anaerobic, corrodes agar (pitting effect).	Endocarditis, human bite infection	
Gardnerella	*vaginalis*	Human vagina	Gram variable, requires enriched media, β-hemolytic	Bacterial vaginosis (in association with anaerobic bacteria). Causes smelly and irritating vaginal discharge.	Metronidazole
Streptobacillus	*moniliformis*	Rats	Gram variable with club-shaped terminal swelling. Requires enriched media	Rat-bite fever	

C10 NEISSERIA

Key Notes

Microbiology	*Neisseria* are fastidious Gram-negative diplococci which grow best in moist environments in the presence of 5–10% CO_2. *N. meningitidis* and *N. gonorrhoeae* infect only humans and are the only pathogenic members of this genus. *N. meningitidis* expresses antigenically variable capsules that divide the species in to serogroups. Serogroups A, B, C and W135 cause the majority of infections.
Pathogenesis	*Neisseria* adhere to and invade human epithelial cells. The meningococcal endotoxin is toxic and is responsible for many of the signs and symptoms of the disease.
Epidemiology	*N. gonorrhoeae* is sexually transmitted and asymptomatic women act as the main reservoir. *N. meningitidis* is a normal nasopharyngeal commensal, and spreads via close kissing contact.
Clinical infections	*N. gonorrhoeae* is the causative agent of gonorrhea and may cause infertility in the long term. *N. meningitidis* causes severe (often fatal) meningitis and/or septicemia. It produces a typical nonfading purpuric skin rash.
Laboratory diagnosis	Gram-negative diplococci seen in clinical specimens (e.g. cervical discharge or cerebrospinal fluid) which yield oxidase-positive colonies are presumptively identified as *Neisseria* species. Biochemical and serological tests are carried out to confirm the identity of the organism. *N. meningitidis* isolates are also routinely serogrouped.
Treatment	Gonorrhea is treated with penicillin or ciprofloxacin. Meningococcal disease is treated with penicillin or a cephalosporin and additional life-supporting interventions.
Prevention and control	In addition to safe sex, screening sexually active women for early gonococcal detection and treatment is key to prevention. Meningococcal disease caused by serogroups A and C is vaccine-preventable. There is no vaccine against serogroup B.

Microbiology   Neisseriaceae are fastidious Gram-negative cocci that are usually seen in pairs under the microscope (hence described as diplococci). Many *Neisseria* species colonize humans; however, only *Neisseria gonorrhoeae* (gonococcus) and *N. meningitidis* (meningococcus) are pathogenic. The organisms are sensitive to low temperature and drying and grow best at 37°C in the presence of humidity and 5–10% CO_2.

Pathogenesis   Several virulence factors are produced by *N. gonorrhoeae* and *N. meningitidis* that
                        enable the organisms to gain access to, reproduce, and persist in various body
                        sites. The organisms produce toxins (**endotoxin and exotoxins**) that bind mole-
                        cules on host cell surfaces. They have pili that enable them to adhere to human
                        epithelial cells and to exchange DNA (virulence genes) between strains. They
                        have the capability to **vary their antigens rapidly** and undermine host defense
                        mechanisms. *N. meningitidis* also has a capsule that enables it to survive the
                        hostile intravascular environment. Different strains of meningococci possess
                        different capsules, which are categorized as serogroups, e.g. A, B, C. Some of
                        these are immunogenic. The majority of cases of meningococcal meningitis in
                        the UK are caused by serogroup B strains.

Epidemiology   Meningococcal and gonococcal infections are **endemic** worldwide. Humans are
                        the only host for these organisms. *N. gonorrhoeae* inhabits the mucous
                        membranes of the genital tract and anal canal and, less commonly, the
                        oropharynx and synovial membranes. The organisms are transmitted between
                        hosts via direct or sexual contact. Transmission to neonates may occur at birth,
                        during passage through an infected birth canal.

                        Gonococci infect mainly sexually active young adults. In women, infection is
                        often asymptomatic, therefore, the true prevalence of gonococcal infection
                        remains unknown. It is thought to infect three million people in the USA alone.
                        Meningococci inhabit mainly the nasopharynx and are transmitted via droplets
                        or close kissing contact. Of the young adult population, 10–30% carry meningo-
                        cocci. Carriage is highest among enclosed populations, such as secondary
                        schools, universities and prisons. *N. meningitidis* infection is most common in
                        sub-Saharan Africa, between The Gambia and Ethiopia (the meningitis belt),
                        where tens of thousands of cases are reported each year. It is also common in
                        countries of the Middle East, Europe (including the UK) and Latin America. The
                        most vulnerable age groups are children aged < 5 years, particularly those < 2
                        years, followed by teenagers and young adults. The disease is more common
                        during autumn and winter seasons.

Clinical infections *Gonorrhea*
                        N. gonorrhoeae is the causative agent of gonorrhea, a sexually transmitted
                        infection. Its primary site of invasion is the genital tract of males and females,
                        although it can infect other body sites.

                        In males, *N. gonorrhoeae* causes acute **urethritis** with **dysuria** and a purulent
                        discharge. Ascending infection may result in **epididymitis, orchitis, prostatitis,
                        periurethral abscess**, and **urethral stricture** and **infertility**.

                        In females, the primary site of infection is the **endocervix**, usually accompa-
                        nied by urethritis. A cervical and/or vaginal discharge is usually present.
                        Dysuria, intramenstrual bleeding, and abdominal and/or pelvic pain are often
                        experienced. If untreated, the bacterium can invade other organs beyond the
                        genitalia (e.g. **salpingitis** leading to scarring and narrowing of the Fallopian
                        tubes, **ectopic pregnancy**, and **sterility**). It may disseminate via the bloodstream
                        to other organs, such as the skin and joints.

                        In newborns, infection of the conjunctiva (**ophthalmia neonatorum**) is the
                        common presenting symptom.

Meningococcal disease

N. meningitidis primarily causes **meningitis and/or septicemia** which can kill rapidly. It is the most common cause of bacterial meningitis in the UK and the only bacterium capable of generating epidemic outbreaks of meningitis. Septicemic patients will suffer extensive disease as a result of endotoxin-generated deranged immunological and clotting mechanisms, which can precipitate **shock and multi-organ failure**. Classically, meningococcal disease is associated with purpuric skin rashes which do not blanch under pressure, caused by subcutaneous hemorrhages.

The bacterium can invade and damage other organs of the body, e.g. the conjunctiva, lungs, joints. Meningococcal disease is one of the few microbiological medical emergencies requiring urgent management.

Laboratory diagnosis

To select *Neisseria* species from other contaminating bacteria (which inhabit the genital tract or the throat), **selective culture media** (containing vancomycin, trimethoprim, and nystatin) are used. Oxidase-positive colonies of Gram-negative diplococci are presumptively identified as *Neisseria* species.

In gonorrhea, direct smears, obtained from the male genital or eye discharge, can be immediately Gram-stained. This may reveal intracellular diplococci in segmented neutrophils. The female genital tract is populated with normal flora that confuse accurate diagnosis by Gram stain.

In meningococcal meningitis and/or septicemia, the cerebrospinal fluid (CSF), blood and throat swabs are urgently examined. The CSF **deposit is Gram-stained and cultured**. The CSF will contain high numbers of white cells (mainly polymorphonuclear cells), high protein levels and reduced sugar levels.

Identification of *Neisseria* species can be confirmed using biochemical reactions. *N. gonorrhoeae* produces acid from glucose, but not from maltose, whereas *N. meningitidis* produces acid from both sugars. Genetic methods of identification (e.g. **PCR**) are also available. *N. meningitidis* isolates are routinely serogrouped.

Treatment

N. gonorrhoeae is sensitive to many antibiotics, including **penicillin** and **ciprofloxacin**. However, resistance to these agents is spreading worldwide and penicillinase-producing *N. gonorrhoeae* are fairly common. Fluoroquinolones are typically effective, although resistance against these drugs is also emerging.

N. meningitidis is also sensitive to several antibiotics, including **penicillin** and **cephalosporins**. Resistance is not a major issue in meningococcal management. Patients with meningococcal septicemia will require additional life-supporting interventions, depending on severity. Nasopharyngeal carriage of meningococci in patients and close kissing associates is eradicated with ciprofloxacin, rifampicin or ceftriaxone.

Prevention and control

No vaccine is available against *N. gonorrhoeae*. Prevention is by practising **safe sex**, e.g. using condoms. Sexually active women who carry the organism asymptomatically act as reservoirs for the organism and may also develop long-term complications if untreated. Therefore, all teenagers and young adult females should be regularly screened for gonococcal carriage. This is not always easy.

There are **vaccines available** against serogroups A, C, W136 and Y meningococci. Serogroup C vaccines are now routinely offered to all infants in the UK. No suitable vaccine is available for serogroup B organisms. Prevention is by early detection and treatment of disease.

C11 LEGIONELLA

Key Notes

Microbiology	*Legionella* are Gram-negative bacteria that are found in both natural and man-made water systems. They live and multiply principally within free-living amoebae. There are > 30 *Legionella* species but most human infections are caused by *Legionella pneumophila* serogroup 1.
Epidemiology	Transmission occurs principally by inhalation of contaminated aerosol droplets from cooling towers and other sources including showers and fountains. Hot and cold water systems may also be a reservoir in large buildings (such as hospitals and hotels). Infections are more common during warmer months and are often related to foreign travel. Infections are more common in men, between 50–70 years. Risk factors include smoking and immunosuppression. Legionnaires' disease is usually a rare cause of pneumonia (only 2–3% of community-acquired pneumonia), but outbreaks may occur. The incubation period is 2–10 days.
Clinical infections	There are two main types of *Legionella* infection: Legionnaires' disease (a type of pneumonia) and Pontiac fever (a milder self-limiting 'flu-like illness). Legionnaires' disease can be severe with an overall mortality rate of 10–15%.
Laboratory diagnosis	*Legionella* bacteria can be grown on specialised agar within 3–10 days from respiratory tract samples (e.g. sputum). Rapid diagnosis of *L.pneumophila* infections can be made by detection of legionella antigen in urine or respiratory tract samples. Antibody responses to legionella can be detected by a variety of serological tests in paired (acute and convalescent) blood samples.
Treatment	Suitable antibiotic therapy is with macrolide or fluoroquinolone antibiotics. Rifampicin can be added for severe infections.
Prevention and control	Water systems should be maintained to minimize the proliferation of legionella, and to reduce the chance of aerosols. Cases of suspected or confirmed Legionnaires' disease should be investigated to determine the likely source of infection.
Related topics	Lower respiratory tract infections (F7)

Microbiology   *Legionella* are Gram-negative bacteria with a typical three-layer cell wall. The chemical structure of the lipid-sugar complexes is somewhat different with shorter sugar chains [lipo-oligosaccharide (LOS)] compared to that of enterobacteria [lipopolysaccharide (LPS)]. *Legionella* LOS contains endotoxin-like activity.

   Legionella are found in both natural and man-made **water systems**. They live and multiply principally within free-living **amoebae**, where they are capable of evading intracellular killing mechanisms. Following phagocytosis they survive within the phagosome and inhibit phago-lysozomal fusion. Man is an accidental host for *Legionella*, but once inside the lung *Legionella* can survive and multiply within alveolar macrophages in a similar manner. *Legionella* are not part of the normal human bacterial flora.

There are > 30 *Legionella* species. Not all of these cause human infections. Some species can be further subdivided into serogroups. Most human infections are caused by *Legionella pneumophila*. There are ≥ 15 serogroups (SG) of *L. pneumophila*, but **L. pneumophila SG1** is the most pathogenic. Further subdivision of particular species/serogroups can be done by monoclonal antibody subtyping or genotypic **typing methods** such as pulsed gel electrophoresis (PFGE). This can be useful during **outbreak investigations** for comparison of strains isolated from human infections, with those obtained from water samples of suspected sources.

Identification of different *Legionella* species involves a variety of phenotypic tests, serotyping or genotypic analysis. Some species exhibit autofluorescence when exposed to long-wavelength UV light.

Epidemiology

Incidence

Legionnaires' disease is usually a rare cause of pneumonia (only 2–3% of community-acquired pneumonia), although outbreaks may occur. Pontiac fever appears to be less common than Legionnaires' disease, but may be under-diagnosed.

Incubation period

The incubation period is usually 2–10 days.

Route of transmission

The main route of transmission is by inhalation of **aerosol droplets** containing *Legionella* bacteria. Droplets of a suitable size can arise from a number of sources. The principal source is from **wet cooling towers** or evaporative condensers used in air-conditioning or other cooling systems. Other sources include showers, decorative fountains, and whirlpool spas.

Hot and cold water systems may also be a reservoir in large buildings (such as hospitals and hotels). Large buildings contain complicated plumbing systems that are prone to colonization and proliferation with *Legionella*. Transmission may occur from showers or other aerosol-generating procedures.

A minority of cases that occur within hospital result first from *Legionella* colonization of the mouth/throat (oropharynx) followed by subsequent aspiration of legionella into the lungs. This is more likely in very immunosuppressed patients on intensive care units.

Seasonal variation

Infections are generally more common during warmer months and are often related to foreign travel. *Legionella* is more likely to multiply within water systems at water temperatures between 20 and 40°C.

Demographics and risk factors

Infections are more common in men between 50–70 years. Risk factors include smoking and immunosuppression (e.g. recipients of organ transplants or patients undergoing cancer chemotherapy). Infections with *Legionella* other than

L.pneumophila SG1 are more often encountered amongst immunosuppressed patients.

Clinical infections   There are two main types of *Legionella* infection: Legionnaires' disease (a type of pneumonia) and Pontiac fever (a milder self-limiting 'flu-like illness).

Legionnaires' disease
Legionnaires' disease usually presents with a slow onset of fever, dry (non-productive) cough and shortness of breath. Other symptoms such as diarrhea and confusion (disorientation) are common. The presence of pneumonia is usually confirmed by chest X-ray. Legionnaires' disease can cause **severe pneumonia**, and some patients are admitted to intensive care units for support of their ventilation (breathing). The overall mortality is 10–15%.

Pontiac fever
Pontiac fever is a milder illness that presents with fever and nonspecific 'flu-like symptoms including myalgia (muscle aching), but without pneumonia. It is self-limiting and usually does not require antibiotic treatment.

A variety of other infections may rarely be caused by *Legionella* (e.g. cellulitis, pericarditis).

Laboratory diagnosis   The laboratory diagnosis of *Legionella* infection can be made by:

(i)   culture;
(ii)   antigen detection;
(iii) serology (antibody detection);
(iv)   genome detection.

Culture
Legionella bacteria can be grown on specialized agar (e.g. BCYE – buffered charcoal yeast extract agar) within 3–10 days from appropriate respiratory tract samples (e.g. sputum or lung washings [broncho-alveolar lavage fluid]). A positive culture is 100% diagnostic as *Legionella* is not a normal commensal of humans. Culture allows for further subtyping and genomic analysis of strains, and for the detection of rare species/serogroups that are missed by antigen/serological detection (see below). The disadvantages of culture are the low sensitivity and delay in results.

Antigen detection
Rapid diagnosis of *L.pneumophila* infections can be made by detection of *Legionella* antigen in urine or respiratory tract samples. **Urine antigen tests** are fairly rapid and have good sensitivity and specificity. These primarily detect infection caused by *L. pneumophila* SG1. Antigen testing of respiratory samples by direct immunofluorescence is less sensitive.

Serology
Antibody responses to *Legionella* can be detected by a variety of serological tests in paired (acute and convalescent) blood samples. A 4-fold rise in antibody titer that is temporally related to symptoms is diagnostic. Antibody responses can be delayed (up to 6 weeks) or absent in some patients. Single high antibody titers are also suggestive of recent infection. IgM antibody is *not* useful in distin-

guishing acute infection as it can be detected for several months. A number of cross-reactions can occur in serological tests, notably with campylobacter.

Genome detection
Amplification and subsequent detection of *Legionella* DNA from clinical samples (e.g. PCR) allows for rapid detection of *Legionella* infection. This time-consuming and expensive technology is not widely used for *Legionella* diagnosis outside of reference laboratories.

Treatment

Suitable antibiotic therapy for Legionnaires' disease is with **macrolides** (e.g. erythromycin or clarithromycin). Alternative treatment is with **fluoro-quinolones** (e.g. ciprofloxacin or levofloxacin). Rifampicin can be added for severe infections. These antibiotics are highly active against *Legionella in vitro* and are able to achieve therapeutic concentrations inside human macrophages (in which the *Legionella* reside).

Prevention and control

Man-made water systems should be maintained to minimize the proliferation of *Legionella*, and to reduce the chance of escape of aerosols. Maintenance of cooling towers includes regular dosing with a biocide, minimizing build-up of organic substances and biofilm, preventing corrosion, and checking the integrity of the aerosol drift eliminator. Cooling towers should be registered with the appropriate local authority and regularly sampled. *Legionella* growth within hot and cold water systems should be minimized by temperature control (i.e. avoidance of the temperature range 20–45°C) or by the addition of biocides. Pipework configurations should avoid 'dead-legs' and outlets should be regularly flushed to prevent the build-up of stagnant water, if they are not in regular use. Once *Legionella* has colonized a water system, it can be difficult to eradicate as the bacteria can live and multiply within the semi-protected environment of a biofilm.

Cases of suspected or confirmed Legionnaires' disease should be investigated to determine the likely source of infection. Local, regional, national and international reporting schemes maintain **surveillance** for linked cases of Legionnaires' disease. Outbreak investigations comprise careful epidemiology and sampling of potential water sources, with microbiological typing of environmental and clinical isolates.

C12 CAMPYLOBACTER AND HELICOBACTER

Key Notes

Microbiology	*Campylobacter* and *Helicobacter* are helical, strictly micro-aerophilic, Gram-negative and oxidase-positive bacteria.
Pathogenesis	These bacteria are motile (possess flagella) and are antigenically variable. *H. pylori* produces potent enzymes and exotoxins, including urease and VacA.
Epidemiology	*C. jejuni* and *H. pylori* are prevalent worldwide. *C. jejuni* colonizes animal gut and is transmitted to humans via meat and other animal products. *H. pylori* infects humans but its route of transmission is not clear.
Clinical manifestations	*C. jejuni* causes food-related gastroenteritis. Occasionally, *Campylobacter* species cause reactive diseases (immune-mediated), including Guillain–Barré syndrome and reactive arthritis. *H. pylori* causes peptic ulcers and stomach cancer and lymphoma.
Diagnosis	Diagnosis is by isolation of the responsible organism from feces (*C. jejuni*) or gastric biopsies (*H. pylori*). The latter is more commonly diagnosed by the urea breath test.
Treatment	*C. jejuni* diarrhea is self-limiting. *H. pylori* infection requires triple therapy, combining two antibiotics (e.g. amoxicillin and clarithromycin) with a proton pump inhibitor.
Prevention	*C. jejuni* prevention is by good kitchen practise and safe handling of cooked and uncooked food. *H. pylori* is not easily prevented.
Curved bacteria	*Campylobacter* and *Helicobacter* are grouped here because they are both curve (helical)-shaped, micro-aerophilic, Gram negative and oxidase positive, and they share many more phenotypic and genetic characteristics.

Campylobacter   *Campylobacter* and **thermophilic** and strictly **micro-aerophilic** bacteria, i.e. grow best at 43°C in the presence of 5–10% of oxygen, 10–15% CO_2, and nitrogen. The organisms are motile, expressing a single **flagellum** at one or both poles. They colonize various **poultry** and domestic animals.

Three main species are pathogenic to humans, *C. jejuni* (mainly from chickens), *C. coli* (mainly from pigs) and *C. lari* (mainly from seagulls). The *Campylobacter* are antigenically highly variable and capable of mutation and horizontal exchange of DNA.

C. jejuni is by far the most common cause of **gastroenteritis** in the UK and is associated with **food-poisoning**. This is manifested by a protracted diarrhea which can be bloody. The organism is also associated with post-infection reactive syndromes which are immunologically precipitated. These include the **Guillain–Barré syndrome** and reactive arthritis.

C. jejuni is a commensal of chickens and heavily contaminates meat and other products. Diagnosis is by isolation of the organism from feces. Treatment is by **fluid replacement** and rest. Antibiotics are not necessary. Prevention is by good kitchen practise and safe handling of cooked and uncooked food.

Helicobacter

Helicobacter pylori is the only pathogenic species of this genus and is increasingly recognized as a causative agent of many diseases related to the stomach and duodenum, including chronic active **gastritis,** chronic **peptic ulcers,** gastric **cancer** and gastric **lymphoma.**

The bacteria have much in common with *Campylobacter*, e.g. they have similar shape and size, are both micro-aerophilic, motile and grow on the same media. *H. pylori* expresses multipolar flagella and a powerful **urease enzyme** which is necessary to neutralize the strong stomach acidity in the microenvironment of the organism. The **vacuolating toxin, VacA**, is a potent toxin released by *H. pylori* and is a key virulence factor.

H. pylori is endemic worldwide and infection is life-long unless eradicated by multiple antibiotics in the presence of a proton-pump inhibitor. Amoxicillin, tetracycline, metronidazole and clarithromycin are all effective, although resistance is spreading, especially in developing countries.

Diagnosis is usually carried out by the **urea breath test**. Here, the patient is offered a drink containing radiolabelled urea which will be split by the bacterial urease enzyme in the stomach. This reaction releases radiolabelled CO_2 which is then detected in the patient's breath. The organism can be isolated from gastric biopsies and antibodies to it are detectable in the serum. Endoscopy is used to visualize ulceration, and demonstration of *H. pylori* in stained biopsy specimens is another method for diagnosis.

C13 VIBRIOS

Key Notes

Microbiology	*Vibrio* species are helical Gram-negative and oxidase-positive bacteria. Vibrios grow best in alkaline media and some species are halophilic, i.e. salt-tolerant. *Vibrio cholerae* is the main pathogenic species, responsible for 'cholera pandemics'. Rare vibrios include *Vibreo parahaemolyticus* (causes gastroenteritis) and *Vibrio vulnificus* (causes severe, often fatal, necrotizing cellulitis and septicemia).
Pathogenesis	*Vibrio cholerae* produces cholera toxin, which is the key virulence factor.
Epidemiology	*Vibrio cholerae* spreads fast via drinking contaminated water, leading to major epidemics.
Clinical manifestations	*Vibrio cholerae* causes cholera, consisting of profuse watery diarrhea.
Diagnosis	Diagnosis is by isolation of the responsible organism from feces
Treatment	Cholera is treated primarily with fluid replacement. Antibiotics (tetracycline) may also help.
Prevention	Cholera is preventable via water treatment, proper sewage disposal and good personal hygiene.

Vibrios
Vibrio cholerae is by far the most common and important pathogenic species of the genus.

Microbiology
Vibrio species are curve (helical)-shaped, Gram-negative and oxidase-positive bacteria. They are actively motile, expressing a single polar flagellum. *Vibrio cholerae* expresses an antigenically diverse endotoxin (O antigen) which forms the basis of its serological classification. Serogroups O1 and O139 cause cholera, whereas all the other serogroups cause milder diarrheal illness, and are hence called the 'noncholera vibrios'.

Serogroup O1, the main cause of several major cholera epidemics, is divided into two biotypes (classic and El Tor) on the basis of biological and biochemical activities. Each of these biotypes can be subdivided into Ogawa, Inaba and Hikojima serotypes.

Pathogenesis
Vibrio cholerae produces a potent exotoxin, **the cholera toxin**, which is the single most important virulence factor responsible for the clinical manifestations of cholera. The toxin recognizes specific receptors on the intestinal epithelial cells, enters the cytoplasm and changes the cellular physiology. It does not kill the enterocytes, but makes them release salt and water into the gut lumen at a much greater speed than can be absorbed – leading to the profuse diarrhea.

Epidemiology

Vibrio cholerae infects humans and **contaminates water**. It remains viable in clean or sea water for days or weeks, but dies readily in polluted water. Spread is fast and efficient via drinking contaminated water, leading to major epidemics or pandemics. The disease is endemic in the Asian subcontinent, South East Asia and other tropical regions.

Clinical manifestations

Vibrio cholerae infection classically leads to **profuse and often fatal watery diarrhea**. The stool consists of rice-colored liquid with minimal or no fecal material. The patient may lose tens of liters of water per day.

Diagnosis

Diagnosis is by recognition of the signs and symptoms of disease and isolation of the organism from feces.

The bacterium grows readily on ordinary laboratory media over a wide temperature range and optimally at an alkaline pH of 8–8.2. Alkaline peptone water or TCBS (thiosulfate–citrate bile sucrose) agar, both with pH of 8.6, are used as selective growth media. They suppress other gut bacteria while promoting growth of *Vibrio cholerae*.

Treatment

The mainstay of management is by fluid replacement. Antibiotics (such as tetracycline) are secondary to rehydration.

Prevention

Prevention is by protection of drinking water, treatment of sewage and good personal and food hygiene.

Vibrio parahaemolyticus is a salt-tolerant (halophilic), sucrose-negative and water-related organism which infects shellfish, mainly in South East Asia. It causes gastroenteritis.

Vibrio vulnificus is also a halophilic shellfish-associated organism associated with severe, often fatal, necrotizing cellulitis and septicemia. It is treated with tetracycline, ciprofloxacin or ceftriaxone.

C14 CLOSTRIDIA

Key Notes

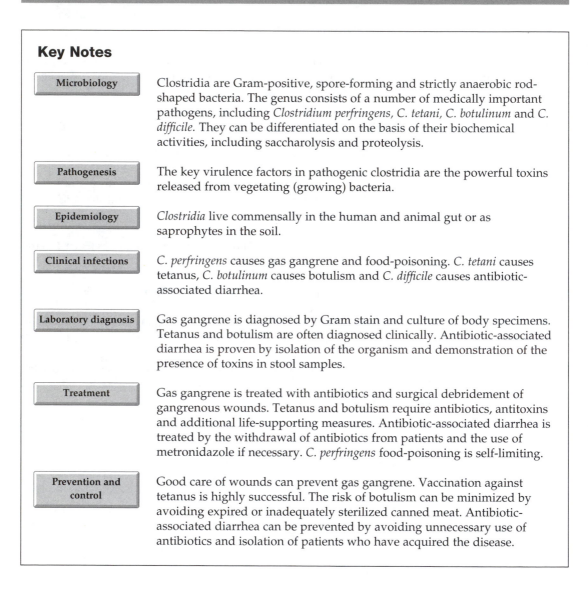

Microbiology	Clostridia are Gram-positive, spore-forming and strictly anaerobic rod-shaped bacteria. The genus consists of a number of medically important pathogens, including *Clostridium perfringens, C. tetani, C. botulinum* and *C. difficile.* They can be differentiated on the basis of their biochemical activities, including saccharolysis and proteolysis.
Pathogenesis	The key virulence factors in pathogenic clostridia are the powerful toxins released from vegetating (growing) bacteria.
Epidemiology	*Clostridia* live commensally in the human and animal gut or as saprophytes in the soil.
Clinical infections	*C. perfringens* causes gas gangrene and food-poisoning. *C. tetani* causes tetanus, *C. botulinum* causes botulism and *C. difficile* causes antibiotic-associated diarrhea.
Laboratory diagnosis	Gas gangrene is diagnosed by Gram stain and culture of body specimens. Tetanus and botulism are often diagnosed clinically. Antibiotic-associated diarrhea is proven by isolation of the organism and demonstration of the presence of toxins in stool samples.
Treatment	Gas gangrene is treated with antibiotics and surgical debridement of gangrenous wounds. Tetanus and botulism require antibiotics, antitoxins and additional life-supporting measures. Antibiotic-associated diarrhea is treated by the withdrawal of antibiotics from patients and the use of metronidazole if necessary. *C. perfringens* food-poisoning is self-limiting.
Prevention and control	Good care of wounds can prevent gas gangrene. Vaccination against tetanus is highly successful. The risk of botulism can be minimized by avoiding expired or inadequately sterilized canned meat. Antibiotic-associated diarrhea can be prevented by avoiding unnecessary use of antibiotics and isolation of patients who have acquired the disease.

Microbiology

The clostridia are a group of strictly anaerobic large (4–6 x 1μm) Gram-positive bacilli. They can be pleomorphic. Some tolerate small amounts of oxygen. They form **spores** which can be centrally or terminally positioned causing bulging within the cell. They are soil **saprophytes** or normal commensals of the human and animal gut. However, they are capable of causing deadly diseases, which are invariably mediated by **potent exotoxins**.

In addition to their colonial and microscopic morphologies, *Clostridium* species can be differentiated on the basis of their biochemical activities. Many species break down sugar (saccharolysis) and/or protein (proteolysis) molecules. These activities can be detected and used for species differentiation. The ability to produce aromatic fatty acid end-products (detected by gas–liquid chromatography) can also be used for species differentiation.

Toxigenicity of isolates is often demonstrated in the laboratory using various appropriate techniques, e.g. observing cytopathic effect on human cells, or neutralization of enzymatic effect on substrates using antibodies.

The clostridia, like all other medically important anaerobic organisms, are sensitive to metronidazole and clindamycin. They are also sensitive to many other antibiotics, including penicillin and erythromycin, but resistant to amino-glycosides, which are therefore added to culture media to enhance the selective isolation of these organisms.

Clostridium perfringens

C. perfringens rarely forms spores under normal laboratory conditions. On blood agar, it produces a characteristic **double zone of β-hemolysis**, a narrow transparent zone and a wide shadowy zone. It is mainly saccharolytic and produces acid and gas from milk, producing a 'stormy clot' in a test tube (litmus milk test).

C. perfringens produces a **number of potent toxins**, the most important of which is the α-toxin (phospholipase C) which causes host cell lysis. This toxin is produced by all isolates of *C. perfringens*. Different strains of the organism produce any of another 11 well-known toxins, including collagenase, proteinase, hyaluronidase and deoxyribunuclease. Based on these toxins, the *C. perfringens* strains are divided into five types, A–E. *C. perfringens* uses these toxins to kill human muscle cells, causing necrosis (**myonecrosis**). Death of these cells creates an even more suitable anaerobic condition where the organism can grow rapidly and release gas, hence the disease is known as '**gas gangrene**'. Infected and discolored blood inside the wound and under the skin turns the infected area to black. Gas gangrene can be caused by other, less common clostridia, including *C. novyi*, *C. septicum*, *C. histolyticum* and *C. sordelli*.

Gas gangrene is treated by surgical debridement of infected areas and the use of high doses of penicillin given intravenously.

Some strains of *C. perfringens* also produce enterotoxins which, when ingested in large numbers, can cause food-poisoning (diarrhea, vomiting and abdominal pain). This is self-limiting and does not require treatment. For prevention, see topic E9.

Clostridium tetani

C. tetani is a straight, slender, anaerobic Gram-positive bacillus with a rounded terminal spore which looks like a 'drumstick' under the microscope. It is Gram positive, but readily loses the Gram stain. The organism is motile, lives in the human and animal gut and contaminates soil, mainly from cattle feces.

C. tetani produces an **extremely potent single toxin** (tetanus toxin) which underlies the pathogenesis of tetanus. Tetanus toxin consists of two components, including the neurotoxic 'tetanospasmin' and the hemolytic tetanolysin. The toxin **prevents muscle relaxation**, leading to persistent contraction of facial and body muscles. Simultaneous contraction of opposite groups of muscles leads to a characteristic grin on the face (*risus sardonicus*) and spastic posture of the limbs and body trunk.

The organism is rarely isolated from the site of bacterial entry, which may not even be visible. Therefore, diagnosis is made mostly on clinical grounds, the features of which are characteristic.

Tetanus is fatal in the absence of rapid and high quality management. The organism is sensitive to penicillin which is administered along with a specific anti-tetanus antibody (**antitoxin**). The infected wound should be debrided and the patient given a booster dose of tetanus vaccine.

Tetanus is now extremely rare in industrialized countries, due to the availability of a safe and highly effective vaccine offered to all children as part of a childhood immunization programme. Booster doses are given to toddlers and young adults.

Clostridium botulinum

C. botulinum strains are motile, anaerobic, Gram-positive bacilli which form subterminal spores. They are ubiquitous and contaminate water and meat-containing canned food, including canned fish, liver pâté and sausages. *C. botulinum* causes botulism which is a severe, **usually fatal, form of food-poisoning**. The bacterium produces **the most powerful toxin** known to man, which is the key virulence factor responsible for the pathogenesis of disease. Its action is the reverse of that of tetanus toxin, i.e. it prevents muscle contraction, leading to flaccid paralysis of important muscles. It inhibits the release of acetylcholine at **motor nerve endings** in the **parasympathetic nervous system**. This potent toxin is relatively heat-resistant (although destroyed by temperatures > 60°C), hence it is considered by the military as a suitable bioweapon. There are seven serotypes of botulinum toxin, named A–G. These act in almost identical ways; however, only types A, B and E cause human botulism.

C. botulinum can be isolated from left-over (suspect) food items. Isolates can be tested in the laboratory for toxin production. Traces of toxin can be found in food items or patient's serum. Laboratory animals are occasionally used to confirm the diagnoses.

Treatment is by removal of undigested food, injecting antitoxins and intensive therapy. Patients will require assistance with breathing and eating due to muscular paralysis.

Botulinum toxin (Botox™) is sometimes used as a medical treatment for spastic paralysis of facial or bladder muscles.

Clostridium difficile

C. difficile is found in the feces of 3–5% of humans, in the gut of several animals and in the environment. The organism produces at least two potent toxins that are responsible for severe and occasionally fatal diarrhea. The organism is harmless in the normal gut where it cannot compete successfully against the resident gut flora. This competitive environment provides a useful barrier (**colonization resistance**) against *C. difficile* and many other pathogens. Administration of broad-spectrum antibiotics in vulnerable patients, e.g. elderly inpatients, removes the competitive barrier in the gut, allowing *C. difficile* to grow and produce toxins. The latter then cause '**antibiotic-associated' colitis**, ranging in severity from mild diarrhea to overwhelming **pseudomembranous colitis.**

Treatment is by withholding antibiotics where possible, replacing body fluids, administration of oral metronidazole or vancomycin and isolating the patient to prevent further spread of disease. Widespread dispersal of *C. difficile* spores in hospital environment can lead to nosocomial infection.

C15 GRAM-NEGATIVE ANAEROBIC BACTERIA

Key Notes

Microbiology	Almost all pathogenic Gram-negative anaerobic bacteria (GNAB) are bacilli. They include the genera *Bacteroides, Prevotella, Porphyromonas, Fusobacterium* and *Liptotrichia*. Each genus is divided into numerous species. GNAB vary in their antibiotic resistance and biochemical reactions.
Pathogenesis	GNAB vary in their pathogenic capabilities. Some, e.g. *Fusobacterium necroforum*, can cause severe disease. They have a typical Gram-negative cell envelope with endotoxin on its outer layer. Some produce potent exotoxins.
Epidemiology	GNAB inhabit the human gut, female genital tract and oropharyngeal area.
Clinical features	GNAB cause abdominal and pelvic (gynecological) sepsis. They form abscesses in periodontal and oropharyngeal as well as other parts of the body including the brain and viscera. They are usually accompanied by other gut-related aerobes (mostly Enterobacteriaceae and occasionally enterococci). *Fusobacterium necroforum* causes a potentially fatal disease called 'necrobacillosis'.
Diagnosis	Isolation of the bacteria provides the only definitive diagnosis. Gas–liquid chromatography can be used to detect the presence of anaerobes in clinical specimens.
Treatment	All GNAB are sensitive to metronidazole, which is the drug of choice. Clindamycin is also effective. Cephalosporins are also used, in combination with metronidazole, to kill accompanying enterobacteria. Surgical drainage of pus is usually necessary.
Prevention	Difficult.

Related topics	Lower respiratory tract infections (F7)	Gastroenteritis and food-poisoning (F8)

Microbiology

The vast majority of the medically important Gram-negative anaerobic bacteria (GNAB) are bacilli. Most were previously all grouped under the single genus of *Bacteroides*. However, after extensive revision, there are now three major genera: *Bacteroides, Prevotella* and *Porphyromonas*.

Each genus is divided into numerous species which display distinct biochemical characteristics. However, in clinical microbiology laboratories, these three genera are still grouped together because they are morphologically similar, cause similar diseases and are all sensitive to metronidazole.

Other important genera include *Fusobacterium* and *Liptotrichia*, each comprising several medically important species.

Strictly anaerobic Gram-negative cocci rarely cause disease. *Veillonella* is the only known genus and is often considered a passenger, i.e. carried along with other more virulent pathogens. It is also sensitive to metronidazole.

The GNAB genera and species vary in their antibiotic resistance, biochemical reactions (e.g. proteolysis and saccharolysis) and production of aromatic fatty acids. These differences have been exploited in the laboratory for their identification. Some produce pigments and others do not.

Pathogenesis

GNAB vary in their pathogenic capabilities. Some, e.g. *Fusobacterium necroforum,* have strong pathogenic potential and cause severe disease. In common with other Gram-negative bacteria, GNAB have outer membranes, thick peptido-glycan layers and inner plasma membranes. They express **endotoxins**, outer membrane proteins and secrete **toxins** such as hemolysin. GNAB also cause suppurative infections and in normally sterile sites, they can cause abscess formation. They produce a characteristically putrid smell.

Epidemiology

GNAB live in humans and animals. Different genera colonize different parts of the human body. These include mainly the gut (all genera but mainly *Bacteroides*), female genital tract (mostly *Prevotella,* also *Porphyromonas)* and the mouth/oropharynx (mostly *Prevotella,* also fusobacteria and liptotrichia). Each gram of feces contains tens of billions of GNAB.

Clinical features

GNAB are important pathogens which are frequently **associated with abdominal and pelvic sepsis**, including gynecological infections, and periodontal diseases. They are usually found along with other aerobic Gram-negative (e.g. Enterobacteriaceae) and Gram-positive (e.g. enterococci) gut bacteria, particularly in cases of fecal peritonitis or mediastinitis subsequent to gut or esophageal perforation, respectively. The GNAB and aerobes are believed to act in synergy. *Bacteroides fragilis* is the most frequently implicated pathogen among the GNAB. The lipopolysaccharide associated with this GNAB is unique in that it lacks endotoxic properties.

GNAB may **cause abscesses** in many parts of the body, including the periodontal area, oropharynx, sinuses, bone, brain and viscera (e.g. liver).

In association with *Borrelia vincenti,* GNAB may precipitate '**Vincent's angina**' which is a painful oropharyngeal infection. *Fusobacterium necroforum* is a particularly serious, potentially fatal, organism causing multiple abscesses which spread rapidly from the throat through the neck and to the rest of the body. This condition is called 'necrobacillosis'.

Diagnosis

Definitive diagnosis is made **by isolation** of the culprit organisms. The Gram stain, anaerobic growth, colonial morphology and sensitivity to metronidazole are often sufficient in the clinical laboratory. **Selective media** containing neomycin, which inhibits Enterobacteriaceae, are often used for their rapid isolation.

Identification to species level is usually unnecessary; however, it can be achieved using antibiotic sensitivity, biochemical activity and gas–liquid chromatography. The latter can also be applied to samples of pus to indicate whether anaerobic organisms are involved.

Treatment   **Metronidazole is the drug of choice** for almost all GNAB. It is usually combined with other antibiotics, such as cephalosporins, to kill co-infecting enterobacteria. Clindamycin is also effective. Certain GNAB produce β-lactamases, therefore penicillin or amoxicillin alone may not be effective without a β-lactam inhibitor. Surgical drainage of pus is usually necessary.

Prevention   Difficult, except for periodontal diseases where good dental hygiene is important.

C16 SPIROCHAETES

Key Notes

Microbiology	The spirochaetes are long, spiral (or helical)-shaped Gram-negative bacteria. They are very slender and weakly refractile, and therefore not easily visible under ordinary light microscopy. They are visualized with dark-ground microscopy, silver staining or immunofluorescence. Three main genera are known to cause disease in humans – *Treponema*, *Borrelia* and *Leptospira*.
Pathogenesis	Not well understood. Spirochaetes can penetrate intact or broken skin and mucous membranes. They are antigenically variable, hence able to evade host immune responses and establish chronic diseases. They have a three-layered cell wall, typical of Gram-negative bacteria, but (uniquely) have axial filaments in their periplasm. They produce secreted proteins which assist with their invasion and dissemination.
Epidemiology	The spirochaetes are prevalent, and cause disease, worldwide. However, within each genus, some species are more geographically restricted than others. *Treponema* infect humans and spread by direct contact. *T. pallidum* is spread via sexual contact or vertically to the fetus. Most *Borrelia* species infect animals and spread to humans via ticks or lice. *B. recurrentis* is an obligate human pathogen, transmitted between people via the body louse. *Leptospira* infect animals and are spread via infected urine.
Clinical manifestations	*T. pallidum* causes syphilis, which manifests itself in several phases, including primary (chancre, painless ulcer), secondary (systemic dissemination of the organism), latent (silent illness) and tertiary syphilis (neurological and cardiovascular manifestations). *T. endemicum* causes nonvenereal syphilis. *T. pertenue* causes yaws. *T. carateum* causes pinta (skin depigmentation). These species are indistinguishable from *T. pallidum* and may represent subspecies that cause skin infections transmitted by direct contact in tropical areas. *B. vincenti* causes Vincent's angina. In addition to *B. recurrentis*, several other *Borrelia* species cause relapsing fever. *B. burgdorferi* causes relapsing fever and Lyme disease. *L. interrogans* causes leptospirosis.
Diagnosis	Bacteria may be seen in clinical specimens but are not usually grown. Detection of antibodies by serology is the main method of diagnosis. Molecular methods of DNA detection (e.g. by PCR) are increasingly available.
Treatment	Most spirochaetes are sensitive to penicillin and tetracycline.
Prevention	Depends on the method of transmission, and is mainly based on avoiding direct contact with infected patients or with vectors.

Spirochaetes   The spirochaetes are a large group of long and spiral (or helical)-shaped anaerobic bacteria. They are **fastidious** and their cell structure is that of Gram-negative bacteria, although some species stain very poorly with Gram stain, if at all.

They have a three-layered cell envelope consisting of an outer membrane, a thin peptidoglycan layer within the periplasmic space and an inner membrane. Uniquely, they have several buried flagella, called 'axial filaments', which run along the axis of the organism within the periplasmic space (*Fig. 1*). The structure and function of these axial filaments give rise to the spiral shape and the rotary movement of the organisms.

They are very slender and weakly refractile, and are therefore not easily visible under ordinary light microscopy. They are **visualized with dark-field microscopy**, silver staining or immunofluorescence.

Three main genera are known to cause disease in humans, including *Treponema*, *Borrelia* and *Leptospira*.

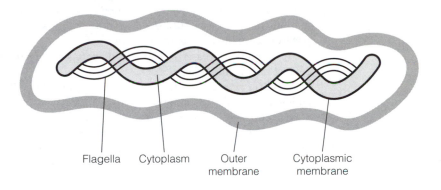

Flagella   Cytoplasm   Outer   Cytoplasmic
                         membrane   membrane

Fig. 1.   The structure of a typical spirochaete. Note the flagella which run along the longitudinal axis of the bacterium, within the periplasmic space.

Treponema   *Treponema pallidum* is the major pathogenic species, followed by *T. carateum*.
infections   *T. pallidum* is divided in to several subspecies, three of which are known to cause disease, including *pallidum, pertenue* and *endemicum*. For ease of reference these are labelled *T. pallidum, T. pertenue* and *T. endemicum*.

All treponemes are morphologically indistinguishable and share much antigenic structure. They have not yet been cultivated in the laboratory, but rabbit testes can be inoculated and used for propagation. Treatment of treponemal infections is with penicillin.

Syphilis

T. pallidum is always pathogenic and causes syphilis, which is a **sexually transmitted disease** of serious consequences. The disease is widespread globally. It has been relatively uncommon in the UK and many Western countries, but is currently increasing.

'**Primary' syphilis** is a painless ulcer (chancre) which develops at the site of first entry of the organism, on the skin or mucous membranes. Untreated, the organism can then disseminate throughout the body and cause '**secondary' syphilis** manifest by a systemic illness and skin rash. The organism will then enter a 'latent' phase during which it may or may not produce symptomatic episodes.

Over the subsequent decades the widely disseminated organism progressively destroys body organs, particularly the central nervous and cardiovascular systems, through a protracted inflammatory process. This is called 'tertiary' syphilis. A serious consequence of syphilis is infection of the fetus during pregnancy (congenital syphilis). Infected babies either die (before or after birth) or suffer serious multi-organ disease.

Primary syphilis may be diagnosed by detection of the organism on dark-ground or fluorescence microscopy; however, the disease is usually diagnosed serologically by detection of antibodies.

Antibodies to the nontreponemal cardiolipin-lecithin-cholesterol can be detected using any one of the following tests, none of which are highly specific to venereal syphilis: VDRL (venereal disease reference laboratory test), RPR (rapid plasma reagin) and ART (automated reagin test). More specific tests for treponemal antigens include TPHA (*T. pallidum* hemagglutination test), FTA-ABS (fluorescent treponemal antibody-adsorption test) and enzyme-linked immunosorbent assay (ELISA). These are more specific tests with fewer false positives than the nontreponemal ones.

Prevention of disease is by practising safe sex, effective management of cases, contact tracing and education.

T. endemicum causes a nonvenereal syphilis-like disease called 'bejel', or endemic syphilis. It is endemic in Africa, West Asia and Australia and mainly affects children in rural areas. Transmission is by direct person-to-person contact. It follows similar phases (primary, secondary and tertiary), like the venereal syphilis, but it does not involve the central nervous system.

T. pertenue causes 'yaws', which is similar to syphilis and follows the same phases, but it is not sexually transmitted. Primary ulceration is usually on the legs and spread is by direct contact. The disease is endemic in rural areas of Africa, South America and South East Asia.

T. carateum causes 'pinta', a skin condition characterized by itchy, scale-producing hypopigmentation and hyperkeratosis of the skin (without ulceration). It disfigures dark skin and is a social stigma. It is not sexually transmitted, spread is by direct contact and the disease occurs mainly in Central and South America.

Borrelia

The *Borrelia* species are long helical bacteria which, unlike most other spirochaetes, stain well with Gram stain (Gram negative). Several species are known to cause disease in humans.

Borrelia vincenti is a commensal inhabitant of the human oropharynx and is strictly anaerobic. In association with other GNAB (e.g. *Leptotrichia buccalis)*, it can cause gingivostomatitis and Vincent's angina. Diagnosis is by Gram stain of oropharyngeal exudates and treatment is with penicillin (and metroinidazole for mixed infections).

Borrelia burgdorferi and several other species cause endemic tick-borne 'relapsing fever'. This disease occurs worldwide and is transmitted from small mammals (e.g. rodents) via the soft-bodied *Ornithodorus* ticks. *Borrelia recurrentis* is responsible for epidemic louse-borne relapsing fever. The organism is an obligate human pathogen and is transmitted between people via the body louse, *Pediculus humanus.*

Relapsing fever is characterized by a febrile generalized illness which relapses after periods of apparent improvement. It is diagnosed by detection of the organism in stained peripheral blood samples or by detection of anti-borrelia antibodies. Treatment is with tetracycline, chloramphenicol, penicillin or erythromycin. Prevention is by avoiding or eradicating the vectors.

Borrelia burgdorferi is also the causative agent of Lyme disease which is transmitted from a range of wild and domestic animals to humans by ixodid ticks. Three main subspecies are well described, including the *sensu stricto* (mainly USA) and those which occur largely in Europe, *afzeli* and *garinii*.

Three stages of Lyme disease are known:

(i)   stage 1: skin rash (erythema chronicum migrans) appears at the site of tick bite;
(ii) stage 2: a systemic illness occurring in some patients weeks or months later when patients suffer cardiac or neurological and musculoskeletal symptoms;
(iii) stage 3: chronic disease, occurring years later when patients present with chronic skin, nervious system or joint abnormalities.

Diagnosis is by clinical assessment, histological biopsies, serology and molecular identification (e.g. by PCR). Treatment is with penicillin, tetracycline or a cephalosporin.

Leptospira

Leptospira are long spirochaetes with numerous coils and hooked ends. They are obligate aerobes and grow best in a moist environment on enriched media at temperatures ~ 30°C. They infect animals and concentrate largely in the kidneys, hence their spread is via infected urine, as well as other body fluids and tissues. The bacteria survive in the environment for days.

Leptospira interrogans is the only pathogenic species in the family Leptospiraceae. It is divided into 23 different serogroups, split into > 200 serovars. Important serovars include *icterohemorrhagiae* (rats), *hardjo* (cattle), *canicola* (dogs and cats) and *pomona* (pigs). For ease of use, the name of the species is often omitted.

Leptospira cause **leptospirosis** or Weil's disease , a febrile illness with systemic upset. Patients may develop bleeding, jaundice and renal impairment. The disease is **biphasic** (bacteremic and immune phases) and is often fatal.

Diagnosis is by detection of the organism in urine (rare), of **antibodies** (serology) or of **DNA** (PCR).

The organism is sensitive to **penicillin** and **tetracycline**. Animals can be vaccinated as a measure to control the disease. Personal protection and hygiene are effective measures.

C17 ATYPICAL BACTERIA

Key Notes

Microbiology	'Atypical bacteria' is a collective term encompassing *Mycoplasma, Ureaplasma, Chlamydia, Chlamydophila, Rickettsia, Ehrlichia, Coxiella* and *Bartonella*. These are considered 'incomplete bacteria' because they lack some essential components present in other bacteria, and are incapable of independent survival (except for *Mycoplasma*). They infect viable eukaryotic cells for reproduction.
Pathogenesis	Most of the atypical bacteria are obligate intracellular pathogens. They vary in terms of their pathogenic capability, target organs and clinical syndromes caused. Details of their pathogenesis are not well understood.
Epidemiology	*Mycoplasma, Ureaplasma, Chlamydia* and *Chlamydophila* are ubiquitous in the human environments. Rickettsiae, Ehrlichiae and *Bartonella* are zoonotic infections transmitted by arthropods. They are geographically limited and are acquired via occupational or recreational activities. *Coxiella burnetti* is also a zoonotic infection but acquired directly from infected animals.
Clinical features	*M. pneumoniae , C. pneumoniae* and *C. psittaci* cause 'atypical pneumonia'. *Chlamydia trachomatis* causes trachoma and sexually transmitted diseases (including genitourinary infections and lymphogranuloma venereum). Rickettsiae species cause febrile illnesses, which vary in severity. Rickettsial infections are often preceded by a local skin lesion (eschar) at the site of arthropod bite. *Coxiella burnetti* causes Q-fever and endocarditis. *Ehrlichiae* species attack and destroy human white or red cells and cause febrile illnesses. *B. henselae* and *B. bacilliformis* cause cat-scratch fever and Oroya fever, respectively.
Diagnosis	Isolation of the organism (rarely carried out), detection of antibodies in serum or detection of DNA in specimens (by PCR).
Treatment	Most of the 'atypical' bacteria are sensitive to erythromycin, tetracycline or both. Penicillin is not effective against any of these organisms.
Prevention	Those species which are spread from animals, including those transmitted by arthropods, are usually acquired via occupational or recreational activities. Protective clothing and awareness are essential.

Atypical bacteria   'Atypical bacteria' is a collective term used to encompass *Mycoplasma, Ureaplasmas, Chlamydia, Chlamydophila,* Rickettsiae, Ehrlichiae, *Coxiella* and *Bartonella*. These are considered 'incomplete bacteria' because they lack some essential components present in all other bacteria.

The free-living mycoplasmas lack a 'true' bacterial cell wall. Their outermost layer is a flexible, triple-layered structure of protein and lipids (in some species containing cholesterol). *Chlamydia* and Rickettsiae are, rather like viruses, obligate intracellular parasites, i.e. unable to grow independently.

Mycoplasmas

Mycoplasmas are fastidious bacteria which are able to grow slowly (over several days) on nutritious soft agar and form 'fried-egg' colonies. They are not stainable with Gram's stain due to lack of cell wall and are not susceptible to penicillins or cephalosporins. They are widespread in the human environment and often contaminate laboratory tissue cultures. *Ureaplasma* is a closely related genus, previously known as 'tiny' or 'T' mycoplasmas.

The commonest species include *M. pneumoniae*, *M. hominis* and *Ureaplasma urealyticum*.

M. pneumoniae causes acute bronchitis, acute exacerbations of chronic bronchitis, bronchiolitis (in children under the age of two) and community acquired 'atypical' pneumonia (Topic F7). Although the organism can be grown in the laboratory, this is not easy and not carried out routinely. Diagnosis is mainly by serological methods, including the detection of specific IgM antibodies (by **latex agglutination** or ELISA) or a rising titre of antibodies (by complement fixation test). Erythromycin is the drug of choice.

M. hominis **and** *U. urealyticum* frequently colonize the genital tract of normal, sexually active men and women. *M. hominis* is associated with pelvic infection in women and post-abortal and post-partum fever. It may also cause 'atypical pneumonia'. *U. urealyticum* is associated with nonspecific urethritis, vaginitis and cervicitis. Treatment is with tetracycline.

Chlamydia and Chlamydophila

Chlamydia and *Chlamydophila* (previously all were called *Chlamydia*) are obligate intracellular parasites with a distinct life cycle that involves the production of small **elementary bodies** inside cells and fully grown infective **reticulate bodies**. Cells are small and have the smallest genome known amongst medically important bacteria (roughly a quarter of the size of *Esch. coli*).

Like many viruses, these organisms require the presence of eukaryotic cells for culture. Several medically important species are known, including *Chlamydia trachomatis*, *Chlamydophila pneumoniae*, *Chlamydophila psittaci* and *Chlamydophila abortus*. All of these species share common antigens which are detected by complement fixation tests. Specific immunofluorescent-labelled monoclonal antibodies are available to distinguish between them.

Chlamydia trachomatis forms iodine-staining, glycogen-containing **inclusion bodies** in eukaryotic cells. *Chlamydia* and *Chlamydophila* species are sensitive to tetracyclines, erythromycin or fluoroquinolones.

Chlamydia trachomatis (*C. trachomatis*) is divided into several serotypes, of which **serotypes A, B, Ba and C cause trachoma**. In this disease, the organisms bind to conjunctival cells to cause 'inclusion conjunctivitis'. Repeated and chronic inflammation may leave scars on the edges of the eyelids. These cause retraction of eyelashes which irritate the cornea and eventually cause corneal abrasion and scarring, leading to blindness.

C. trachomatis is the commonest cause of blindness in developing countries. A total of 600×10^6 cases are estimated to occur worldwide with $10–20 \times 10^6$ cases of blindness. Transmission is by direct contact, via contaminated fingers, or inanimate objects such as towels, or indirectly by flies. Treatment is by topical application of tetracycline and/or systemic use of erythromycin. Prevention is by improved personal hygiene

Serotypes D–K of *C. trachomatis* are considered one of the most common sexually transmitted diseases worldwide, causing **genitourinary infections**. In men, these give rise to urethritis, epididymitis, proctitis and conjunctivitis. Systemic spread can lead to **Reiter's syndrome.** In women, these serotypes cause urethritis, cervicitis, bartholinitis, salpingitis and follicular conjunctivitis. Salpingitis may cause narrowing of the Fallopian tubes and result in ectopic pregnancy and infertility. The bacteria can spread systemically and cause peri-hepatitis, arthritis and dermatitis. Neonates may acquire infection following passage through an infected birth canal, giving rise to **conjunctivitis** which is sometimes accompanied by interstitial pneumonitis. Treatment is by ocular tetracycline and/or systemic erythromycin.

Serotypes L1, L2 and L3 *C. trachomatis* cause another, distinct, **sexually transmitted disease**, lymphogranuloma venereum. This predominantly occurs in Africa, Asia and South America and its primary lesion includes an ulcerating papule (after 1–4 weeks of exposure) at the site of inoculation. This leads to fever, headache, myalgia, swelling of draining lymph nodes in the groin (inguinal buboes), and possible dissemination to other lymphoid tissues, giving rise to fever, pain, hepatitis, pneumonitis or meningoencephalitis. Diagnosis is by cell culture and treatment is with systemic tetracycline.

Chlamydophila psittaci and *Chlamydophila pneumoniae* cause severe 'atypical pneumonia'. *Chlamydophila psittaci* (psittacosis) is acquired from domestic, tropical and wild birds. *Chlamydophila pneumoniae* is endemic (not related to avian contact), transmitted from person to person and has a long incubation period. It is also associated with pharyngitis, persistent cough and sinusitis. Symptoms include fever, headache and myalgia. Chest X-rays show single lesions rather than patchy shadowing.

Diagnosis is by isolation of the organism, detection of its antigen from clinical specimens or detection of antibodies in patients' sera. Tetracycline or erythromycin can be used for treatment.

Chlamydophila abortus infects domestic animals (mainly pregnant sheep and goats) and is occasionally transmitted to pregnant farm workers. It can infect the fetus and cause abortion or stillbirth. Treatment is by intravenous erythromycin.

Rickettsiae

Rickettsiae are small Gram-negative bacteria which are **obligate intracellular parasites**. They grow in cell culture or experimental animals. They are not routinely cultured in the laboratory, and diagnosis is by serological detection of antibodies.

There are numerous rickettsial species which **cause similar febrile diseases** (with or without skin rash) and are **transmitted by arthropods** (*Table 1*). The skin bite (port of entry, also called eschar) often shows early signs of inflammation and precedes the systemic illness. Natural reservoirs include domestic or wild animals and humans (*Table 1*). Rickettsiae are sensitive to tetracycline, erythromycin, chloramphenicol and fluoroquinolones which are used for treatment.

Ehrlichiae

Ehrlichiae species are tick-borne Rickettsiae that infect wild or domestic animals and occasionally humans. The organisms attack white blood cells and cause systemic upset of varying severity. Diagnosis is largely by serology (antibody detection) and treatment is with doxycycline.

Table 1.   Species of Rickettsiae and Coxiella, their clinical syndromes, vectors, reservoirs and geographical distributions

Organism	Disease	Arthropod vector	Vertebrate reservoir	Clinical severity	Geographical distribution
Spotted fevers					
Rick. rickettsia	Rocky Mountain spotted fever	Tick	Dogs, rodents	+	Rocky Mountain states, Eastern USA
Rick. akai	Rickettsial pox	Mite	Mice	–	Asia, Far East, Africa, USA
Rick. conorii	Mediterranean Spotted fever	Tick	Dogs	+	Mediterranean
Typhus					
Rick. prowazeki	Epidemic typhus	Louse	Human	+	Africa, South America
Rick. typhi	Endemic typhus	Flea	Rodents	–	worldwide
Rick. tsutsugamushi	Scrub typhus	Mite	Rodents	+ +	Far East
B. quintana	Trench fever	Louse	Human	+	Asia, Africa, C and S America
Q- fever					
Cox. burnetii	Q fever	None	Sheep, goats	+	worldwide

Coxiella burnetii (Q Fever)

Coxiella are very similar to Rickettsiae but are not transmitted to humans by arthropods, being resistant to drying, heat and sunlight (i.e. robust enough for airborne spread). *C. burnetii* preferentially infects the lungs without causing any skin rash. It is a zoonotic disease infecting mainly farmers, abattoir workers and veterinarians who handle animal placentas and tissue fluids. When inhaled, *C. burnetii* multiplies in the alveoli and after 3 weeks of incubation disease manifests as fever, severe headache, respiratory symptoms, atypical pneumonia and possibly hepatitis. This condition is called Q-fever (from query) and resolves within 2 weeks or leads to chronic infection, e.g. endocarditis. Q-fever is diagnosed serologically by detection of a 4-fold rise in complement fixation titer. The organism has two antigenic forms: phases 1 and 2. Antibodies against the former only appear in chronic Q-fever, whereas the antiphase 2 antibodies are present in acute and chronic disease. The condition is treated with tetracycline and in cases of endocarditis, heart valve replacement may be necessary.

Bartonella

Three *Bartonella* species are known to cause disease in humans. *B. bacilliformis* is transmitted between humans by the sandfly. It causes Oroya fever, a potentially fatal febrile illness associated with severe hemolytic anemia, which is largely restricted to South America. This organism also causes verruga peruna, which is a benign skin eruption.

B. henselae is the causative agent of 'cat scratch fever' (febrile illness acquired from cats) and 'bacillary angiomatosis' which is a hemangiomatous skin eruption that may cause extensive skin damage in patients with AIDS.

B. quintana is the causative agent of trench fever and occasionally causes bacillary angiomatosis. This is transmitted by lice and is most prevalent in the poor and the homeless. *Bartonella* infection is diagnosed by serology (antibody detection) and treated with chloramphenicol or doxycycline.

C18 BRANCHING GRAM-POSITIVE BACTERIA

Key Notes

Microbiology
Actinomyces and *Nocardia* are branching, nonmotile and slow-growing Gram-positive bacilli. Only a few species cause rare diseases in man.

Actinomyces
Most actinomycetes are microaerophilic soil organisms but some live commensally in human and animal oropharynx and gut.
Actinomyces israelii is the most common human pathogenic species. In tissue, colonies develop to form yellowish 'sulfur granules', which are found in pus discharged through draining sinuses. The typical sites of the disease are: cervicofacial; abdominal and, rarely, thoracic. *A. israelii* often colonizes intrauterine contraceptive devices.
The organism is sensitive to penicillin, erythromycin and tetracycline.

Nocardia
Nocardia are weakly acid-fast and grow slowly under aerobic conditions, forming star-shaped colonies. They cause chronic granulomatous suppurative infections.
Nocardia asteroides affects lungs mainly, but can cause multiple brain abscesses.
N. modurae and *N. brasiliensis* may cause '*madura foot*' or 'mycetoma' a subcutaneous tissue infection of the foot, occurring in tropical areas.
Nocardia are sensitive to sulfonamides and co-trimoxazole, but universally resistant to penicillins. Long period of therapy (e.g. 6 months) may be necessary.

Branching Gram-positive bacteria
Actinomyces and *Nocardia* are branching Gram-positive bacilli. They are nonmotile and largely saprophytic soil organisms and only a few species cause rare diseases in man. *Actinomyces* are microaerophilic or anaerobic whereas *Nocardia* are aerobic bacteria.

Actinomyces
Most actinomycetes are soil organisms but some live commensally in human and animal oropharynx and gut.
Species are identified by colonial appearances (some are pigmented), ability to grow aerobically and biochemical tests.
Actinomyces israelii is the most common human pathogenic species. It is a nonmotile, nonsporing, Gram-positive bacillus which grows in filaments that readily break up into rods and may show branching. Unlike *Nocardia*, *Actinomyces* are not acid-fast. In tissues, colonies develop to form diagnostic yellowish 'sulfur granules', which are visible to the naked eye and which are found in pus discharged through draining sinuses.
The organism is microaerophilic and requires long incubation on semi-solid media. Furthermore, clinical samples (e.g. pus) often contain other faster-growing bacteria which dominate the culture. Therefore, isolation of *Actinomyces*

in a routine microbiology laboratory is not easy. However, identification of Gram-positive branching filaments in a sulfur granule is diagnostic. These granules should be washed, crushed and cultured.

The source of Actinomycosis is endogenous (i.e. from the patient's own mouth or gut) and results in a chronic granulomatous infection with abscess formation, involving profuse pus discharging through draining sinuses. Infection often starts after local trauma, e.g. the extraction of carious teeth or appendectomy. The typical sites of the disease are: cervicofacial; abdominal (usually ileocecal) and, rarely, thoracic (affecting the lung).

The organism is sensitive to penicillin, erythromycin and tetracycline. Treatment may take a long time (several weeks) and surgical debridement of sinuses and wounds may be required.

A. israelii often colonizes intrauterine contraceptive devices. If ignored (e.g. for months or years) this may cause invasive disease, possibly in association with other organisms.

Other actinomycetes, namely *A. naeslundii*, *A. viscosu*, *A. odontolyticus* and *A. pyogenes*, inhabit the oropharynx of humans and are associated with dental plaque and caries.

Nocardia

Nocardia are similar to actinomycetes, i.e. nonmotile branching Gram-positive bacilli. However, some species are weakly acid-fast (can be stained with Zeihl–Neelson stain). They grow slowly (over 5–14 days) under aerobic conditions. Their colonies are characterisitic, forming rosette or star shapes which are initially white, but turn yellow and then pink/red.

Nocardia cause chronic granulomatous suppurative infections. *Nocardia asteroides* affects lungs mainly, and sometimes can cause multiple brain abscesses (usually secondary to lung infection).

N. asteroides can cause opportunistic infection, including pulmonary nocardiosis, in immunocompromised patients, such as transplant recipients.

N. modurae and *N. brasiliensis* may cause '**madura foot**' or '**mycetoma**' which is a subcutaneous tissue infection of the foot, occurring in tropical areas. The organism becomes inoculated after minor injuries (e.g. walking with bare feet) causing chronic discharging sinuses.

Diagnosis is by demonstrating acid-fast (branching) bacilli in typical lesions. Nocardia are sensitive to sulfonamides and co-trimoxazole, but resistant to practically all other antibiotics. Treatment may have to be continued for 6 months.

D1 YEASTS

Key Notes

Candida species	The principal human pathogen is *C.albicans*, which is found as part of the normal human flora. Opportunistic infection is often related to antibiotic treatment, and is more common in diabetes, during pregnancy and in immunosuppressed patients. Infections range from common superficial mucosal infections (e.g. vaginal thrush) to less common, deeper invasive infections. Laboratory diagnosis is principally by microscopy and culture on selective agar. *C.albicans* is easily differentiated from other *Candida* species by the germ-tube test. Treatment is with topical, oral or intravenous antifungal drugs, including imidazoles, polyenes, flucytosine and caspofungin. Antifungal drugs may also be used as prophylaxis against *Candida* infection in certain immunosuppressed patients.
Cryptococcus neoformans	*C.neoformans* is an encapsulated yeast that is found in the environment, particularly in pigeon or other bird droppings. It is an opportunistic pathogen of immunosuppressed patients, and is a particular cause of meningitis in patients with HIV infection. Laboratory diagnosis is by microscopy and fungal culture of cerebrospinal fluid or other specimens, or by detection of cryptococcal antigen. Treatment is with a prolonged course of systemic antifungal drugs (amphotericin B, flucytosine, imidazoles). Prophylaxis (e.g. with fluconazole) is used to prevent further infections.
Other yeast infections	*Malassezia furfur* is a lipophilic yeast that is part of the normal skin flora, but it can also cause a common, mild superficial skin infection called pityriasis versicolor. *Trichosporum beigelii* causes a mild superficial infection of the hair in tropical countries (white piedra). Both these yeasts are rare causes of fungemia in immunosuppressed patients.
Related topics	Antifungal drugs (E8) Infection in immunocompromised patients (F14)

Yeasts are **unicellular** fungi that reproduce by **budding**. The cells are generally round or oval, but under certain conditions they can become elongated and join together to form a **pseudomycelium**. The medically important yeasts are candida and cryptococcus. *Malassezia* and *Trichosporum* are also briefly described.

Candida species

Microbiology
Candida are relatively common opportunistic pathogens of humans. The principal pathogen is **Candida albicans**. Other species that occasionally cause infection are *C.tropicalis*, *C.parapsilosis*, *C.glabrata* and *C.krusei*. Some of theses species (notably *C.glabrata* and *C.krusei*) can be resistant to certain antifungal drugs (e.g. azoles).

Epidemiology

Small numbers of *Candida* species may be found in the mouth, gastrointestinal tract and genital tract without causing infection. Much larger numbers may be found during or after broad-spectrum antibiotic therapy, with a risk of superficial or systemic candidiasis following such treatment.

Candida is principally an **opportunistic infection**, with most infections occurring in debilitated or immunosuppressed patients, those on broad-spectrum antibiotics, or patients at the extremes of age. Infections are more common in patients with diabetes mellitus, and during pregnancy. Treatment with steroids, including inhaled steroids (e.g. for asthma) can predispose to candidiasis. Cutaneous candidiasis occurs particularly in moist areas of skin (e.g. between skinfolds in obese patients). Bloodstream infection (candidemia) is sometimes seen in patients receiving total parenteral nutrition (TPN) via a central venous catheter. Patients who abuse drugs intravenously are at particular risk of candida endocarditis and endophthalmitis.

Most infections are endogenous (*Candida* is often part of the normal human flora). Occasionally cross-infection can occur, especially in healthcare settings. Whilst *C.albicans* is the main pathogen, other species are gaining increasing clinical importance in some healthcare settings (e.g. intensive care units, transplant centers).

Clinical infections

Candida can cause a variety of both superficial and invasive infections (*Table 1*). Recurrent oropharyngeal candidiasis is infrequently encountered in normal healthy adults, and should raise the possibility of underlying disease or immunosuppression. Deep infections are almost invariably found in debilitated or immunosuppressed patients.

Laboratory diagnosis

Candida can often be identified presumptively in microbiology specimens by microscopic visualization of yeast cells. *Candida* can be readily grown on most types of agar in 48–72 h. Selective agar (e.g. Sabouraud agar) is used to inhibit bacterial growth. Some specialized (chromogenic) agars are available to allow for easy differentiation of the main *Candida* species.

A simple, rapid test for the identification of *C.albicans* is the **germ-tube test**. Most strains of *C.albicans* produce true hyphae when incubated for 2–3 h in human serum, whereas other species do not. Identification of nonalbicans *Candida* to species level is by more time-consuming biochemical tests.

In vitro antifungal sensitivity tests are not routinely available in most diagnostic microbiology laboratories, although they are an important component of the management of some patients with deep or invasive candidiasis, particularly the nonalbicans candida infections where imidazole resistance is more common.

Treatment and prevention

The main treatments for candida infections are outlined in *Table 1*. Most superficial infections can be treated with topical antifungal drugs, although oral systemic therapy may be required for highly immunosuppressed patients, for recurrent infections, and for nail infections (where prolonged treatment is required). For deep or invasive infections, parenteral (intravenous) therapy may be required.

Table 1. The main infections caused by Candida

	Infection	Clinical features	Treatment
Superficial candidiasis	Oropharyngeal candidiasis (oral thrush)	White plaques on buccal mucosa, gums (including denture stomatitis), tongue (glossitis) and throat. Can cause painful ulceration and dysphagia (pain on eating)	Topical suspension of nystatin or amphotericin B. Systemic antifungal drugs for severe infection
	Vaginal candidiasis (vaginal thrush)	Vulval and vaginal itching (pruritus), vaginal discharge	Topical treatment with imidazole creams/pessaries e.g clotrimazole, or topical nystatin. Oral fluconazole (single dose) is also used.
	Cutaneous candidiasis	Pruritic rash with erythema, vesicles, pustules and sometimes fissuring of the skin. Nappy rash.	Topical creams +/– steroids
	Penile candidiasis	Soreness of the glans penis (balanitis)	Topical creams. Investigate sexual contacts.
	Paronychia and onychia	Nailfold infection (paronychia) of the finger(s) causing pain and swelling. This may progress to affect the nail (onychia)	Prolonged oral treatment with itraconazole +/– topical creams. Removal or avoidance of predisposing factors (e.g. repeated immersion of hands in water)
Deep candidiasis	Esophageal candidiasis	Painful dysphagia with retrosternal chest pain, usually in the context of HIV or cancer chemotherapy	
	Urinary tract infection	Dysuria and frequency	
	Hepatosplenic candidiasis	Abdominal pain and abnormal liver function tests following dissemination of candida during a neutropenic episode	Systemic antifungal drugs. Prolonged antifungal therapy may required. Surgical intervention is usually necessary for endophthalmitis and endocarditis
	Endophthalmitis	Blurred vision and ocular pain, mainly seen in intravenous drug users	
	Candidemia, endocarditis	Persistent fever, not responding to antibiotics. Dissemination to skin and other organs may occur	

Topical drugs include **imidazoles** (e.g. clotrimazole) and **polyenes** (e.g. nystatin, amphotericin B).

Oral systemic drugs for candidiasis are the imidazoles (fluconazole, itraconazole and voriconazole). The polyenes can be used as oral suspensions for the treatment of oropharyngeal candidiasis, but polyenes are not absorbed systemically. Intravenous drugs available for deep/invasive infections are **amphotericin B** (including lipid formulations), flucytosine, caspofungin and imidazoles (Topic E8).

Prophylaxis (e.g. with fluconazole) is used to prevent infections in certain high-risk patient groups, although this predisposes to colonization or infection with some nonalbicans strains.

Cryptococcus neoformans

Cryptococcus neoformans is an **encapsulated yeast**. Humans are commonly exposed to this organism from the environment, the most abundant source of which is in dried pigeon droppings or soil contaminated with bird droppings. The fungus is usually inhaled and the lungs are the principal portal of entry. *Cryptococcus* is an opportunistic pathogen, and most infections occur in immunocompromised patients, particularly those with **HIV infection**.

The principal clinical manifestation of cryptococcosis is **meningitis** (Topic F3). This may be either acute or chronic. Some patients also have pulmonary symptoms or signs. Clinical infection can also occur in the skin, eyes and musculoskeletal system, usually as a result of hematogenous spread.

Laboratory diagnosis of cryptococcal infection is usually by detection of the organism in cerebrospinal fluid (CSF) or blood. Encapsulated yeast cells can be visualized microscopically in CSF, particularly when mounted in India ink. Cryptococci can be cultured from CSF and blood cultures, although cultures may take up to 10 days. Detection of cryptococcal antigen (e.g. by latex agglutination) can also be used to rapidly detect cryptococci in CSF or blood. The level of detectable antigen (antigen titer) can be used to monitor response to therapy.

Cryptococcal meningitis carries a relatively high mortality. Treatment is with high-dose systemic antifungal drugs. A combination of either amphotericin B or fluconazole with flucytosine is usually given initially for 4–6 weeks. Recurrent infection or relapse is common in HIV patients and maintenance therapy with oral fluconazole may be given as prophylaxis against second episodes in HIV patients with low CD4 counts.

Other yeast infections

Malassezia furfur

This is a **lipophilic yeast** that is part of the **normal skin flora**. It is also the cause of a common, mild superficial skin infection called **pityriasis versicolor**. This is a potentially disfiguring skin condition with the development of numerous brown scaly patches, and sometimes hypopigmentation of the skin. Treatment is with topical antifungal creams or oral itraconazole.

Malassezia can also cause central venous catheter-associated fungemia, particularly in patients receiving fat emulsions as part of total parenteral nutrition (TPN).

Trichosporum beigelii

Trichosporum causes a mild superficial infection of the hair in tropical countries (white piedra). However it is also a rare but important opportunistic pathogen in immunosuppressive disease (e.g. leukemia) causing disseminated infection.

D2 MOLDS

Key Notes

Aspergillus

Aspergillus is a common environmental mold, but an uncommon cause of human infection. Clinical infections mostly affect the lungs (following inhalation of *Aspergillus* spores), causing either allergic aspergillosis (producing asthma-like symptoms – Farmer's lung), aspergilloma (a fungus ball that often develops in pre-existing lung cavities), or invasive aspergillosis (in immunosuppressed patients). Laboratory diagnosis is by detection of *Aspergillus* antibodies or antigen in blood, culture from clinical specimens, or detection of fungal hyphae by histology. However, clinical and radiological diagnosis is important because laboratory investigations are insensitive. Allergic aspergillosis can be treated with inhaled steroids. Surgical resection of an aspergilloma may be required. Treatment of invasive aspergillosis is with high-dose systemic antifungal drugs. Prophylactic antifungal drugs and an ultraclean air supply may be used to reduce the risk of invasive aspergillosis in at-risk patients.

Mucorales

Mucorales are a group of environmental molds that cause mucormycosis, a rare infection of humans. Infection may follow inhalation of spores, or direct inoculation. Risk factors for infection include diabetes mellitus, immunosuppression and extensive burns. The commonest type of infection is rhinocerebral mucormycosis, which starts in the paranasal sinuses, but then spreads rapidly into adjacent tissues, including the eye, palate and brain. Diagnosis is by clinical, radiological and histological features (showing broad nonseptate hyphae within tissue). Occasionally the fungus can be grown from a clinical specimen. Treatment is by surgical resection of necrotic tissue and high-dose antifungal chemotherapy, but mortality is high.

Dermatophytes

Dermatophytes are a group of filamentous fungi that commonly cause superficial infection of skin, nails and hair, known as ringworm or tinea (skin and hair), or onychomycosis (nails). Clinical diagnosis can be confirmed by microscopy and fungal culture of skin scrapings, hair or nail clippings. Treatment is usually with topical antifungal preparations, but systemic antifungal drug therapy may be required for hair and nail infections.

Other mold infections

Other molds that occasionally cause human infection include *Fusarium* species and *Penicillium marneffei*. Mycetoma is a chronic suppurative condition caused by higher bacteria (e.g. *Actinomyces*) and certain other fungal pathogens (e.g. *Madurella*) found in the tropics.

Related topics

Antifungal drugs (E8)
Skin and soft tissue infections (F1)

Infection in immunocompromised
patients (F14)

Molds

Molds (or filamentous fungi) are a diverse group of saprophytic eukaryotic organisms that, in their vegetative state, consist of a network of fungal hyphae or filaments (the **mycelium**). Many genera of molds are found in the environment, but only a few are capable of causing human disease. The important filamentous fungi that cause human infection are:

* *Aspergillus*;
* Mucorales;
* Dermatophytes.

Aspergillus

Microbiology

The genus *Aspergillus* contains several species that can cause human infection, including *A. fumigatus*, *A. niger* and *A. flavus*. These fungi are abundant in the environment (soil, dust, and decaying organic matter). Airborne spores are common and most infections follow inhalation of *Aspergillus* spores.

Epidemiology

Risk factors for invasive aspergillosis are principally **immunosuppression** (e.g. bone marrow transplant recipients or prolonged neutropenia). Outbreaks of hospital-acquired aspergillosis can occur when building demolition or construction occurs in the vicinity of high-risk units such as bone marrow transplant wards, as a large number of *Aspergillus* spores can be released into the air during such work.

Clinical infections

The lungs are the principal site of aspergillosis. There are several distinct forms of **pulmonary aspergillosis**.

(i) Allergic aspergillosis.
    This is a relatively uncommon condition that can occur in otherwise healthy individuals and is due to an allergic response (hypersensitivity reaction) to the inhalation of *Aspergillus* spores. This causes asthma-like symptoms and the formation of mucus plugs which may contain *Aspergillus* mycelium. Farmer's lung is an allergic pneumonitis associated with exposure to mold-contaminated grains.

(ii) Aspergilloma.
    An aspergilloma is a fungus ball that usually occurs in patients with pre-existing lung cavities (e.g. as a result of previous tuberculosis or bronchiectasis). Patients may be asymptomatic, or have chronic cough and malaise. The most serious complication is hemoptysis (coughing up blood) due to the predilection of *Aspergillus* to invade blood vessels.

(iii) Invasive pulmonary aspergillosis.
    This is a rapidly progressive and life-threatening infection in highly immunosuppressed patients. Patients develop persistent fever, shortness of breath and pleuritic chest pain. Invasive pulmonary aspergillosis can cause widespread lung necrosis and hemoptysis. Some patients will develop hematogenous seeding to other organs including the brain (cerebral aspergillosis), eye (endophthalmitis), skin (with characteristic black necrotic lesions) and liver.

Aspergillosis can also affect the paranasal sinuses in immunosuppresed patients causing a syndrome similar to rhinocerebral mucormycosis (see below). *Aspergillus* can also cause fungal keratitis and occasionally onychomycosis. Other rarer infections include endocarditis (prosthetic valve) and osteomyelitis.

Diagnosis
Allergic pulmonary aspergillus may be diagnosed by the detection of high levels of *Aspergillus* antibodies (**precipitins**) in serum. *Aspergillus* antibodies are usually elevated in patients with aspergillomas.

The diagnosis of invasive aspergillosis in immunosuppressed patients is difficult. Cultures of blood, sputum or other samples are frequently negative, and these patients do not usually mount an antibody response. Clinical features are often nonspecific, and antifungal treatment may have to be given on an empiric basis to at-risk patients with a persistent fever that is nonresponsive to antibiotics.

Radiological investigations may suggest invasive aspergillosis, particularly high resolution CT scan of the lungs which is more sensitive than a plain chest X-ray. The detection of *Aspergillus* **antigen** in blood is more sensitive than culture techniques, but is prone to both false-negative and false-positive results. Definitive diagnosis can only be made by histological detection or culture of the organism from deep tissue biopsies or sterile body fluids.

Treatment and prevention
Allergic pulmonary aspergillosis may be treated with **steroids** if symptoms do not resolve. Antifungal drugs are not used.

Aspergillomas may not require treatment, but sometimes **surgical excision** is required to manage hemoptysis. Direct installation of antifungal drugs into the aspergilloma is occasionally attempted.

Invasive aspergillosis is treated by high-dose systemic **antifungal therapy** (lipid derivatives of amphotericin B, caspofungin or itraconazole) but mortality is high.

Prophylaxis against invasive aspergillosis in at-risk patients includes isolation facilities with **specialized ventilation systems** and use of prophylactic **antifungal drugs** during periods of highest risk (e.g. itraconazole). Isolation rooms can be provided with clean air [using high-efficiency particulate air (**HEPA**) filters] at a positive pressure relative to the outside. This is particularly important when construction, demolition or other building work is occurring in the vicinity of at risk units.

Mucorales

The mucorales consist of a group of molds that can cause a variety of infections called **mucormycosis** (or zygomycosis). The commonest species are *Rhizopus*, *Rhizomucor* and *Absidia*. These fungi are ubiquitous in the environment, particularly soil and decomposing organic matter. Infection usually follows inhalation of airborne spores, but direct inoculation can also occur.

Mucormycosis is an **opportunistic infection**, and is seldom seen in otherwise healthy individuals. Risk factors for infection are **diabetes mellitus, hematological malignancies** and **extensive burns**. The fungi are able to invade blood vessels (as can *Aspergillus*) causing tissue infarction and **necrosis**.

The commonest clinical infection is **rhinocerebral mucormycosis**. This starts in the paranasal sinuses, but infection then spreads to involve the palate, the orbit, the face and the brain. Mortality is very high. Rapidly progressive necrotic lesions are the hallmark of the disease, which can be difficult to diagnose in the early stages.

Mucormycosis can also affect the lungs, gastrointestinal tract and skin. Cutaneous mucormycosis may occur in burns patients and also neonates. Outbreaks caused by contaminated dressings or splints have been described.

Diagnosis is made by a combination of clinical and radiological features with **histological examination** of a tissue biopsy showing the characteristic **broad nonseptate hyphae** (this differentiates the mucorales from *Aspergillus*). Histology is more important than culture of the organism from clinical specimens.

Treatment is by a combination of **surgical debridement** (which may need to be extensive) and **high dose** intravenous **amphotericin B** (usually a lipid preparation). Unfortunately many patients do not survive.

Dermatophytes

Dermatophytes are a group of filamentous fungi that cause superficial infection of keratinized structures (skin, nails and hair). There are three genera of dermatophytes:

- *Trichophyton*;
- *Epidermophyton*;
- *Microsporum*.

Of > 40 different species, only a few are common causes of human infection. The natural reservoir of dermatophytes can be humans (**anthropophilic**), animals (**zoophilic**), or soil (**geophilic**). Common species causing human infection include *E. floccosum*, *T. rubrum*, *T. mentagrophytes*, *T. tonsurans* (anthropophilic) and *M. canis* (zoophilic). These are very common infectious agents found throughout the world, although some other types of dermatophytes are found in restricted geographical areas.

Clinical infection is known as **ringworm** or **tinea** and can affect the skin or hair. The different clinical manifestations and treatment are briefly described in Topic F1. Infection of nails is called **onychomycosis** (or onychia). This is a chronic condition causing unsightly discoloration of nails (usually on the feet). Eventually the nail becomes friable (dry and brittle) and may separate from the nail bed. Treatment is difficult and requires long courses (several months) of oral antifungal drugs (griseofulvin, terbinafine or itraconazole).

Occasionally other nondermatophyte molds may cause onychomycosis (e.g. *Scopulariopsis brevicaulis*). These respond less well to drug treatment and surgical removal of the nail may be required.

Other mold infections

Mycetoma

Mycetoma is a chronic suppurative infection of skin, soft tissue and bone, found in tropical and subtropical countries. Mycetoma may be caused by higher bacteria (e.g. *Actinomyces*, *Nocardia*) (Topic C18) or fungal pathogens (e.g. *Madurella*). Infection usually occurs in the feet (madura foot). Fungal mycetoma is very difficult to treat medically, once chronic infection is established.

Fusarium *infections*

Fusarium species are common soil organisms. They can cause onychomycosis as well as fungal keratitis. Occasionally disseminated infection occurs in immunosuppressed patients, and respiratory infection is the usual manifestation.

Penicillium marneffei

Molds of the genus *Pencillium* are extremely common in the environment, and are often found as culture contaminants in the laboratory. Only one species, *P. marneffei*, causes significant human infection. This mold is limited to **South East Asia** where it causes an invasive infection of the lungs, skin, liver and spleen. **Disseminated infection** is more common in patients with **HIV**. Treatment is with amphotericin B.

D3 Dimorphic fungi and Pneumocystis Jiroveci (carinii)

Key Notes

Dimorphic fungi	Dimorphic fungi comprise a diverse group of fungi characterized by mycelial growth at 22°C, but development of yeast forms at 35–37°C. Sporotrichosis is an infection of subcutaneous tissue with *Sporothrix shenckii*, usually on the arms or hands after a contaminated inoculation injury. *Histoplasma capsulatum* (causing the infection histoplasmosis) is principally found in North, Central and South America. The natural reservoir is soil contaminated by bird or bat droppings. Histoplasmosis is usually an asymptomatic or mild, self-limiting chest infection, but chronic pulmonary infection (resembling tuberculosis) and disseminated infection (in immunosuppressed patients) can occur. *Coccidioides immitis* is a soil fungus found in South Western USA and some parts of Central and South America. Inhalation of airborne spores can cause asymptomatic or mild pulmonary infection, but occasionally chronic pulmonary or disseminated disease occur. Other geographically restricted dimorphic fungi with similar clinical features include *Blastomyces dermatitidis* (North, Central and South America) and *Paracoccidioides brasiliensis* (South America).
Pneumocystis carinii	*Pneumocystis Jiroveci* is found worldwide, but the natural reservoir is unknown. Humans are exposed via the airborne route, but clinical infection generally occurs only in patients with depressed cell-mediated immunity (e.g. HIV infection). *P. Jiroveci* principally causes pneumonia (PCP) and, in patients with HIV infection, this is an AIDS-defining illness. Other affected patients include bone marrow and solid-organ transplant recipients. Laboratory diagnosis is by microscopic examination of induced sputum or broncho-alveolar lavage fluid, using specialized stains (e.g. immunofluorescence). Treatment is principally with high dose trimethoprim-sulfamethoxazole. Chemoprophylaxis is used to prevent PCP in the post-transplant period, and in HIV patients with low CD4 counts.
Related topics	Antifungal drugs (E8) Infection in immunocompromised patients (F14)

Dimorphic fungi   Dimorphic fungi comprise a diverse group characterized by mycelial (filamentous) growth at 22°C, but development of yeast forms at 35–37°C in the human body. There are a number of dimorphic fungi that can cause human infection including:

- *Sporothrix shenckii* (sporotrichosis);
- *Histoplasma capsulatum* (histoplasmosis);

- *Coccidioides immitis* (coccidioidomycosis);
- *Paracoccidioides brasiliensis* (paracoccidioidomycosis);
- *Blastomyces dermatitidis* (blastomycosis).

Sporotrichosis

Sporothrix schenckii is a common environmental fungus found worldwide. Sporotrichosis is an infection of subcutaneous tissue that usually occurs within 3 months of a contaminated inoculation injury, usually on the arms or hands. A painless nodule develops at the site of injury, and then further lesions develop along the lymphatic channels. Treatment is with oral itraconazole or saturated potassium iodide solution.

Histoplasmosis

Histoplasma capsulatum is a dimorphic fungus principally found in North, Central and South America. The natural reservoir of *H. capsulatum* is soil contaminated by bird or bat droppings, or caves where bats are roosting. Outbreaks may occur when such sites are disturbed, releasing large numbers of airborne spores.

In most humans, exposure to *Histoplasma* causes asymptomatic infection, or only a mild, self-limiting chest infection. In some patients, however, a more chronic pulmonary infection occurs that resembles tuberculosis. Occasionally disseminated histoplasmosis develops, with infection involving lungs, liver, spleen, skin and central nervous system. This is a particular problem in immunosuppressed patients, particularly those with HIV infection.

Diagnosis is made by microscopy of stained smears of blood, sputum or other body fluid, fungal culture (incubated for 4–6 weeks), and serological tests for both histoplasma antibodies and antigen. Treatment of invasive or chronic pulmonary histoplasmosis is with amphotericin B or itraconazole. Long-term suppressive therapy in HIV patients may be required.

Coccidioido-mycosis

Coccidioides immitis is a fungus found only in soil in South Western USA and some parts of Central and South America. Infection occurs after inhalation of airborne spores. Like histoplasmosis, coccidioidomycosis often causes asymptomatic or mild acute pulmonary infection, although sometimes rashes (including erythema nodosa) can occur. Chronic pulmonary and disseminated disease is less common, occurring particularly in HIV or other immunosuppressed patients. Diagnosis and treatment are similar to histoplasmosis.

Other geographically restricted dimorphic fungi with similar clinical features include *Blastomyces dermatitidis* (North, Central and South America) and *Paracoccidioides brasiliensis* (South America).

Pneumocystis Jiroveci

Pneumocystis Jiroveci was originally classified as a protozoan and known as carinii, but is now classed as a fungus on the basis of genetic studies. However, like some protozoa, the organism exists in two main forms: trophozoites and cysts.

P. Jiroveci is found in a worldwide distribution, but the natural reservoir is unknown. Based on sero-epidemiological studies, most humans have been exposed to the organism, but infection generally occurs only in patients with depressed cell-mediated immunity.

Clinical infection is seen most commonly in patients with HIV infection, or recipients of bone marrow or solid-organ transplants. The commonest type of infection is Pneumocystis pneumonia (or PCP). This is one of the commonest **AIDS-defining illnesses in HIV patients**. Symptoms include fever, progressive

breathlessness and cough. Chest X-rays usually show diffuse fine bilateral infil-
trates.

Diagnosis is often made on clinical and radiological grounds in the context of
HIV infection or transplantation. Laboratory diagnosis can be confirmed by
microscopic examination of induced sputum or broncho-alveolar lavage fluid
(obtained at bronchoscopy), using specialized stains (e.g. immunofluorescence
with specific monoclonal antibodies).

The mainstay of treatment is with high-dose trimethoprim-sulfamethoxazole
(co-trimoxazole) for 2–3 weeks. Steroids are sometimes given as adjunctive
therapy. Alternative treatments include pentamidine and atovaquone.
Chemoprophylaxis (oral co-trimoxazole or nebulized pentamidine) is used to
prevent PCP in recipients of bone marrow or solid-organ transplants, and in
HIV patients with low CD4 counts or after a first attack of PCP (to prevent
recurrence).

D4 PROTOZOA

Key Notes

Definition	Protozoa are unicellular eukaryotic organisms distributed worldwide. They vary in size, shape and lifestyle and reproduce sexually and/or asexually.
Gut protozoa	• *Entamoeba histolytica*, a water-borne pathogen, infects the human intestine causing amoebic dysentery or liver abscess. • *Giardia lamblia* causes malabsorption and diarrhea. • *Cryptosporidium parvum* is a complex pathogen causing protracted diarrhea mainly in children and immunocompromised adults.
Bloodstream protozoa	• *Plasmodium falciparum*, *P. vivax*, *P. malariae* and *P. ovale* are transmitted by female Anopheles mosquitoes and cause malaria. • *Trypanosoma brucei rhodesiense* and *T. gambiense* are transmitted by the tsetse fly and cause African sleeping sickness. *T. cruzi* is transmitted by the triatomine bug causing Chagas, disease in South America • *Leishmania* species are transmitted by sandflies causing leishmaniasis. *L. donovani* causes visceral leishmaniasis (kala azar) in the Old and New Worlds. *L. tropica* causes cutaneous leishmaniasis (oriental sore) in the Old World, whereas *L. mexicana and L. braziliensis* cause mucocutaneous leishmaniasis in the New World. • *Toxoplasma gondii* lives in domestic cats whose feces contaminate fingers and infect humans. It causes severe disease mainly in the fetus, immunocompromised adults (with AIDS) and rarely young children.
Other protozoa	• *Trichomonas vaginalis* is sexually transmitted, causing vaginal discharge in females and urethritis in males.

Protozoa   Protozoa are **unicellular eukaryotic** organisms distributed worldwide in most habitats. They reproduce sexually (by fusion of male and female gametocytes) and/or asexually (by binary fission). Most species are free-living, but some are pathogenic, causing infections that range from asymptomatic to life-threatening diseases.

Protozoa **vary in size, shape and lifestyle** and are classified largely on the basis of their microscopic morphology. The stages of parasitic protozoa that actively feed and multiply are called trophozoites; in some protozoa, other terms are used for the various stages of the life cycle. Some protozoa surround themselves with protective membranes (forming cysts) during exposure to harsh environmental conditions.

Medically important protozoa infect various parts of human body. These are summarized in *Table 1*. Some protozoa infect more than one organ *de novo* or as part of their development where they go through different reproductive stages.

Table 1.   Medically important protoza

Protozoa infecting various organs	
Skin	*Leishmania*
Eye	*Acanthamoeba*
Gut	*Entamoeba* (and invasion to liver), *Giardia*, *Cryptosporidium*, *Isospora*
Genito-urinary tract	*Trichomonas*
Bloodstream	Plasmodia, *Trypanosoma*
Spleen	*Leishmania*
Liver	*Leishmania*, *Entamoeba*
Muscle	*Trypanosoma*
CNS	*Trypanosoma*, *Naegleria*, *Toxoplasma*, Plasmodia

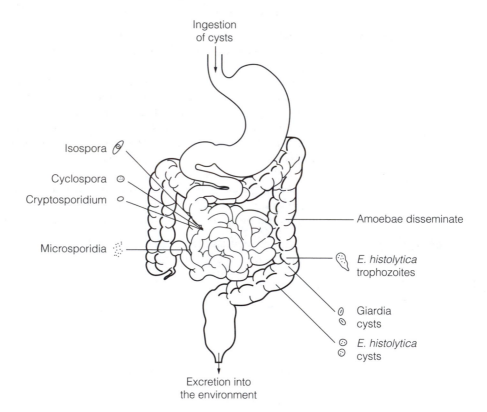

Fig. 1.   Intestinal protozoa.

Gut protozoa (Figs. 1 and 2)

Amoebiasis

Entamoeba histolytica is a **water-borne** protozoon, which divides asexually in the human intestine where it feeds on gut cells. Man is the definitive host and the parasite's life cycle is simple (*Fig. 3*). **Cysts** are shed by humans via the feces and new infection is via ingestion of sewage-contaminated water or on contaminated fruit, vegetables etc. The cyst opens up into growing cells (vegetates) in the distal part of the small intestine and colon where they feed and multiply. Enta.

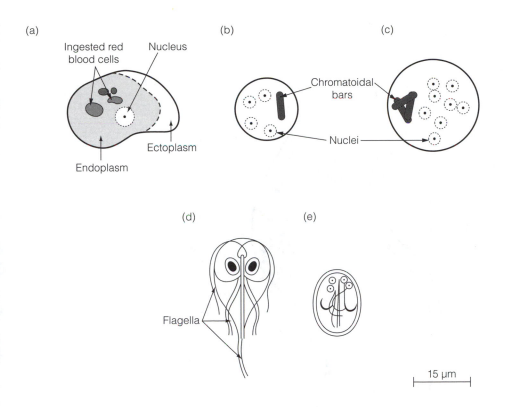

Fig. 2. Schematic illustration of some of the more common intestinal parasites. (a) Entamoeba Histolytica tropozoite, (b) E. histolytic cyst, (c) Entamoeba coli cyst, (d) Giardia lamblia tropozoite, (e) G. lamblia cyst

histolytica readily forms cysts which are passed out with feces. The growing amoeba is highly virulent and destroys all types of human cells. It is capable of invading the epithelial and subepithelial layers of the large intestine including blood vessels, causing '**amoebic dysentery**', which is characterized by loose stool mixed with mucus and blood. Occasionally, invading amoebae are carried via the bloodstream to the liver or other organs where they cause **amoebic abscesses** which can be life-threatening. The organism is sensitive to metronidazole which is the treatment of choice.

Giardia lamblia
G. lamblia is another gut protozoon with a simple life cycle, similar to that of *Enta. histolytica* but it is much less pathogenic. It does not invade host cells but colonizes and damages the small intestine where it causes **malabsorption and diarrhea**. Diagnosis is by microscopy of feces to visualize the parasite's trophozoites or cysts. It is sensitive to metronidazole.

Cryptosporidium parvum
This protozoon has a coccidian life cycle, i.e. reproduces sexually and asexually and transforms into several complicated phases during its life cycle. The parasite is shed in the feces of humans and animals and contaminates drinking water or swimming pools. Ingested oocysts vegetate in the intestine, reproducing inside gut cells. This results in the death of host cells leading to **protracted**

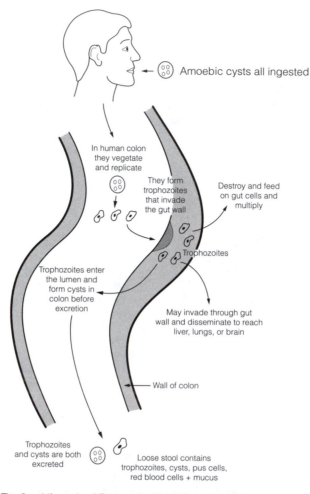

Amoebic cysts all ingested

In human colon
they vegetate
and replicate

They form
trophozoites
that invade
the gut wall

Destroy and feed
on gut cells and
multiply

Trophozoites

Trophozoites enter
the lumen and
form cysts in
colon before
excretion

May invade through gut
wall and disseminate to reach
liver, lungs, or brain

Wall of colon

Trophozoites
and cysts are both
excreted

Loose stool contains
trophozoites, cysts, pus cells,
red blood cells + mucus

Fig. 3.   Life cycle of Entamoeba histolytica.

diarrhea, mainly **in children and immunocompromised** adults. In patients with
AIDS the disease can be fatal. Diagnosis is by microscopy of feces, stained with
auramine or modified Ziehl–Neelson method. There is no effective treatment.
Filtration to remove parasites is critical and municipal sanitation lapses have
been associated with epidemics of cryptosporidiosis.

Others

Balantidium coli is similar to *Enta. histolytica* but causes milder disease. *Cyclospora
cayetanensis* is similar to cryptosporidium and causes watery diarrhea. *Isospora
belli* is another coccidian protozoon with similar lifecycle to that of
Cryptosporidium. It is also important as an infective agent among the immunocompromised and in AIDS patients and there is no effective treatment.

**Bloodstream
protozoa**

Malaria

Malaria is one of the most serious, global and common protozoal infections
caused by one of four species of the genus *Plasmodium*, namely *P. falciparum*, *P.
vivax*, *P. malariae* and *P. ovale*. Forty per cent of the world's population are at risk

of infection and it is a major cause of death in many parts of tropical and subtropical Africa, Asia and South America. *P. falciparum* is capable of causing cerebral malaria, the most serious form of the disease, which is fatal if not treated rapidly (also termed malignant malaria).

Man is the definitive host and the parasite is transmitted by **female *Anopheles*** mosquitoes (*Fig. 4*). These inoculate sporozoites as they bite during feeding. The sporozoites are carried via the bloodstream to the liver where they invade and

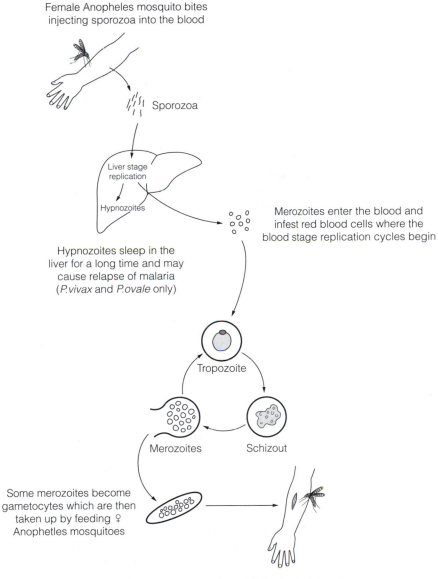

Female Anopheles mosquito bites injecting sporozoa into the blood

Sporozoa

Liver stage replication

Hypnozoites

Merozoites enter the blood and infest red blood cells where the blood stage replication cycles begin

Hypnozoites sleep in the liver for a long time and may cause relapse of malaria (*P.vivax* and *P.ovale* only)

Tropozoite

Merozoites

Schizout

Some merozoites become gametocytes which are then taken up by feeding ♀ Anophetles mosquitoes

♀ and ♂ Gametocytes mate inside the insect and replicate into sporozoa

Fig. 4.   Malaria.

multiply within hepatocytes before targeting red blood cells for further reproduction (*Fig. 4*). *P. vivax* and *P. ovale* may lay dormant for long periods (i.e. symptom-free) inside liver cells before they reactivate and cause 'relapse' of disease. Cyclical destructions of red blood cells are responsible for the classical symptoms of malaria, namely regular periodic fever and chills. This occurs every 2 days except in case of *P. malariae* (every 3 days). Destruction of red blood cells causes chronic anemia. *P. falciparum* is capable of plugging capillaries of the brain, placenta and other organs. It is the only species that is capable of causing 'cerebral malaria' which is a severe and often fatal form of disease. Pregnant women and people without a spleen are at greatest risk of severe disease.

Inside red cells, the parasite either divides rapidly or transforms into male and female 'gametocytes' which are taken up by new mosquitoes and initiate a new round of 'sexual reproduction'. The latter produces infective parasites, ready to infect a human at the next feed.

Diagnosis is by demonstrating the blood parasite in stained thick and thin blood films. The characteristic features of the various species of malaria parasite inside red blood cells are diagnostic. Serological and genetic methods of diagnosis are also available.

Treatment is by antimalarial drugs, including chloroquine which is the drug of choice. This is a safe and cheap drug; however, chloroquine-resistant *P. falciparum* is now widespread, therefore alternative drugs (e.g. quinine, mefloquine or artemisinin) are used in areas where this species is most common. Chloroquine with or without proguanil (or mefloquine) is also used for prophylaxis against malaria. These, combined with the use of insect repellents, protective clothing and sleeping under mosquito netting, are highly effective measures. Timing of prophylaxis both prior to travel to endemic regions and extending beyond return should be stressed.

Trypanosomiasis

Trypanosomes (and Leishmania, see below) may exist in several forms (*Fig. 5*) depending on their species and environmental conditions. Trypanosomis is a zoonotic disease transmitted by insects which feed on human blood. The trypanosomes are taken up by the insect during a blood meal, multiply and undergo developmental cycles in the gut of the insect before infecting another human host.

The African trypanosomes leave the gut and head for the salivary gland of the 'tsetse' fly from where they develop further and pass to the next host (humans) via the saliva during feeding. Initially, the parasite proliferates at the bite site in the skin, forming a characteristic lesion (called chancre). The protozoa then invade the bloodstream causing parasitemia which is accompanied by clinical symptoms, including fever and malaise. Subsequently, the parasite invades the central nervous system leading to the 'African sleeping sickness' syndrome. Two forms of the disease are known: the East African form (Rhodesian sleeping sickness caused by *Tryp. brucei rhodesiense*) is acute whereas the West African form (Gambian sleeping sickness caused by *T. gambiense*) is more chronic. A total of 25 000 cases are reported per year. Wild and domestic animals act as reservoir hosts of disease.

Tryp. cruzi causes Chagas' disease in northern parts of South America and is transmitted by the triatomine bug. The parasite passes out via the bug's feces and enters either via the bite site or mucosal area (eyes or mouth), reaches the bloodstream and multiplies as amastigotes in muscle or glial cells. Infected human cells

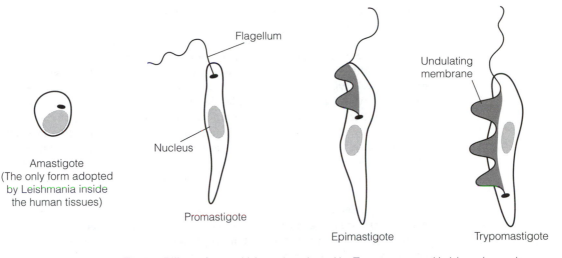

Amastigote
(The only form adopted
by Leishmania inside
the human tissues)

Flagellum

Nucleus

Promastigote

Undulating
membrane

Epimastigote

Trypomastigote

Fig. 5. Different forms which can be adopted by Trypanosoma and Leishmania species.

inflate into pseudocysts before they rupture and release infective protozoa which start the cycle again. This process of muscle cell destruction weakens the integrity and power of the involved organ leading to distension of esophagus (**mega-esophagus**) and colon (**megacolon**) and cause enlargement and disruption of heart function leading to sudden death. An estimated $16-18 \times 10^6$ are thought to be infected. The reservoirs of the parasite are opossums and other forest animals. Domestic animals are a more common reservoir (dogs, cats, rats, mice and pigs).

Diagnosis is by demonstration of the parasite, its DNA or antibodies in blood or CSF of patients. Treatment is difficult and involves the use of toxic drugs.

Leishmaniasis

Leishmaniasis is another zoonotic protozoal infection caused by four main species of the genus *Leishmania*. *L. donovani* causes visceral leishmaniasis (kala azar) in the Old and New Worlds. *L. tropica* causes cutaneous leishmaniasis (oriental sore) in the Old World, whereas *L. mexicana* and *L. braziliensis* cause mucocutaneous leishmaniasis in the New World.

Dogs and rodents are particularly important reservoirs for the parasite which is **transmitted by sandflies**. The organisms are inoculated when the sandfly bites and are taken up by **skin or liver macrophages** where they multiply as small round parasites (amastigotes) and kill the host cell.

Diagnosis is mainly based on clinical symptoms and demonstration of the parasite (*Fig. 5*) in tissue biopsies of skin (in **cutaneous leishmaniasis**) or bone marrow and liver (in **kala azar**). Treatment is difficult and involves the use of toxic drugs such as pentostam or amphotericin B.

Toxoplasmosis

Toxoplasmosis, caused by *Toxoplasma gondii*, is a **zoonotic** disease that occurs *in utero*, in immunocompromised adults (with AIDS) and rarely young children. The main reservoir and definitive host is the **domestic cat** and other felines where full sexual and asexual coccidian lifecycles take place, very similar to *Cryptosporidium parvum*. Most people have been exposed to this organism early in life, especially in some cultures where consumption of under-cooked meats is a part of the culture (e.g. France).

Infective oocysts from **cat feces** contaminate hands or food, reach the human intestine where they open up and **penetrate gut epithelial cells**. After several cycles of multiplication, the parasite may enter the circulation and **infect various nucleated cell types**; finally tissue cysts are formed in brain and muscle, which contain many slowly dividing toxoplasmas and last for years, unless new cycles of proliferation are initiated in **immunocompromised patients**. In cattle and sheep, the parasite follows the same fate as in humans, and ingestion of viable toxoplasmas in under-cooked meat is another source of infection for humans. The parasite can **cross the placenta** to cause serious brain damage to the unborn fetus, particularly in early pregnancy. A common manifestation of congenital toxoplasmos is choroido-retinitis which may lead to blindness.

Primary infection in immunocompetent adults is almost always asymptomatic. In children, it can cause severe clinical symptoms mimicking 'infectious mononucleosis' with swelling of **lymph glands**. The most important factor in pathogenicity in adults is the immune status of the infected individual. In the immunocompromised, dormant tissue cysts in the brain and heart may **reactivate** and cause severe disease.

Diagnosis is by **detection of rising titer of specific antibodies** to *T. gondii*. Brain infection may be diagnosed by high-resolution imaging, e.g. CT or MRI scans.

Treatment is difficult and requires combinations of drugs such as pyrimethamine and sulfonamide. Reactivation of *Toxoplasma* cysts can be prevented by administration of co-trimoxazole.

Children and pregnant women should be advised to avoid exposure to kittens, cat litter, gardening and under-cooked meat.

Trichomoniasis

Trichomoniasis is caused by *Trichomonas vaginalis*, a weak noninvasive pathogen. It is a common STD and causes vaginal discharge in females and urethritis in males. The parasite is sensitive to metronidazole.

D5 HELMINTHS AND PARASITIC ARTHROPODS

Key Notes

Helminths are parasitic worms; they are eukaryotic multicellular organisms and often have a complex life cycle.

Nematodes	They are the commonest helminths. Some parasitize the gut and others enter the bloodstream or other body organs.

The gut nematodes are diagnosed by identification of their characteristic eggs or larvae in stool samples (by microscopy). They include **common roundworm** (*Ascaris lumbricoides*), **pinworm** (*Enterobius vermicularis*), **hookworms** (*Ancylostoma duodenale* and *Necator americanus*) *Strongyloides stercoralis* and **whipworm** (*Trichuris trichiura*).

Tissue or blood nematodes are transmitted by biting insects and cause systemic infections. The filariae include *Wuchereria bancrofti* (causes elephantiasis), *Onchocerca volvulus* (causes 'river blindness') and *Loa loa* (causes 'Calabar swellings').

Trematodes	Trematodes have a complex life cycle involving a definitive host and two intermediate hosts. Blood flukes cause bilharzia and include *Schistosoma haematobium* (urinary form), *S. mansoni* (intestinal) and *S. japonicum* (intestinal). These infest major abdominal veins and lay characteristic eggs. They induce aggressive host responses and finally cause fibrosis.

Other important trematodes include the liver fluke (*Fasciola hepatica*); Chinese liver fluke (*Clonorchis sinensis*); and lung fluke (*Paragonimus westermani*).

Cestodes (tapeworms)	Tapeworms pass through two hosts to complete their life cycle. Hydatid worm, (*Echinococcus granulosus*) is a parasite of the dog and uses sheep as the intermediary host. Man is an accidental intermediate host for the larva which forms 'hydatid cysts'. Beef tapeworm (*Taenia saginata*) and pork tapeworm (*T. solium*) infest human intestine and are the longest worms. Fish tapeworm (*Diphyllobothrium latum*) occurs in lakeland areas where fish is eaten raw.

Parasitic arthropods	Insects and other arthropods (ticks and mites) are implicated in disease in three ways: as vectors of infection; as ectoparasites and as allergens.

Helminths

Helminths are parasitic worms; they are eukaryotic, multicellular organisms and often have a complex life cycle. Helminths of medical importance fall into three main groups: nematodes (roundworms) and two types of platyhelminths (flatworms) – trematodes (flukes) and cestodes (tapeworms). The common helminths are mentioned here and illustrated in *Fig. 1*.

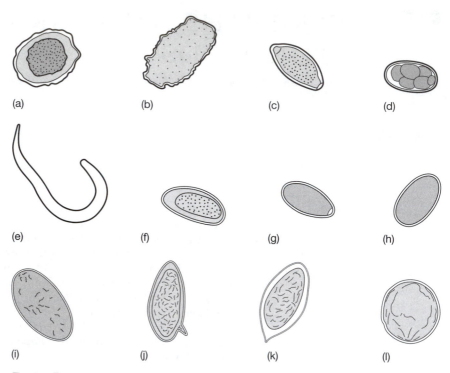

Fig. 1.   Eggs of intestinal helminths: (a) Ascaris lumbricoides (fertile eggs); (b) A. lumbricoides (infertile egg); (c) Trichuris trichiura; (d) hookworm; (e) Strongyloides stercoralis (larva); (f) Enterobius vermicularis. Eggs of trematodes: (g) Clonorchis sinensis; (h) Fasciola hepatica; (i) Paragonimus westermani; (j) Schistosoma mansoni; (k) S. haematobium; (l) S. japonicum.

Nematodes

Nematodes include a large number of pathogenic worms, ranging in size from a few millimeters to tens of centimeters long. Their lifestyle ranges from those that require a single host to those that require the presence of two mammalian species. Some parasitize the gut and others enter the bloodstream or other body organs.

Gut worms

Common roundworm (*Ascaris lumbricoides*) is the commonest cause of helminthiasis. Well over a billion people harbor this worm, chiefly in developing countries. The adult worm is ~ 20 cm long, living in the small intestine and laying large numbers of eggs which are transmitted feco-orally to a new host. The eggs hatch in the small intestine of new host, and the larvae migrate to the liver and lungs before finding their way back to the small intestine. During this migration (which takes ~ 6 weeks), larvae cause mild hepatitis, pneumonitis and allergic symptoms. Adult worms may cause malnutrition or intestinal obstruction. Diagnosis is by identifying the eggs in fecal samples or the adult worms when passed via stool. The worm is sensitive to mebendazole.

Pinworm or threadworm (*Enterobius vermicularis*) is a common cause of pruritus ani in children worldwide, in both industrialized and in developing countries. The adult worm is 1 cm long, living in the cecum or colon. Female worms lay eggs on the perianal area. The eggs can be picked up with Sellotape and examined under the microscope. Treatment is with mebendazole.

The most common, medically important, **hookworms** are *Ancylostoma duodenale* and *Necator americanus*. Adult worms reside in the small intestine where they feed and lay eggs. The eggs are transmitted feco-orally to a new host. Hookworms may cause anemia as they ingest blood and cause bleeding lesions in the lining of the gut. Diagnosis is by stool microscopy and treatment is with mebendazole.

Strongyloides stercoralis is related to hookworms. Adult females lay eggs containing infective larvae which hatch in the intestine. These are seen in the stool and are diagnostic. The parasite leads a free-living phase in soil where adult males and females reproduce. Infective larvae may penetrate the skin and migrate (larva migrans) via the lungs and reach the intestine. Treatment is by using mebendazole. Long-term latent infections are known to reactivate in patients with immunosuppressive conditions.

Whipworms (*Trichuris trichiura*) are common worms living in the colon and lay 'tea-tray' eggs (diagnostic) which do not hatch until they reach the intestine of a new host. They may cause rectal prolapse in children. Treatment is with mebendazole.

Tissue or blood nematodes

These include microscopic **'filarial' worms (microfilaria)** which are transmitted by biting insects and cause systemic infections (microfilariasis). The filariae include *Wuchereria bancrofti*, the causative parasite of elephantiasis, found in many tropical countries; *Onchocerca volvulus*, which is responsible for 'river blindness' in West Africa, but also occurs in parts of Central and South America; and *Loa loa*, which causes transient subcutaneous reactions ('Calabar swellings') in parts of tropical Africa. Diagnosis is by demonstrating the larvae or adult worms in tissue or blood. Treatment is with some toxic drugs, including diethylcarbamazine.

Trichinella spiralis is a tissue nematode in humans who are infected accidentally, usually by eating under-cooked pork products. Infective larvae mature into adult male and female worms and produce numerous larvae that penetrate the gut wall and migrate to skeletal muscles. Most of the symptoms of trichinosis are associated with the migration of larvae. Treatment is unsatisfactory, although mebendazole is thought to be effective.

Trematodes

Trematodes generally have a complex life cycle involving a snail and a second intermediate host as well as the definitive host.

Blood flukes (schistosomes, bilharzias) are the most important trematodes. *Schistosoma haematobium* (predominantly in Africa and Middle East) infest the major veins of the urinary bladder and cause hematuria and hepatosplenomegaly (bilharzia). **S.** *mansoni* (Africa) and **S.** *japonicum* (Far East) infest the mesenteric veins of the rectum and give rise to loose stools and hepatosplenomegaly. Deposition of eggs in tissues causes granuloma formation, fibrosis and some times malignancy. Adult worms lay eggs with characteristic spikes which are passed out via urine (*S. haematobium*) or feces (*S. mansoni* and *S. japonicum*). These are diagnostic when seen under the microscope. The eggs must find their specific intermediate host snails in water before transforming into infective parasites (called cercariae) which penetrate the skin upon contact (causing swimmer's itch) and migrate via the bloodstream into the liver where they develop into adult worms. The mature male parasite embraces the longer and slimmer female in its highly adapted copulatory canal, before migrating further to reach the major veins of the bladder or rectum. Treatment is with praziquantel.

Other important trematodes include the **sheep liver fluke** (*Fasciola hepatica*); **Chinese liver fluke** (*Clonorchis sinensis*); and **lung fluke** (*Paragonimus westermani*). The distribution of fluke infections is governed by the ecology of the intermediate hosts.

Cestodes (tapeworms)

Tapeworms have a simpler life cycle than the flukes, but transmission generally involves an intermediate host. The most important cestode is the **hydatid worm**, *Echinococcus granulosus*, which is a parasite of the dog. Rearing animals (e.g. sheep) are the intermediate hosts harboring the larval forms, which give rise to hydatid cysts, usually in the liver. Man is an accidental intermediate host. The cyst grows inexorably and death may ensue unless it is removed surgically. Hydatid disease occurs mainly in sheep-rearing areas, including the UK, where the cycle is maintained between sheep and sheep-dogs. Diagnosis is critical because rupture of hydatic cysts can cause anaphylaxis and shock in the host.

Other important cestodes include the **beef tapeworm**, *Taenia saginata* and **pork tapeworm**, *Taenia solium* which infest human intestine. These worms may reach a size of >10 m, but are generally innocuous, although the larval stage of *Taenia solium* may lodge in the brain and give rise to an epileptiform disease, cerebral cysticercosis. An important source of infection is consumption of undercooked meat. Treatment of cysticercosis can include surgical excision. Niclosamide and praziquantel are effective treatments.

The **fish tapeworm**, *Diphyllobothrium latum* occurs in lakeland areas where fish is eaten raw. It has been held responsible for pernicious anemia, because of competition for dietary vitamin B_{12}.

Parasitic arthropods

Insects and other arthropods (ticks and mites) are implicated in disease in three ways: as vectors of infection; as ectoparasites; and as allergens.

Arthropod vectors

Various diseases caused by helminths (e.g. filariasis), protozoa (e.g. malaria, trypanosomiasis and leishmaniasis), bacteria (e.g. plague, typhus and Lyme disease) and viruses (e.g. yellow fever and other arbovirus diseases) are transmitted by the bite of arthropods, which form an essential component of the life cycle. The geographical distribution of such diseases is governed by that of the arthropod vector. Containment of disease is associated with vector control and/or reservoir animal control or vaccination.

Ectoparasites

Some lice (*Pediculus humanus* and *Phthiris pubis*), fleas (*Pulex irritans* and *Tunga penetrans*) and mites (*Sarcoptes scabiei* and *Demodex folliculorum*) spend much or all of their life attached to the human host. Bed bugs (*Cimex lectularius* and *Cimex hemipterus*), reduviid bugs (*Triatoma* species, etc.) and many species of ticks require a periodic blood meal and parasitize man.

With some biting flies (*Cordylobia anthropophaga* and *Dermatobia hominis*), the larval stage penetrates skin (including human skin) as an obligatory part of the life cycle. The condition is known as cutaneous myiasis.

Allergens

The common house dust mite *(Dermatophagoides pteronyssimus)* is a potent allergen and has been incriminated in asthma. Other mites may also give rise to hypersensitivity reactions. Stinging and biting insects may also occasionally elicit a serious allergic response and the hairs of some caterpillars may cause severe urticaria.

E1 LABORATORY DIAGNOSIS OF VIRUS INFECTIONS

Key Notes

Diagnosis of a virus infection may be made: (i) by demonstrating that clinical material derived from the patient contains a virus; or (ii) by demonstrating the presence of an immune response to the virus in the patient's serum.

Techniques for the detection of virus

(i) Virus isolation in tissue culture.
(i) Clinical material from the patient is inoculated into cultured cells, which are then observed for the development of a cytopathic effect (CPE). Development of CPE may be slow. Diagnosis can be speeded up by detection of expressed viral antigens within infected cells before a CPE has appeared, e.g. by hemadsorption or by the detection of early antigen fluorescent foci test. Not all viruses grow in routine cell culture.

(ii) Electron microscopy (EM).
(ii) Visualization of viral particles by EM requires $> 10^6$ particles per ml of fluid. This pertains in the feces of patients with viral gastroenteritis and in virally induced vesicle fluid (e.g. HSV- or VZV-induced vesicles). EM is a rapid diagnostic technique.

(iii) Antigen detection.
(iii) Virally infected cells express viral antigens. This technique requires cells from the patient, e.g. from a nasopharyngeal aspirate. Multiple spots of dried cells are each stained with a different antiviral monoclonal antibody tagged to a fluorescent dye. Only those cells stained with a monoclonal directed against the infecting virus will fluoresce. This is also a rapid technique.

(iv) Genome detection.
(iv) Genome amplification techniques, such as the polymerase chain reaction (PCR) assay, are exquisitely sensitive. They are routinely used in the diagnosis and management of HIV and HCV infection, and are beginning to replace many of the older assay formats for detection of an ever-increasing list of viruses. Drawbacks include expense, and their propensity to generate false-positive results if not performed properly.

Diagnosis by serological methods

Serological diagnosis of a recent infection can be made by: (i) demonstrating a rise in titer of virus-specific IgG in serum samples taken over a short period of time; (ii) demonstrating the presence of virus-specific IgM in a serum sample; or (iii) for some viruses, demonstrating the presence of viral antigens in a serum sample. Antibodies to viruses, or viral antigens in serum can be detected by a variety of techniques, the most commonly used being the enzyme-linked immunosorbent assay.

Laboratory diagnosis of virus infections

Diagnosis of a virus infection may be made in two ways. First, it may be possible to identify the virus, or parts thereof, in material derived from the patient. Second, it may be possible to demonstrate a host immune response (usually antibody) to the virus, which is evidence that the patient has been infected with that virus. Depending on the nature of the immune response, it may be possible to determine whether the infection took place recently (and therefore may account for the patient's current illness), or sometime in the past. Not all the techniques described below are applicable in all circumstances, and the actual approach used to diagnose a particular infection will depend on the infection concerned.

Techniques for the detection of virus

There are four different approaches to demonstrating the presence of virus – or of subcomponents of the virus – in clinical material derived from the patient.

(i) Virus isolation in tissue culture

Principle
Viruses are obligate intracellular parasites (Topic A3). They do not grow on inanimate culture media but require susceptible living cells. Diagnostic virus laboratories maintain various types of cells in culture into which clinical specimens can be inoculated. As viruses replicate within cells, the cells may undergo morphological changes (e.g. swelling due to alteration of membrane permeabilities, shrinkage due to cell death). These changes, visible under the light microscope, are referred to as a **cytopathic effect** (CPE, see Topic A4). The inoculated cell cultures are therefore examined at intervals to determine whether a CPE has developed. The occurrence of a CPE indicates that the original clinical material did indeed contain a virus.

Practical points
A variety of clinical specimens may be used for virus isolation, e.g. throat swab, nasopharyngeal aspirate, feces, urine, CSF, eye swab, vesicle fluid. Swabs are broken off into viral transport medium (isotonic fluid plus antibiotics to inhibit bacterial overgrowth), as viruses will not survive on dry swabs. In the laboratory, the container is vortexed to release cells from the swab into the transport medium, which is then inoculated onto the cell sheets.

Not all viruses will grow in all cell types. In order to maximize the chances of isolating a range of clinically important viruses, most laboratories will run two or three different cell culture types.

Viruses routinely isolated in cell culture include herpes simplex viruses (HSV), varicella-zoster virus, cytomegalovirus, adenoviruses, enteroviruses, respiratory syncytial virus (RSV), rhinoviruses, influenza and parainfluenza viruses. Different viruses produce different CPE on different cell sheets. With experience, it is possible to state with reasonable certainty which virus is responsible for a given CPE.

Virus isolation by cell culture is much slower than bacterial isolation on inanimate media. The fastest-growing virus, HSV, may produce a CPE after 24 h, but most viruses require several days (or even weeks) to do so. This is a considerable disadvantage. Some viruses cannot be grown at all using standard laboratory cell cultures, e.g. hepatitis B virus, HIV, EBV.

In order to speed up the detection of certain slow-growing viruses, various adaptations of cell culture have been devised. These include:

- *Haemadsorption.* Some viruses (e.g. influenza) possess a hemagglutinin, which will be expressed on the surface of infected cells. Thus, red cells added to the cell culture tubes become adherent to the cell sheet. This enables the presence of a virus to be detected *before* a CPE has developed.
- The detection of early antigen fluorescent foci test (DEAFF test). This is used to accelerate the diagnosis of cytomegalovirus (CMV) and adenovirus infections. Virus-infected cells express antigens (known as early antigens, as they are expressed early in the viral replication cycle) in their cell nucleus long before any CPE is visible. These antigens can be demonstrated by staining the cells with appropriate fluorescently labelled monoclonal antibodies directed against the early antigens.

(ii) Electron microscopy

Direct visualization of viral particles can be done with an electron microscope (EM). Paradoxically, the EM is *not* a particularly sensitive instrument – the huge magnification produced (e.g. x 45 000) means that there needs to be a large concentration of viral particles in the sample under test (e.g. 10^6/ml) in order for the microscopist to see them in a reasonable length of time. This limits the practical usefulness of EM to those situations in which virus is likely to be present in high titers. These include:

- Viral gastroenteritis: the causative viruses (rotaviruses, adenoviruses, noroviruses, caliciviruses, see Topic B14) are excreted in feces in very large quantities.
- Vesicle fluid: virus-induced vesicles (e.g. herpes simplex, varicella-zoster) also contain very high titers of virus.

One major advantage of EM is that it is a rapid diagnostic technique. A positive diagnosis can be made within an hour of the sample arriving in the laboratory.

(iii) Antigen detection

It is possible to prove that virus is present within a clinical sample by demonstrating the presence of 'subcomponents' of the virus. There are only two such 'subcomponents' – viral proteins (known as antigens), and viral genomes (i.e. viral nucleic acid). All virally infected cells will express virally encoded antigens on the cell surface or within the cell. These cells can therefore be stained with fluorescently labelled monoclonal antibodies directed against the viral antigens, and the cells will then fluoresce under UV light.

This technique can be used on any cellular material derived from the patient within which the virus may be replicating. In practise, this is best applied to the diagnosis of respiratory tract infections. Cellular material is obtained by aspirating epithelial cells with a syringe and fine tubing inserted into the nose (a nasopharyngeal aspirate). The cells are spotted onto glass slides, allowed to dry, and fixed with acetone. Each slide is then stained with a different fluorescently-labelled monoclonal antibody, unbound antibody is washed off, and the cells examined under UV light. The only cells to fluoresce will be those stained with the appropriate monoclonal antibody.

This is also a rapid diagnostic technique, as a result may be available with in 2 h of the sample reaching the laboratory. One further advantage, which is true also of EM, is that this technique is not dependent on the presence of viable virus.

(iv) Genome detection

Principle
Any cell that is infected with a virus must contain copies of the genome of that virus. There are various technologies available for the detection of specific viral DNA or RNA sequences within clinical material. **Hybridization techniques,** using labelled DNA or RNA probes, depend on the principle that complementary strands of nucleic acid (whether DNA or RNA) will spontaneously hybridize to each other under appropriate temperature and salt conditions. However, these are now being largely replaced with **genome amplification assays,** e.g. the polymerase chain reaction (PCR). This is designed to allow specific sequences of nucleic acids to be amplified by several orders of magnitude (e.g. 10^6-fold), such that detection of the amplified product is relatively straightforward. Thus it is possible to demonstrate the presence of a single infected cell within the midst of 10^6 uninfected ones.

Genome amplification assays have a number of potential advantages, including exquisite sensitivity, speed, and they can be structured in such a way as to give quantitative results (i.e. *viral load* measurement). However, there are potential drawbacks – they are relatively expensive to perform, and if not conducted carefully, samples may become contaminated with amplified product, thereby generating false-positive results. They have become routine in the diagnosis and management of patients infected with hepatitis C virus and HIV, and there is no doubt that such assays will play an increasingly wide role in virus diagnosis in the future.

Diagnosis by serological methods

The diagnosis of viral infection by demonstration of antibodies or viral antigens in serum is known as 'serology'.

Principles
There are three ways of diagnosing viral infections by serological techniques.

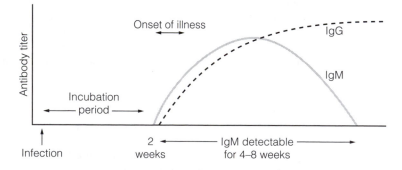

Fig. 1.   Diagram of IgG and IgM responses to infection.

(i)   Measuring virus-specific IgG (e.g. anti-influenza A antibodies). Following a recent infection, the levels – or **titer** – of virus-specific IgG will rise as the immune response develops (*Fig. 1*). Thus, the demonstration of a rise in virus-specific IgG titers in a pair of sera taken at the time of illness (the acute sample) and 7–14 days later (convalescent sample) is indicative of a recent infection. The problem with this approach is that one has to wait until the convalescent sample is available before the diagnosis can be made.

(ii) Measuring virus-specific IgM. As this class of immunoglobulin only appears transiently after an infection (*Fig. 1*), the presence of specific IgM in a single serum sample (e.g. IgM anti-rubella) is indicative of a recent infection. The big advantage of this approach is that the diagnosis can be made using the acute sample only.

(iii) In some virus infections, it is possible to detect the presence of viral antigens themselves, rather than the antibody response to the antigens, in serum. The best example of this is in hepatitis B virus infection (HBV), where the diagnosis is made by the demonstration of HBV surface antigen (HBsAg) in serum (Topic B3).

Practical points

There is a variety of methodologies available for detecting and measuring virus-specific antibodies. All these techniques involve the interaction of antigen with specific antibody, followed by some sort of system to detect this interaction. Two of the more commonly used techniques are the enzyme-linked immunosorbent assay (ELISA) – and the complement fixation test (CFT). The ELISA is a very flexible type of assay – it can be adapted to allow antigen detection rather than antibody detection, e.g. for diagnosis of hepatitis B virus infection, and also can be constructed so as to distinguish between virus-specific IgM and IgG.

E2 LABORATORY DIAGNOSIS OF BACTERIAL INFECTIONS

Key Notes

Specimens	Some specimens (e.g. blood cultures) are normally sterile, and are collected using an aseptic technique, but most specimens are taken from nonsterile sites, or collected in a nonaseptic manner (e.g. throat swabs, urine), producing mixed cultures of potential pathogens, normal flora and contaminants. Specimens should be collected before starting antibiotic therapy. They must be correctly labelled and accompanied by relevant clinical information.
High-risk samples	Certain specimens may contain pathogens that are a hazard to laboratory staff, and these must be labelled as 'high-risk' in order to insure appropriate safety measures are taken.
Transport to the laboratory	Rapid transport is necessary to preserve the viability of delicate organisms, and to minimize the multiplication of bacteria within the specimen (semiquantitative culture techniques are often used). Bacterial multiplication can be reduced by storage at 4°C, if delays occur. Blood cultures, however, should be placed directly in an incubator (37°C).
Microscopy	Wet microscopy is usually used to record the presence of white blood cells. Gram staining is the commonest method to microscopically visualize and differentiate bacteria by their shape (cocci or bacilli) and staining properties (Gram positive or Gram negative). Specialized stains are available for certain types of bacteria.
Culture	Most clinical specimens are cultured on agar plates. Bacteria form characteristic colonies after incubation for 18–48 h. Selective plates are available for specific types of bacteria. Indicator systems can produce characteristic color changes with different bacteria. Semiquantitative cultures are performed with some specimens (e.g. urine) especially when low numbers of bacteria are not clinically relevant. Bacteria can also be grown in liquid media (broth). These are more sensitive, but subculture onto agar is usually required for bacterial identification, and they are prone to contamination.
Bacterial identification	Methods used to identify bacteria include Gram stain, colonial morphology and growth patterns on agar plates, rapid bench tests, and biochemical or antigenic tests. Identification can be made to genus, species or subtype level, according to clinical and epidemiological circumstances. Typing methods include serotyping, phage typing and genotyping. Molecular methods can also be used to identify bacteria, through DNA sequencing. Identification of bacterial toxins is also used to diagnose some types of infection.

Other diagnostic methods	These include antigen detection tests, antibody tests, detection of bacterial DNA (e.g. by PCR) and histology.
Interpretation and reporting	Many bacterial investigations require interpretation, especially cultures yielding mixed growth. There must be good liaison between healthcare workers and the microbiology laboratory to insure that appropriate investigations are performed, results are correctly interpreted, and clinically relevant reports produced.
Related topics	Antibiotic susceptibility testing (E3)

This topic summarizes the laboratory diagnosis of bacterial infections. Similar diagnostic methods are used for the diagnosis of fungal, protozoal and helminthic infections (Section D). The laboratory diagnosis of viral infections is summarized in Topic E1. Antibiotic susceptibility testing is covered in Topic E3.

Specimens

Many different specimens can be sent for microbiological examination from patients with suspected bacterial infection. Common specimens include urine, feces, swabs from wounds, throat swabs, vaginal swabs, sputum, and blood cultures. Less common, but important specimens include cerebrospinal fluid (CSF), pleural fluid, joint aspirates, tissue, bone and prosthetic material (e.g. line tips).

Some types of specimen are normally **sterile** (e.g. blood cultures, joint aspirates). These samples are usually obtained via a percutaneous route with needle and syringe, using appropriate skin disinfection and an aseptic technique. The culture of bacteria from such specimens is usually indicative of definite infection – unless organisms normally resident on the skin are detected – '**skin contaminants**'. However, these skin bacteria (e.g. *Staphylococcus epidermidis*) can also cause biofilm-related infection of prosthetic devices (e.g. prosthetic hip joints, central venous catheters), making the interpretation of positive culture results difficult.

By contrast, many microbiological specimens are taken from **nonsterile sites** (e.g. from mucosal surfaces with a **resident microbial flora,** such as vaginal or throat swabs), or are collected through a nonsterile site (e.g. urine passed through the urethra). Such samples often contain bacteria of no clinical relevance in addition to possible pathogens, making the interpretation of culture results more difficult. To further complicate matters, **potential pathogens** may also be found as '**contaminants**' (e.g. *Streptococcus pneumoniae* colonization of the pharynx may contaminate a sputum sample), giving rise to possible **false-positive results**. In general it is preferable to send samples from sterile sites if available.

It is good practise to obtain the samples for bacteriological culture **before antibiotic therapy is started**, to maximize the sensitivity of the investigations and reduce false-negative results. Similarly, samples of tissue or pus are preferred over swabs, to maximize the recovery of bacteria in the laboratory. However, in certain emergency situations (e.g. suspected invasive meningococcal disease), antibiotics should be administered immediately.

Specimens must be accurately labelled and accompanied by a properly completed **request form**, indicating the precise nature of the specimen, the date collected, relevant clinical information, the investigations required, and details of antibiotic therapy. This allows the laboratory to perform the correct range of tests, and helps in the interpretation of results and reporting.

As well as clinical specimens, medical microbiology laboratories may also process samples of food, water and other environmental samples (e.g. air sampling from operating theaters) as part of infection control procedures.

High-risk samples Certain bacterial infections are a particular hazard to laboratory staff, and specimens that might contain these pathogens should be labelled as 'high risk' to allow for **additional safety measures** if necessary. This includes blood cultures from suspected **typhoid** (*Salmonella typhi*) or **brucellosis** (*Brucella* species), or samples that might contain *Mycobacterium tuberculosis*.

Transport to the laboratory Most specimens are sent to the laboratory in sterile universal containers, but swabs should be placed in a suitable charcoal-based transport medium (unless they are to be plated onto agar at the bedside). Specimens should be **transported rapidly** to the laboratory, or otherwise stored at 4°C.

Rapid transport is necessary in order to:

(i) preserve the viability of the more 'delicate' bacteria, such as *Streptococcus pneumoniae* or *Haemophilus influenzae* (delays in processing can cause false-negative culture results);

(ii) minimize the multiplication of bacteria (e.g. coliforms) within specimens before they reach the laboratory. This is particularly important with urine and other specimens that utilize a **semiquantitative culture** technique, as delays in transport can give rise to falsely high bacterial counts when the specimen is processed.

If delays are inevitable, then storage at 4°C minimizes bacterial multiplication. Alternative methods for urine samples include:

(i) sterile containers with **boric acid** added to inhibit bacterial multiplication;

(ii) the **'dip-slide'** method. A special agar slide is dipped briefly into freshly voided urine, and then the slide (not the urine) is forwarded to the laboratory for incubation and reading of culture results.

Many laboratories employ **automated systems** for processing blood cultures. Specialized **blood culture bottles** are employed. These are inoculated with blood, using careful aseptic technique, and then either transported rapidly to the laboratory, or placed within an **incubator** (at 37°C).

Microscopy Some specimens are routinely examined microscopically. Microscopy is a relatively **rapid technique** (in contrast to culture techniques) and may give an early indication as to the nature of the infection.

'**Wet**' **microscopy** (i.e. not using a specific bacterial stain) is performed to determine the number of white blood cells in a liquid specimen (e.g. urine or cerebrospinal fluid). Wet microscopy can also be used to visualize yeast cells (indicative of possible candida infection), or protozoa such as *Trichomonas vaginalis* (in a vaginal swab). Wet microscopy of stool samples (using an iodine-containing stain) can detect a wide range of protozoal cysts (e.g. *Giardia*, *Entamoeba*), or helminthic ova (e.g. *Ascaris*, hookworm). However, wet microscopy is not particularly suitable for visualizing or differentiating bacteria.

The commonest method for visualizing bacteria within a specimen is with the **Gram stain**. This is most suitable for use with normally sterile specimens (e.g. CSF, joint aspirates), pus and tissue. A Gram stain helps with the visualization of bacteria, and gives an indication of the type of bacteria present, based on the shape of the bacteria (cocci or bacilli) and the staining properties (Gram positive: purple; Gram negative: pink/red). A Gram stain also helps to identify mixtures of bacteria, helps to determine the appropriate range of agar plates to be used for subsequent culture, and helps with the interpretation of culture results.

Although a rapid technique, Gram stain is not especially sensitive. For liquid specimens (e.g. CSF), the sample is first centrifuged (to concentrate any bacterial cells in the deposit), and Gram stain and culture is performed from the deposit after the supernatant is decanted. This helps increase the sensitivity of both microscopy and culture.

Not all bacteria are stained well by the Gram stain, and a variety of **specialized stains** are available. The commonest is the **Ziehl–Neelson (ZN) stain** which is used to demonstrate the presence of **mycobacteria**. This staining technique is based on the ability of mycobacteria to retain fuchsin stain in the presence of both acid and alcohol (they are '**acid–alcohol-fast bacilli**' or **AAFB**). Mycobacteria can also be visualized using the fluorescent dye auramine and a **fluorescence microscope**. Direct immunofluorescence is employed to detect certain pathogens (e.g. *Legionella*, *Pneumocystis*) using specific antibodies conjugated to a fluorescent dye.

Another microscopic technique is dark-ground (or dark field) microscopy. This is mainly used to detect the thin spirochaetal cells of *Treponema pallidum* (syphilis).

Culture

Culture on solid media

The principal method for the detection of bacteria from clinical specimens is by culture on **agar plates** (i.e. solid media). Bacteria (and fungi) will grow on the surface of agar plates to produce distinct **colonies**. Different bacteria produce different but characteristic colonies, allowing for early presumptive identification and easy identification of mixed cultures.

There are many different types of agar plates. Agar is used as the gelling agent to which is added a variety of nutrients (e.g. blood, peptone and sugars) and other factors (e.g. buffers, salts and indicators).

Some types of agar are **nonselective** (e.g. blood agar, nutrient agar) and these will grow a wide variety of bacteria. Some agar plates (e.g. MacConkey agar) are more **selective** (in this case through the addition of bile salts selecting for the 'bile-tolerant' bacteria found in the large intestine such as *Escherichia coli* and *Enterococcus faecalis*). MacConkey agar also contains lactose and an **indicator system** that identifies lactose-fermenting coliforms (e.g. *Escherichia coli*, *Klebsiella*) from lactose-nonfermenting coliforms (e.g. *Morganella*, *Salmonella*). Agar plates can be made even more selective by the addition of antibiotics or other **inhibitory substances**, and sophisticated indicator systems can allow for the easy detection of defined bacteria from mixed populations.

Colonies of most bacteria are usually large enough to identify after 18–24 hours of incubation (usually at 37°C), but for some bacteria longer incubation times are required (from 2 days to several weeks). Agar plates may be incubated in **air**, in air with added **carbon dioxide** (5%), **anaerobically** (without oxygen) or **micro-aerophilically** (a trace of oxygen) according to the requirements of the different types of bacteria that may be present in specimens.

Semi-quantitative culture techniques can be employed with agar plates using the principle that each colony grows from a single 'colony-forming unit'. For example, a defined quantity of a specimen (e.g. 2 µl of urine) is plated onto agar in a standardized manner, and an assessment is made of the number of colonies produced. Sputum specimens are often diluted (e.g. 1 in 1000) before plating onto agar, in order to detect only those bacteria present in significantly large numbers.

Culture in liquid media

Bacteria can also be grown in liquid media (broth). Like agar plates, broth cultures may be nonselective or selective. Bacterial growth is easy to detect as the clear liquid turns **turbid**, usually within 24–48 h, but incubation may need to be extended to 14 days or more.

The advantage of broth culture is that it is significantly **more sensitive** than direct culture on agar. The disadvantage is that, by itself, it is not easy to determine the type of bacteria present or whether a mixed growth has occurred, and in most cases the broth must be **subcultured** onto solid agar plates. This causes an additional delay in culture results. Broth cultures are also prone to **contamination**.

Broth enrichment media are used when high sensitivity is required (e.g. for detection of bacteria from CSF, or to detect small numbers of *Salmonella* in a stool sample containing many millions of other bacteria).

Automated blood culture systems utilize liquid culture. Bacterial growth may be detected by a variety of methods (e.g. detection of bacterial CO_2 production). Automated liquid culture systems are also available for the culture of mycobacteria, and similar technology can be used to automate sensitivity testing (Topic E3).

Bacterial identification

A number of methods are used to identify bacteria that have been grown in the laboratory (*Fig. 1*), including Gram-stain appearance, colonial morphology, growth patterns on selective and nonselective agar, atmospheric growth requirements, rapid bench tests (e.g. catalase, oxidase, coagulase), biochemical tests (e.g. sugar fermentations), and antigenic tests (e.g. serogrouping).

The range of methods utilized and the degree of identification performed depends on the nature of the specimen, the clinical context, the presumptive bacterial identification and whether additional epidemiological information is required, such as during outbreaks. For instance, it may be acceptable to identify an organism only to the **genus** (family) level (e.g. 'coliform' or 'coagulase-negative staphylococcus'). However, for important infections (e.g. endocarditis) identification is usually done to **species** level (e.g. *Enterobacter cloacae* or *Staphylococcus epidermidis*).

For some bacteria it is important to further subdivide the species in order to establish the **serogroup** or **serotype** present. A good example of this is with *Neisseria meningitidis*, when the main serogroups (A, B, C, W135 and Y) can usually be determined by simple latex agglutination.

Many **typing methods** are available including serotyping, phage typing and a variety of genotyping techniques [e.g. pulsed field gel electrophoresis (PFGE), random amplified polymorphic determinants (RAPD)], but these are generally expensive and time-consuming, and are often available only in reference or larger laboratories. Typing methods are valuable tools in the investigation of outbreaks, and in interpreting the epidemiology of infectious disease.

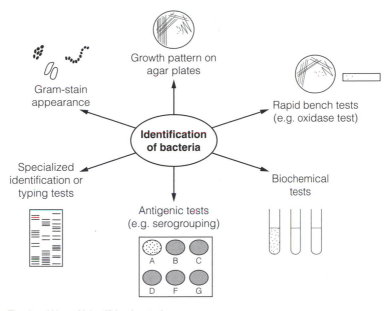

Fig. 1.   Way of Identifying bacteria.

For some infections, it is not just the identification of the organism that is important, but whether it has the ability to produce **specific toxins**. The best example of this is **diphtheria**, which is caused by **toxigenic strains** of *Corynebacterium diphtheriae* (Topic C4). The toxin may be detected by immuno-precipitation (the Elek test) or by detection of the presence of the toxin gene (DNA) by PCR. Another important example is the detection of *Clostridium difficile* **toxins**. This is used to make the diagnosis of *Clostridium difficile*-associated diarrhea, and the toxin can be detected directly in stool samples, without the need to culture the bacteria.

Some bacteria are difficult to identify by conventional tests, but with the increasing availability of automatic **DNA sequencers** and bacterial DNA data-bases, it is now possible to identify bacteria by comparing DNA sequences with other known bacterial sequences. This technique can also be used to identify **nonculturable** bacteria from specimens, so long as the DNA can be extracted and amplified.

Other diagnostic methods

In addition to microscopy and culture, there are several other methodologies employed for the diagnosis of bacterial infections. These are:

(i)   antigen detection tests;
(ii) antibody tests;
(iii) detection of bacterial DNA (e.g. by PCR);
(iv) histological examination.

Methods (i), (ii), and (iii) are described in more detail in Topic E1, but the same principles apply to bacteria, fungi, protozoa and helminths.

Some important examples of the use of these methods are listed in *Table 1*.

Table 1.   Examples of the diagnosis of bacterial infections by noncultural methods

Infection	Organism	Specimen(s)	Detection methods
Bacterial meningitis	*Neisseria meningitidis*	CSF	Latex agglutination, PCR
		Whole blood	PCR
		Serum	Antibody tests
	Streptococcus pneumoniae	CSF	Immunochromatography
		Whole blood	PCR
Legionnnaires' disease	*Legionella pneumophila*	Urine	EIA, immunochromatography
		Respiratory specimens	Direct immunofluorescence
Tuberculosis	*Mycobacterium tuberculosis*	Sputum	Microscopy (ZN or auramine stain), PCR
		Lymph node	Histological appearance (caseating granulomas, AAFB)
Syphilis	*Treponema pallidum*	Exudate from chancre	Dark-ground microscopy
		Serum	Serological tests (e.g. TPPA, EIA)

CSF, cerebrospinal fluid;
PCR, polymerase chain reaction;
EIA, enzyme immunoassay;
ZN, Ziehl-Neelson stain;
AAFB, acid–alcohol-fast bacilli;
TPPA, *Treponema pallidum* particle agglutination.

Interpretation and reporting

The results of many bacteriological investigations require a degree of interpretation, particularly culture results. Many cultures yield mixed growth comprising potential pathogens, commensal bacteria (normal flora), and sometimes contaminants. Not all bacteria identified are automatically reported. For instance, a swab from a wound growing coagulase-negative staphylococci and diptheroids may be reported as 'mixed skin organisms only' or 'no significant pathogens identified'.

The following are some of the factors that are considered in interpreting bacteriological culture results:

• type of specimen;
• any delays in processing;
• types of bacteria recovered;
• knowledge of the normal human flora at different sites;
• clinical information provided on the request form;
• details of recent antibiotic therapy.

There must be **good liaison** between **healthcare workers** and the microbiology **laboratory**, in order to insure that the most appropriate investigations are performed, results are interpreted correctly, and clinically relevant bacteriological reports are produced.

E3 ANTIBIOTIC SUSCEPTIBILITY TESTING

Key Notes

***In vitro* sensitivity tests**	Bacterial pathogens are tested for their susceptibility to antibiotics to guide antibiotic treatment. Sensitivity tests are generally performed from single pure bacterial colonies on an agar plate. Direct sensitivity tests are set up directly from specimens or liquid cultures, producing quicker, but less standardized results.
Disk sensitivity tests	Antibiotic diffuses out of a disk placed on the surface of the agar. If bacteria are sensitive to the antibiotic, then a zone of growth inhibition forms around the disk after incubation. The zone size depends on several factors and two methods are available to control this process: comparative disk testing (where both a test and control organism are tested on the same plate), and standardized disk testing.
Breakpoint sensitivity tests	Antibiotic is incorporated into the agar at a uniform concentration and bacteria inoculated onto the agar surface. Only bacteria resistant to the antibiotic at the breakpoint concentration will then grow. Using a multipoint inoculator, many bacterial strains can be tested simultaneously on each agar plate.
Minimum inhibitory concentration (MIC)	The MIC is the minimum (lowest) concentration of an antibiotic that will inhibit the growth of a bacterial strain. This can be determined by several methods including macro- and microdilution tests, extended breakpoint sensitivity tests, and E-test strips. Determination of MIC is important in the management of certain infections (e.g. endocarditis).
Minimum bactericidal concentration (MBC)	The MBC is the lowest concentration of the antibiotic that will kill a bacterial strain. The MBC is less clinically relevant than the MIC, as MBC tests are harder to standardize.
Detection of bacterial resistance mechanisms	Various bacterial resistance mechanisms (e.g. β-lactamase production, antibiotic resistance genes) can be detected in the laboratory, providing a quick method of predicting *in vitro* sensitivity results.
Automated sensitivity tests	Automated systems can reduce the technical time required to perform sensitivity tests. These systems often utilize liquid culture, producing faster results than conventional agar-based tests.
Clinical relevance of *in vitro* antibiotic sensitivity tests	*In vitro* sensitivity test results should only be used as a guide to treatment, and the results do not always correlate with clinical response. The success of antibiotic treatment can be affected by many factors including immune responses, pharmacological factors and other biological variables, and the presence of biofilms.

Related topics	Laboratory diagnosis of bacteriological infections (E2)	Antibiotics (2): clinical use and antibiotic resistance (E7)
	Antibiotics (1): classification, mode of action and pharmacology (E6)	

In vitro sensitivity tests

In order to guide the appropriate antibiotic treatment of bacterial infections, bacterial pathogens isolated from clinical specimens are usually tested against a selection of antibiotics to assess their degree of susceptibility. This is usually done with bacteria that have been grown on solid media. Sensitivity tests are performed from **single pure colonies** and require a further 18–24 h of incubation. Thus while culture results may be available within 24 h of receipt of a specimen, sensitivity results usually take an additional day.

In some situations, **direct sensitivity tests** are performed, either from the specimen itself (e.g. urine) or from a liquid broth with bacterial growth (e.g. blood culture bottle). In this case, sensitivity tests are set up at the same time as the specimen is subcultured to agar plates. Although this speeds up the process, there are several disadvantages:

(i) it is difficult to ensure the correct **inoculum** (the number of bacteria spread onto the agar surface);
(ii) the inoculum may be **mixed** (more than one type of bacteria), making the results difficult to interpret and requiring the test to be repeated;
(iii) the selection of antibiotics tested may be inappropriate for the bacterium subsequently grown.

Several different methods are available for assessing the susceptibility of bacteria to antibiotics.

Disk sensitivity tests

Disk sensitivity tests are performed on agar plates. A small disk of filter paper, pre-impregnated with a defined quantity of antibiotic, is placed on the surface of an agar plate that has already been inoculated with a suspension of bacteria. The antibiotic diffuses out of the disk into the agar, along a concentration gradient, as the plates are incubated (for 18–24 h). If the bacterial strain is sensitive to the antibiotic, then a **zone of inhibition** (no growth) occurs around the disk (*Fig. 1*).

The diameter of the zone depends on a number of factors including:

(i) the quantity of antibiotic within the disk;
(ii) the degree of susceptibility of the bacteria to the antibiotic;

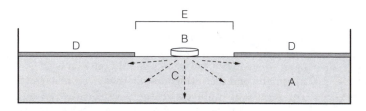

Fig. 1. Disk sensitivity test. A – agar; B – antibiotic disc; C – antibiotic diffuses into agar along concentration gradient; D – bacterial growth on surface of agar after 18 hours of incubation; E – zone (diameter) of inhibition.

(iii) the physicochemical properties of the antibiotic;

(iv) the depth (in mm) of the agar plate;

(v) the concentration of bacteria in the inoculum (semiconfluent growth is required).

There are two methods employed to determine the sensitivity pattern. The **comparative disk test (Stokes' method)** uses both a test organism and a control organism on the same plate (*Fig. 2*). The control organism is of defined sensitivity to the antibiotics being tested, and this method allows a direct comparison of the diameter of the zones of inhibition between the test and control organisms.

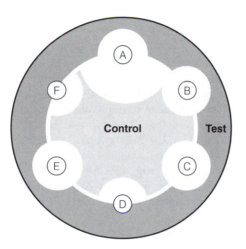

Fig. 2. Schematic representation of comparative disk sensitivity test (Stokes' method). Control – control bacterial strain (known sensitivity to antibiotics); Test – bacterial strain under test; A-F – six different antibiotic disks. In this figure, the test organism is sensitive to antibiotics B, C & E, but resistant to antibiotics A (>3 mm reduction in zone diameter compared to control), D & F.

Standardized disk testing uses carefully standardized agar plates and inocula. A standardized inoculum of the test organism is plated out across the whole surface of the agar plate (control organisms are tested on a separate plate). The diameter of the zones of inhibition are measured in mm, and the organism reported as sensitive or resistant based on defined cut-off points (for example <18 mm = resistant).

One disadvantage of disk testing is that it is usually only possible to have a maximum of six different antibiotic disks on a standard agar plate.

Breakpoint sensitivity tests

Breakpoint sensitivity tests use a different principle to disk testing. A defined concentration of antibiotic (the 'breakpoint') is incorporated into the agar during production of the agar plates. Bacteria are then inoculated onto a small part of the surface of the agar (usually with a 'multipoint' inoculator that allows many different strains to be tested on the same plate) and the agar plate incubated for 18–24 h. Bacteria that are resistant to the antibiotic (at the breakpoint concentration) will grow, whilst those that are sensitive will not. A control agar plate with no added antibiotic is used to check for viable bacterial growth (*Fig. 3*).

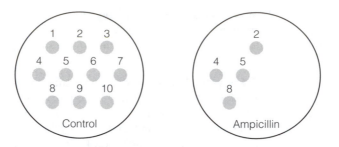

Fig. 3.   Breakpoint sensitivity tests. Control – sensitivity agar plate with no added antibiotic; Ampicillin – sensitivity agar plate incorporating ampicillin at a defined (uniform) concentration 1 to 10 – different bacterial strains inoculated onto the surface of the agar plates with a multipoint inoculator. In this example, all 10 strains have grown on the control plate. Strains 1, 3, 6, 7, 9 & 10 are sensitive to ampicillin. Strains 2, 4, 5 & 8 are resistant.

Using a **multipoint inoculator**, > 30 different strains of bacteria can be tested against a wide range of different antibiotics in one batch. This process is less technically time-consuming than the equivalent number of disk tests.

Sometimes the same antibiotic is used at two different concentrations in separate agar plates (e.g. 1 and 4 mg/l). Using these low and high breakpoint concentrations, bacteria can be classified as 'sensitive', 'intermediate' or 'resistant'. By including a whole range of concentrations of the same antibiotic in separate plates, the minimum inhibitory concentration of the antibiotic can be determined for each of the strains being tested (see below).

Minimum inhibitory concentration (MIC)

The MIC is the minimum (lowest) concentration of an antibiotic that will inhibit the growth of a bacterial strain.

Conventionally, this is determined using a series of doubling dilutions of the antibiotic in liquid culture medium, to produce a range of concentrations in test tubes (macrodilution) or in a microtiter tray (microdilution). After inoculation of the test strain into each antibiotic concentration, bacterial growth is determined by visible turbidity after 18–24 h of incubation (*Fig. 4*). The MIC is the lowest concentration of antibiotic with no visible bacterial growth.

MIC tests can also be done by **extended breakpoint sensitivity tests** (see above). These methods are technically time-consuming and relatively expensive. An alternative method is by use of commercially available **E-test** strips. These are specialized antibiotic-impregnated strips which, like disk testing, are placed on the surface of inoculated agar plates. During incubation, antibiotic diffuses into the agar forming a zone of inhibition. There is a manufactured concentration gradient within the strip, and numerical gradations are marked along the edge of the strip to reflect this. The MIC is determined by measuring the point at which the edge of the zone of inhibition crosses the E-test strip (*Fig. 5*).

Antibiotic MIC tests are usually performed only in certain situations in a clinical bacteriology laboratory. They are most commonly used when a very precise assessment of the *in vitro* susceptibility of a bacterial strain is required, for instance in the treatment of pneumococcal meningitis (Topic F3) or streptococcal endocarditis (Topic F13).

MIC tests are also used to assess the overall degree of activity of antibiotics against different strains of the same bacterial species, particularly when evaluating or developing new antimicrobial agents. A simple way of describing the

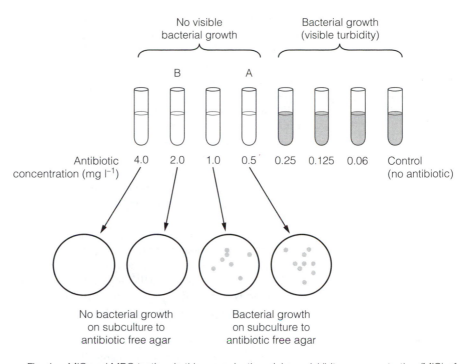

Fig. 4.   MIC and MBC testing. In this example, the minimum inhibitory concentration (MIC) of the antibiotic is 0.5 mg/l (Tube A). The minimum bactericidal concentration (MBC) is 2.0 mg/l (tube B).

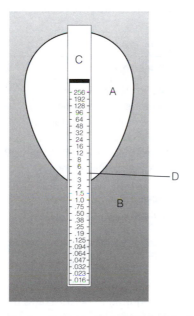

Fig. 5.   Determination of MIC by E-test. A – zone of inhibition; B – bacterial growth; C – E-test strip; D – the MIC is the point at which the edge of the zone crosses the E-test strip – in this example it is 3 mg/l.

relative activity of an antibiotic against a group of organisms, is by using the terms MIC_{50} and MIC_{90}. These are the lowest concentrations of the antibiotic that inhibit 50 and 90% of the bacterial strains tested, respectively.

Minimum bactericidal concentration (MBC)

The MBC is the lowest concentration of the antibiotic that will 'kill' a bacterial strain. The definition of 'killing' is a 99.9% (3 $\log_{10}$) reduction in viable bacteria.

The MBC test is an extension of an MIC test (*Fig. 4*). The simplest method for determining the MBC is to perform a subculture from antibiotic concentrations with no visible growth in the MIC test onto antibiotic-free agar. This will determine whether the bacteria have been inhibited from growing but are still viable, or whether they have been killed.

Some antibiotics are highly **bactericidal**. In this case the MIC and MBC are usually very similar. **Bacteristatic** antibiotics on the other hand have much higher MBC than MIC. Occasionally a bacterial strain may have a high MBC but low MIC with a normally bactericidal antibiotic (e.g. penicillin). This is described as bacterial '**tolerance**' to the antibiotic.

MBC tests are very difficult to standardize and are often not entirely reproducible. The clinical relevance of MBC tests and the demonstration of tolerance is less clear than with MIC determinations, but they are occasionally performed to guide antibiotic therapy in some difficult cases of infection.

Detection of bacterial resistance mechanisms

An alternative method for guiding appropriate antibiotic therapy is through the detection of bacterial resistance mechanisms. These can be used to predict the results of conventional sensitivity tests, especially when a specific resistance mechanism is detected. Often these tests do not require overnight incubation, and thus the results may be available at an earlier stage to guide treatment.

Some common examples of bacterial resistance detection used in clinical laboratories are given in *Table 1*. This type of approach is likely to become increasingly used, especially as molecular techniques become more widely available.

Table 1. Examples of the detection of bacterial resistance mechanisms

Resistance mechanism	Organism	Method of detection	Comment
β-Lactamase production	*Haemophilus influenzae*	Rapid 'stick' test (hydrolysis of nitrocefin)	Predicts resistance to ampicillin and amoxicillin
	Neisseria gonorrhoeae	Rapid 'stick' test (hydrolysis of nitrocefin)	Predicts high level resistance to penicillin
Methicillin resistance (altered PBP2')	*Staphylococcus aureus*	Latex agglutination for PBP2' Detection of *mec A* gene by PCR	Predicts resistance to β-lactam antibiotics (MRSA)
Rifampicin resistance	*Mycobacterium tuberculosis*	Detection of *rpo B* gene mutations by PCR	Detects 95% of rifampicin-resistant *M.tuberculosis* strains

PBP, penicillin binding protein; MRSA, methicillin-resistant *Staph. aureus*.

Automated sensitivity tests

There are a variety of commercially available automated systems available to help reduce the technical time required to perform and record routine sensitivity tests. For example, the results of disk sensitivity tests and breakpoint sensitivity tests can be read using a camera interfaced to a computer system. Other systems utilize **liquid cultures**, and detect the effect of antibiotics on the rate of bacterial growth through measurement of turbidity (**nephelometry**) or the production of CO_2. These automated systems can significantly shorten the necessary incubation time, with the possibility of some results being available within the same working day. They can also significantly reduce the time taken to produce sensitivity results for slow-growing organisms, notably *Mycobacterium tuberculosis*.

Clinical relevance of *in vitro* antibiotic sensitivity tests

It must be remembered that *in vitro* sensitivity tests are only a **guide** to the appropriate antibiotic treatment. A laboratory report indicating that organism A is resistant to antibiotic B does not necessarily mean that antibiotic B will not work, and *vice versa*. Whilst *in vitro* tests are designed to try and reflect the *in vivo* situation (e.g. through utilization of appropriate breakpoints to reflect antibiotic pharmacokinetic parameters), they can never take account of all the human and bacterial biological variables.

There are a number of factors to consider when interpreting laboratory reports that include sensitivity test results:

(i) Many infections will resolve spontaneously (assuming a normal immune system). If a patient has already responded clinically to a certain antibiotic treatment, then it is not always necessary to change the antibiotic if the laboratory report indicates that the organism isolated is 'resistant'.

(ii) The organism identified on the laboratory report may not be the primary pathogen.

(iii) An organism reported with sensitivity results does not always require treatment. A good example of this is with catheter-specimens of urine. Treatment is generally required only if the patient is symptomatic (i.e. treat the patient not the result!).

(iv) An antibiotic with apparent *in vitro* activity may not work clinically, as there are many pharmacokinetic and other factors to consider in choosing the most appropriate antibiotic therapy (Topics E6 and E7).

E4 ANTIVIRAL AGENTS

Key Notes

The key details relating to **mechanism of action, clinical indications, resistance mechanisms** and **toxicity** of all currently licensed antiviral agents (except those for the treatment of HIV infection, see Topic E5) are shown in *Table 1*.

(1) Anti-HSV and VZV agents

Aciclovir
Basis of selectivity: Requires triphosphorylation, the first stage of which is performed by a viral enzyme, thymidine kinase. The drug therefore becomes activated and concentrated only in virally infected cells. Aciclovir triphosphate has a much higher affinity for viral DNA polymerase than for cellular DNA polymerase.
Analogues: Valaciclovir is a valine ester of aciclovir which is better absorbed orally. Penciclovir is a structural analogue with the same mechanism of action. It is also administered orally in the form of an ester, famciclovir.
Spectrum of activity: Aciclovir and analogues only work against those viruses able to phosphorylate them, i.e. herpes simplex and varicella-zoster viruses.

(2) Anti-CMV agents

Ganciclovir
Basis of selectivity: Similar to aciclovir. Requires triphosphorylation, the first step being mediated by a viral enzyme, the UL97 protein. However, normal cells can also phosphorylate ganciclovir, resulting in greater toxicity than with aciclovir.
Analogues: Valganciclovir, the valine ester, is much better absorbed orally.
Spectrum of activity: CMV, as well as HSV and VZV.

Cidofovir
Basis of selectivity: Cidofovir triphosphate has much greater affinity for viral DNA polymerases than cellular DNA polymerase.
Spectrum of activity: Much broader than aciclovir, with potential activity against all DNA viruses.

Foscarnet
Basis of selectivity: Much greater affinity for viral DNA polymerases than cellular DNA polymerase.
Spectrum of activity: All herpesviruses.

(3) Antirespiratory virus agents

Ribavirin
Spectrum of activity: In vitro has a broad spectrum, but in clinical practise is used only for treatment of severe RSV infection (inhalation of aerosolized drug), Lassa fever, and chronic hepatitis C virus infection.

Amantadine
Spectrum of activity: Extremely narrow – influenza A viruses only.

Zanamavir and oseltamivir
Spectrum of activity: All influenza viruses.

(4) Antihepatitis virus agents

Interferon
Interferons (IFN) are naturally occurring protein molecules produced by cells in response to antigenic or viral stimuli. There are several IFN (α, β, and γ), produced by different cells.
Spectrum of activity: Theoretically broad, but in practise limited. IFN-α is useful in the treatment of some patients with chronic hepatitis B virus infection, where the mode of action is through immunomodulation. In chronic hepatitis C virus infection, IFN conjugated to polyethylene glycol (PEG-IFN) plus ribavirin combination therapy results in sustained virus clearance in ~ 50% of patients.

Lamivudine
Spectrum of activity: Chronic HBV infection (also HIV, see Topic E5).

Adefovir
Spectrum of activity: Chronic HBV infection.

Related topics   Viruses and virus replication (A3)   Antiretroviral agents (E5)

Antiviral agents

Viruses are obligate intracellular parasites (Topic A3). Viral replication is intimately bound up with normal cellular processes, and therefore it is difficult to develop antiviral drugs which do not also poison the host cell. However, a certain amount of success has been achieved, and a handful of antiviral agents are now in regular use (*Table 1*). Most, but not all, act at the step of macromolecular synthesis in the viral replication cycle (step 4, see Topic A3), particularly viral nucleic acid production. Currently licensed antiviral drugs are described in this topic. Drugs which inhibit the replication of HIV (antiretrovirals) are considered in the next topic. The agents are grouped together on the basis of the main viruses against which they act.

(1) Anti-HSV and VZV agents

Aciclovir
Aciclovir was the first antiviral drug to be developed with selective toxicity, i.e. it inhibits viral replication whilst exhibiting virtually no toxic side-effects on host cells. It is acycloguanosine (*Fig. 1*), an analogue of the purine nucleoside, guanosine, in which the deoxyribose moiety has lost its cyclic configuration. Aciclovir itself is inactive. It must first be phosphorylated to the triphosphate form. Although the second and third phosphate groups are added by cellular enzymes, the first phosphorylation step is accomplished by a **viral** enzyme, thymidine kinase (TK); normal cells are unable to perform this phosphorylation step, therefore the drug only becomes activated in infected cells. Aciclovir diffuses from the extracellular space into cells, until the intracellular and extracellular concentrations are the same. However, in virally infected cells, as the

Table 1. Antiviral drugs

Activity	Agents	Mode of action	Clinical indications	Resistance	Toxicity
Anti-HSV and -VZV	Aciclovir	Nucleoside analogues. Drug-triphosphate is	Primary HSV – mucosal, keratitis	TK⁻ variants	Virtually Nonexistent
	Valaciclovir (prodrug)		Recurrent HSV – prophylaxis	TK mutants	
	Penciclovir	viral DNA polymerase inhibitor and chain terminator	Herpes encephalitis, disseminated disease	DNA pol mutants	
	Famciclovir (prodrug)		Primary and recurrent VZV infection		
Anti-CMV	Ganciclovir	Nucleoside analogue. Ganciclovir triphosphate is viral DNA polymerase inhibitor	Serious CMV infection, aciclovir-resistant HSV	UL97 mutants, DNA pol mutants	Bone marrow
	Foscarnet	Pyrophosphate analogue. Viral DNA polymerase inhibitor	Serious CMV infection, aciclovir-resistant HSV	DNA pol mutants	Nephrotoxic
	Cidofovir	Nucleotide analogue. Triphosphate is viral DNA polymerase inhibitor	CMV retinitis	DNA pol mutants	Nephrotoxic
Antirespiratory viruses (1) RSV	Ribavirin	Nucleoside analogue. Interferes with mRNA capping	RSV infection in at-risk babies. Also effective in Lassa fever and in chronic HCV infection (with IFN)	Not described	Haemolytic anemia
(2) Influenza	Amantadine	Blocks M protein ion channel essential for uncoating of influenza A virus	Influenza (as treatment or as prophylaxis)	Mutations in M protein arise easily	Central nervous system stimulation
	Zanamivir Oseltamivir	Neuraminidase inhibitors	Influenza (as treatment or as prophylaxis)	Uncommon	Uncommon
Antiviral hepatitis	Interferon-alpha	Immunomodulatory action in HBV infection. ?Antiviral action in HCV infection	Chronic HBV infection. Chronic HCV infection (in combination with ribavirin)	Not known	Headache, fever myalgia, depression
	Lamivudine	RRT inhibitor	Chronic HBV infection	Mutations in *RT* gene	Not common
	Adefovir	DNA polymerase inhibitor	Chronic HBV infection	Unusual. Mutations in *RT* gene	Nephrotoxic

Footnotes: HSV, herpes simplex virus; VZV, varicella-zoster virus; CMV, cytomegalo virus; RSV, respiratory syncytial virus; HBV, hepatitis B virus; TK = Thymidine kinase; TK⁻ = No TK gene present; DNA pol = DNA polymerase; *UL97* = Gene whose product phosphorylates ganciclovir; RT = Reverse transcriptase; IFN = Interferon

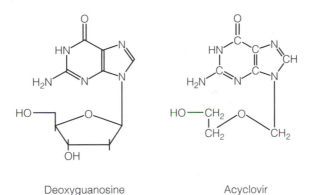

Deoxyguanosine   Acyclovir

Fig. 1.   Deoxyguanosine (left) and acyclovir (right).

aciclovir molecule becomes phosphorylated, thereby shifting the equilibrium aciclovir ↔ aciclovir monophosphate to the right, the concentration of free aciclovir within the cell will drop, causing more drug to diffuse into the cell. Thus, the drug becomes specifically concentrated in infected cells. The triphosphate form of aciclovir inhibits viral DNA replication by two mechanisms: as an analogue of guanosine triphosphate it is incorporated into the growing DNA chain and causes **chain termination**, since aciclovir triphosphate lacks the 3'-hydroxyl group in the deoxyribose ring necessary to form the 5'-3' phosphodiester linkage with the next base. Aciclovir triphosphate also directly inhibits viral DNA polymerase, to a considerably greater extent than it acts on normal cellular DNA polymerase.

Whilst aciclovir is undoubtedly a successful antiviral drug, almost without toxicity to the host, on the downside, it has a very narrow spectrum of activity – only working against those viruses able to perform the first phosphorylation step, which in clinical practise means herpes simplex and varicella-zoster viruses.

Resistance
Virus may become resistant to aciclovir by a number of mechanisms:

(i)   **Loss of the gene encoding TK.** Such strains, known as TK⁻ mutants, will be inherently resistant to aciclovir, as they will be unable to perform the first phosphorylation step.
(ii) **Mutations in the TK gene.** These may reduce the affinity of the *TK* for aciclovir, and thereby prevent the first phosphorylation step.
(iii) **Mutations in the viral DNA polymerase gene.** Such DNA pol mutants may be unable to bind aciclovir triphosphate, and thereby escape the inhibitory properties of the drug.

All these mechanisms occur both in the laboratory and in nature. However, loss of the TK gene, or mutations within it, is associated with a considerable decrease in virulence, and such strains are not a clinical problem. Fully virulent DNA pol mutants are, however, a real clinical problem. They emerge particularly in immunocompromised patients (e.g. with HIV infection), in whom recurrent HSV infections cause frequent and extensive disease necessitating prolonged treatment with aciclovir.

Analogues of aciclovir

Valaciclovir, the L-valyl ester of aciclovir, is an oral prodrug that is better absorbed orally than aciclovir – and once in the bloodstream is hydrolyzed to aciclovir. The mode of action of **penciclovir** (*Fig. 2*) is very similar to aciclovir – it requires phosphorylation, initially by thymidine kinase, and the triphosphate derivative selectively inhibits viral DNA polymerase. It is also marketed as an oral prodrug – **famciclovir**, which is metabolized into penciclovir after oral administration.

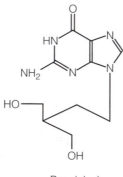

Penciclovir

Fig. 2.   Penciclovir.

(2) Anti-CMV agents

It is particularly disappointing that the otherwise excellent drugs described above have little activity against cytomegalovirus (CMV), which although also a herpesvirus, does not encode a thymidine kinase enzyme. The drugs described below were specifically developed with the aim of finding an effective anti-CMV agent.

Ganciclovir

Ganciclovir (*Fig. 3*), also a nucleoside analogue, is a derivative of aciclovir that shows useful activity against CMV. It must be activated by phosphorylation, but in contrast to aciclovir, it can be phosphorylated by CMV, which uses a different virally encoded phosphotransferase enzyme, the product of the CMV gene

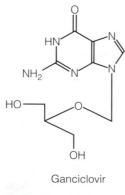

Ganciclovir

Fig. 3.   Ganciclovir.

UL97. Subsequent phosphorylation to the triphosphate generates a compound which acts as a viral DNA polymerase inhibitor.

Unfortunately, cellular enzymes can also phosphorylate ganciclovir, so that active drug is generated in uninfected cells, leading to unwanted toxicity. Bone marrow suppression is the most frequent toxic effect, resulting in neutropenia and thrombocytopenia. Ganciclovir is poorly absorbed (<10%) when given orally, and has to be administered intravenously, although the valyl ester, valganciclovir, is much better absorbed orally (40%). The anti-CMV activity of ganciclovir can be sight- or life-saving in immunosuppressed patients with severe CMV infections (Topic B7).

Resistance to ganciclovir arises through point mutations in the *UL97* gene resulting in an inability of the virus to phosphorylate the drug. Such mutants arise fairly commonly in clinical practise, but fortunately they remain sensitive to foscarnet (see below). Mutations in the CMV DNA polymerase gene may also confer resistance to ganciclovir.

Cidofovir

This **acyclic nucleoside phosphonate** or **nucleotide analogue** (*Fig. 4*) already possesses a phosphate-mimetic group. It therefore needs only two, not three, phosphorylation steps to reach its active form, and these steps are performed by cellular enzymes. Thus, active drug may arise in both infected and uninfected cells, but a selective antiviral activity is maintained due to the higher affinity of the diphosphorylated form for viral DNA polymerases than for the corresponding cellular enzymes. In this form, the drug acts as a chain terminator and competitive inhibitor.

Cidofovir is the first member of this class of drugs licensed for use. It is an acyclic cytosine analogue. Adefovir, an acyclic adenine analogue, is another member of this class. These agents have a broader spectrum of activity than aciclovir and derivatives, potentially against all DNA viruses, as there is no need for the first virally performed phosphorylation step. *In vitro*, cidofovir is active against all herpesviruses, as well as adeno-, polyoma-, papilloma- and poxviruses, although at present it is only used in the treatment of CMV infections. Adefovir has additional activity against reverse transcriptase (which, after all, is a modified form of a DNA polymerase enzyme), and therefore inhibits replication of both HIV (Topic E5) and hepatitis B virus (see below).

Cidofovir is nephrotoxic in a dose-dependent manner, and must be administered with probenecid to prevent irreversible renal damage. Resistance to

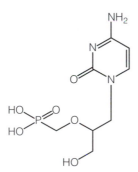

Fig. 4.   Cidofovir (HPMPC).

members of this class of drugs may arise through mutations in viral DNA polymerase enzymes. Clinical experience with cidofovir is not yet extensive enough to know how much of a problem virus resistance will become.

Foscarnet

Unlike the drugs discussed so far, foscarnet is a **pyrophosphate analogue** (*Fig. 5*). It does not require phosphorylation. It forms complexes with DNA polymerases, and prevents cleavage of pyrophosphate from nucleoside triphosphates, resulting in inhibition of further DNA synthesis. It shows some selective toxicity for viral rather than host cell enzymes, and has activity against all the herpesviruses, including CMV (*Table 1*). Like ganciclovir, oral bioavailability is poor, and it has to be administered by intravenous injection. It is nephrotoxic, and can cause acute renal failure.

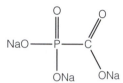

Fig. 5.   Foscarnet.

(3) Antirespiratory virus agents

Ribavirin

Ribavirin (*Fig. 6*) is structurally related to guanosine. Like other **nucleoside analogues** it has to be activated intracellularly by phosphorylation. Its precise mode of action is unknown, but in the triphosphate form it inhibits viral protein synthesis, apparently by interfering with 'cap' formation at the 5′ end of mRNA. As all viruses must generate mRNA, a drug acting in this way should have a broad spectrum of activity, against both DNA and RNA viruses, and this is indeed the case, at least *in vitro*, for ribavirin. In clinical trials, however, in a wide range of virus infections, it has generally been disappointing. Clinical use is now restricted to infections caused by respiratory syncytial virus, Lassa fever virus, and HCV, the latter given in combination with IFN. In severe lower respiratory tract infection in young babies caused by RSV, the compound is administered by inhalation of an aerosolized solution, whereas in the other indications it is given orally. The most commonly encountered serious side-effect of prolonged ribavirin usage is that of macrocytic anemia.

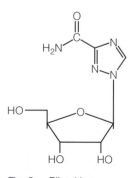

Fig. 6.   Ribavirin.

Amantadine

This extraordinary molecule (*Fig. 7*) has an extremely narrow antiviral spectrum, restricted to influenza A virus – other influenza viruses are unaffected at therapeutically achievable concentrations. It blocks a viral matrix protein ion channel, thereby interfering with uncoating of the viral nucleic acid within the infected cell (i.e. step 3 in the virus replication cycle, see Topic A3). Resistance readily arises by mutation in the matrix protein, which is then unable to bind the drug.

Amantadine binds to *N*-methyl-D-aspartate receptors in the brain, and therefore has dopaminergic effects. These effects can be therapeutically useful (amantadine is also licensed in the UK as an antiparkinsonian drug), but also account for the side-effects of the drug: restlessness, insomnia, agitation, and confusion. Unfortunately, amantadine is poorly tolerated, especially by the elderly, one of the groups who stand to benefit most from an effective anti-influenza drug.

Fig. 7.   Amantadine.

Zanamavir and oseltamivir

These are the first of a new class of anti-influenza compounds with a novel mode of action – selective inhibition of viral neuraminidase (NA, see Topic B11). NA acts to cleave off sialic acid residues from the cell surface and thereby enable budding virus particles, which would otherwise remain attached to the cell from which they have just emerged, to break away and leave an infected cell. The **neuraminidase inhibitors** are purposefully designed drugs. The crystal structure of influenza NA was solved, and small molecules constructed to fit into the enzymatic active site. Zanamavir and oseltamivir inhibit the release and propagation of infectious influenza viruses from the epithelial cells of the respiratory tract, and are efficacious in clinical trials. Unlike amantadine, these drugs act against both influenza A and B viruses, and also inhibit all the known neuraminidase subtypes of influenza A viruses.

(4) Antihepatitis virus agents

Interferons

Interferons (IFN) are glycosylated proteins produced *in vivo* in response to viral or antigenic challenge (Topic A2). There are three types of human IFN: IFN-α, produced by many cell types; IFN-β, produced by fibroblasts; and IFN-γ (immune interferon), produced by T lymphocytes.

Interferons have a wide range of biological effects. They activate several different biochemical pathways within a cell, resulting in resistance to virus infection, e.g. activation of a ribonuclease which digests viral RNA, or phosphorylation and inhibition of initiation factor, a protein which assists in the initiation of transcription of mRNA into protein, thereby effectively preventing production of viral proteins. The importance of each of these pathways differs between different IFN, and indeed between different cells stimulated by the same IFN.

Interferons also have immunomodulatory effects on cells of the immune system. Again, these vary in detail between different IFN, but they include stimulation of natural killer cells, induction or suppression of antibody production, stimulation of T-cell activity, and stimulation of expression of HLA class I and II molecules on the surface of cells. Finally, IFN also affect cell proliferation, which has led to their successful use in the management of certain malignant tumors.

Clinical experience with IFN, given by subcutaneous or intramuscular injection, in a wide range of virus infections has generally been disappointing. Clinical use is now restricted to IFN-α in the treatment of chronic viral hepatitis. In chronic hepatitis B virus infection, the mode of action of IFN-α is through its immunomodulatory rather than its antiviral activity, enhancing the ability of the host cytotoxic T-cells to recognize and kill virus-infected hepatocytes. The mode of action in chronic hepatitis C virus infection is different, but not fully understood. Response rates to IFN-α are considerably enhanced by polymerizing the molecule with polyethylene glycol (PEG)–IFN and also by combination therapy with ribavirin. PEG–IFN increases the half-life of the IFN in the bloodstream, thereby maintaining therapeutic levels of the molecule for much longer than with standard IFN. Overall response rates to combination PEG–IFN and ribavirin therapy in chronic HCV infection are ~ 50%. In both chronic HBV and HCV infection, therapy needs to be given for ≥ 6 months.

Unexpectedly, clinical trials revealed that patients receiving IFN-α experience 'flu-like side-effects: fever, headache, and myalgia. This led to the realization that individuals suffering from influenza complain of 'flu-like symptoms precisely because of the induction of IFN by the virus. Most patients become tolerant to these effects after the first few doses.

Lamivudine

This is a reverse transcriptase inhibitor, developed for the treatment of HIV infection, and is therefore considered in more detail in the next topic. However, the realization that HBV replication involves a reverse transcriptase step led to its use in the treatment of chronic HBV infection. Lamivudine is effective in halting HBV replication and lowering HBV DNA levels. However, resistance arises through point mutations in the viral polymerase enzyme, and once the drug is stopped, viral replication begins again.

Adefovir

This is a nucleotide analogue, as described above. It has activity against HBV DNA polymerase, and is now licensed for use in the treatment of chronic HBV infection. It is effective in reducing HBV DNA levels, and thus far, emergence of resistance appears to be a rare occurrence.

E5 ANTIRETROVIRAL AGENTS

Key Notes

The antiretroviral agents	There are four classes of antiretroviral agents licensed for clinical use – nucleoside analogue and nonnucleoside analogue reverse transcriptase inhibitors (NRTI and NNRTI), protease inhibitors (PI), and fusion inhibitors (FI). These drugs are listed, together with their modes of action and the basis for their selectivity, in *Table 1*.
Adverse effects of antiretrovirals	The most important of these listed in *Table 2*. Long-term effects of PI and possibly NRTI on glucose and lipid metabolism may lead to serious morbidity such as diabetes mellitus and ischemic heart disease.
Resistance to antiretrovirals	Resistance arises to all antiretroviral drugs through the emergence of viruses with mutations in the genes encoding the target proteins RT, protease, and gp41 (*Table 1*). Some mutations confer cross-resistance to a number of agents, other mutations are unique to a single agent. Resistance-conferring mutations within RT differ between the NRTI and NNRTI. Some mutations confer resistance to one drug but simultaneously increase sensitivity to another, e.g. M184V leads to lamivudine resistance but increases AZT sensitivity.
Anti-HIV drugs in development	The next class of drugs likely to emerge will be the integrase inhibitors. Other drugs, e.g. hydroxyurea, or interleukin-2, may also have a role in the management of an HIV-infected patient.
Clinical use of anti-retroviral agents	The aim of therapy is complete inhibition of viral replication. This is associated with restoration of circulating CD4-positive T cells and immune reconstitution. Difficult decisions arise as to when to start therapy, and with which combination of drugs. Therapy is monitored by serial testing of the viral load in peripheral blood, the aim being to suppress this to undetectable levels. Antiretrovirals may also be used for post-exposure prophylaxis.
Related topics	Viruses and virus replication (A3) Human immunodeficiency viruses (B1)

The antivetroviral agents

The first human immunodeficiency virus (HIV) was identified and characterized as a retrovirus in 1983. Long-term infection with HIV leads to inexorable destruction of the host immune system, and the development of the acquired immunodeficiency syndrome (AIDS) (Topic B1). A detailed understanding of the replication cycle of the virus has led to the identification of a number of potential targets for antiretroviral drugs. The outlook now for an HIV-infected patient is markedly different from 1983. Unfortunately, however, the ultimate aim of 'curing' a patient of HIV infection remains an unattainable goal.

There are four classes of antiretroviral agents currently in clinical usage (*Table 1*), which act against three targets in the viral life cycle: reverse transcriptase, HIV protease, and the viral entry fusion process.

Table 1.   Antiretroviral agents, main properties

Drug class	Mode of action	Selectivity of action	Resistance
NRTI e.g. AZT, ddl, ddC, d4T, 3TC, abacavir	Drug triphosphate acts as chain terminator and RT inhibitor	A greater affinity for RT than cellular DNA polymerases	Accumulation of specific point mutations in RT gene
NNRTI e.g. efavirenz, nevirapine	Bind directly to RT of HIV-1 (not HIV-2)	A greater affinity for RT than cellular DNA polymerases	Accumulation of specific point mutations in RT gene (distinct from NRTI mutations)
PI e.g. ritonavir, indinavir	Bind to HIV protease	Higher affinity for HIV protease than cellular proteases	Accumulation of specific point mutations in gene encoding protease
FI e.g. enfuvirtide	Prevent fusion of gp41 to cell membrane		Mutations within gp41

NRTI = nucleoside analogue reverse transcriptase inhibitors;
NNRTI = nonnucleoside analogue reverse transcriptase inhibitors;
PI = protease inhibitors;
FI = fusion inhibitors;
AZT, zidovudine; ddl, didanosine; ddC, zalcitabine; d4T, stavudine; 3TC, lamivudine.

(1) Nucleoside analogue reverse transcriptase inhibitors (NRTI)

Reverse transcriptase (RT) was the first and most obvious target identified for potential antiretroviral drugs, as reverse transcription (synthesis of DNA from an RNA template) is an essential step in HIV replication, mediated by a virally encoded RT enzyme, but does not occur in host cells. Two classes of RT inhibitors have been developed – nucleoside analogues, which bind at the substrate-binding site, and nonnucleoside analogues, which bind elsewhere.

Azidothymidine (AZT) is a thymidine analogue in which the 3′ hydroxyl group has been replaced by an azido (N_3) group (*Fig. 1*). This makes it a 2′–3′ dideoxy nucleoside analogue (lacking hydroxyl groups at both the 2′ and 3′ positions of the deoxyribose ring). AZT is activated by phosphorylation to the triphosphate form, all three steps being performed by cellular enzymes. Tight binding of AZT triphosphate to RT results in inhibition of RT function, and as with aciclovir triphosphate (Topic E4), incorporation of AZT triphosphate into a growing DNA chain results in chain termination. To date, five further drugs in this class, with similar structures, have been licensed for use (*Table 1*). They all require triphosphorylation, and in their triphosphate form they act as chain terminators and competitive inhibitors of HIV-derived RT.

(2) Nonnucleoside analogue reverse transcriptase inhibitors

More than 20 chemical classes of compounds have been shown to bind to RT at sites other than the substrate-binding site (where the NRTI bind), and collectively these drugs are known as nonnucleoside analogue RT inhibitors (NNRTI). Thus far very few of these have made it to the clinic (*Table 1*). They inhibit the RT of HIV-1 but not that of HIV-2. *Figure 2* shows the structure of one of these

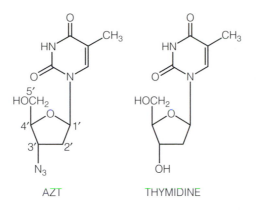

Fig. 1.   Azidothymidine-nucleoside analogue RTI.

agents, nevirapine. Many other drugs in this category are at various stages of clinical and preclinical testing.

(3) Protease inhibitors

When HIV replicates, it produces polycistronic mRNA, which is translated into a series of polyproteins. These are subsequently cleaved into individual protein molecules by a virally encoded aspartyl protease enzyme. The HIV protease was crystallized, its three-dimensional structure elucidated, and small molecules purposefully designed to bind in its enzymatic pocket, as a result of which there are now at least six HIV protease inhibitors currently licensed for use and more are in advanced stages of clinical trials (*Table 1*). Most of these compounds are structurally related.

(4) Fusion inhibitors

This is the latest class of antiretroviral drugs to reach the clinic, with enfuvirtide achieving a license for clinical use in 2003. Enfuvirtide acts after binding of viral gp120 to the CD4 molecule on host target cells. It is a synthetic peptide which interferes with the conformational changes required for viral–host cell membrane fusion and subsequent injection of the virus into the host cell.

Adverse effects of antiretrovirals

All of the above-mentioned antiretroviral agents have important side-effects (*Table 2*). Patients generally become tolerant to the 'flu-like symptoms associated with first use of the NRTI, but development of the more serious side-effects in *Table 2* may preclude the use of some agents in individual patients.

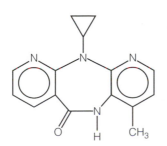

Fig. 2.   Nevirapine – nonnucleoside analogue RTI.

Overall, the most worrying adverse effects of antiretroviral agents are those on glucose and lipid metabolism. Whilst these are particularly associated with the PI, it is possible that the NRTI also act here. The frequency and pathophysiology of these reactions are unclear, but they may predispose to a long-term risk of complications such as diabetes mellitus and ischemic heart disease.

Table 2. Antiretrovirals: side-effects

NRTI	Headache, gastrointestinal upset (all)
	Bone marrow suppression (AZT)
	Peripheral neuropathy (ddl, ddC, d4T)
	Pancreatitis (ddl, ddC)
	Elevation of liver enzymes (d4T)
	Hypersensitivity reactions (abacavir)
NNRTI	Rash
	Hepatitis (nevirapine)
PI	Kidney stones (particularly indinavir)
	Gastrointestinal upset (particularly indinavir)
	Interference with glucose and lipid metabolism
	Lipodystrophy, hyperlipidemia
	Diabetes mellitus
	Ischemic heart disease

For abbreviations, see *Table 1.*

Resistance to anti-HIV drugs

Isolates of HIV derived from patients who have been taking AZT for ≥ 6 months are invariably less sensitive to AZT *in vitro* than isolates taken from the same patient at the initiation of therapy. This arises from mutations in the gene coding for RT, leading to reduced binding of AZT triphosphate. Resistance to AZT arises as a series of sequential point mutations within the RT gene, each additional mutation resulting in an increase in the dose of AZT necessary to inhibit the virus.

Resistance to other NRTI similarly arises through mutations in the RT gene. In general, cross-resistance between AZT and the other NRTI is not a problem, as the positions of the mutations conferring resistance to AZT differ from those giving rise to, for instance, ddI or ddC resistance. In contrast, resistance mutations to ddI and ddC are similar, and also overlap with those causing 3TC resistance. On the plus side, whilst the principle mutation conferring resistance to 3TC is at position 184 in the *RT* gene, paradoxically this mutation causes an increase in sensitivity to AZT in previously AZT-resistant virus. Yet further mutations in the *RT* gene confer resistance to the NNRTI. Some, but not all, lead to cross-resistance among the different NNRTI. Knowledge of the different resistance mutations for each agent allows the rational development of combinations of NRTI, or of NRTI plus NNRTI for therapeutic use, e.g. there is considerable benefit from combining AZT with lamivudine, and also with an NNRTI.

As with the RT inhibitors, resistance to the PI arises through point mutations in the gene coding for the target protein. A range of mutations in the protease gene has been described, but cross-resistance between drugs of this class is common. Continued use of a given protease inhibitor will lead to the accumulation of mutations, with a concomitant increase to high-level resistance.

Anti-HIV drugs in development

Despite the clinical successes achieved by use of the above agents, there is a desperate need for new drugs with different sites of action, and different toxicity profiles in the fight against HIV infection. Many such agents are under development and investigation. These include new NRTI and NNRTI; new nucleotide analogue RT inhibitors (Topic E4), e.g. tenofovir, and new PI which hopefully will have activity against known protease-resistant strains. The proof-of-concept provided by enfuvirtide that the very early stages of viral attachment and entry into target cells can be inhibited by small molecules will hopefully result in the emergence of other fusion inhibitors.

One new target in the viral life cycle of great interest is the virally encoded integrase enzyme. This is responsible for integration of the DNA provirus into host cell chromosomes. Inhibition of this process would have obvious benefits in preventing virus from becoming latent in resting T cells. Integrase inhibitors are likely to be the next class of antiretroviral drugs.

Optimal therapy for patients with HIV infection may involve treatment with other drugs which are not necessarily antiviral. Thus, there are encouraging reports of the use of hydroxyurea, which blocks cellular activation necessary for viral replication in resting CD4-positive T cells. Similarly, immunomodulatory agents such as interleukin-2 also have their advocates.

Clinical use of antiretroviral agents

With the availability of potent antiretroviral drugs, the main issues in the management of HIV infection have become:

(i)   When should antiretroviral therapy be started? This decision depends on a number of factors, such as the clinical status of the patient, the viral load and CD4 counts, and the willingness of the patient to commit to a potentially complicated regimen of drug-taking which may have a range of side-effects.

(ii) Which drugs should be used? The important principle here is that patients should always be treated with a combination of antiretroviral agents, never just a single agent. This is to minimize the risk of emergence of drug-resistant viruses. With the use of a combination of drugs, an individual virus would have to acquire a whole series of resistance mutations in different genes simultaneously. Drugs from different classes of antiretrovirals should be combined, e.g. an NRTI, plus an NNRTI, plus a PI. Such combinations are referred to as highly active antiretroviral therapy (HAART). Compliance with therapy is vital, thus it may be necessary to try different drugs and combinations to find a regimen which the patient is best able to tolerate.

(iii) How should therapy be monitored? HAART aims to suppress completely all virus replication, and therefore the success of therapy is judged by its effect on the viral load. Ideally, viral load in peripheral blood should become undetectable after initiation of therapy, and remain so. In the absence of replication, the virus cannot spontaneously mutate and thereby acquire resistance. Also, if the virus is not replicating, it cannot kill infected cells. As viral load is suppressed, the CD4 count rises, with consequent restoration of immune function. Re-emergence of virus in peripheral blood means either that the patient is not taking the tablets, or that drug-resistant virus has emerged.

Note that antiretroviral drugs may also be used as post-exposure prophylaxis, e.g. for a healthcare worker who has suffered a needlestick injury from a known HIV-positive source patient. A combination of antiretrovirals is administered as soon as possible after the event, and continued for a period of 4 weeks.

E6 ANTIBIOTICS (1): CLASSIFICATION, MODE OF ACTION AND PHARMACOLOGY

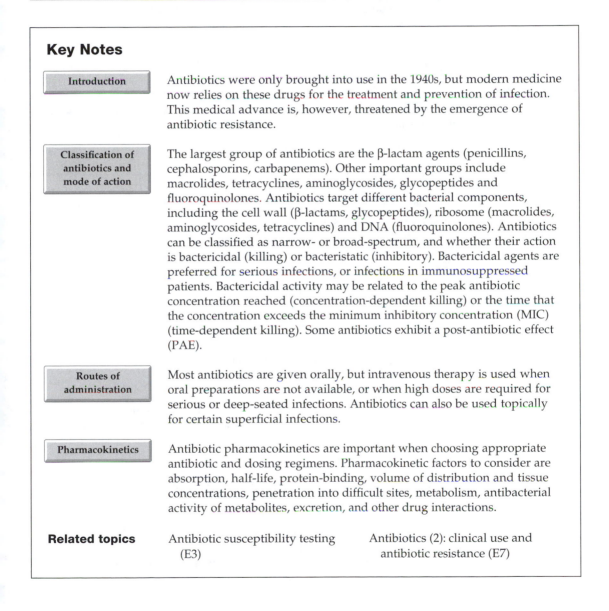

Key Notes

Introduction	Antibiotics were only brought into use in the 1940s, but modern medicine now relies on these drugs for the treatment and prevention of infection. This medical advance is, however, threatened by the emergence of antibiotic resistance.
Classification of antibiotics and mode of action	The largest group of antibiotics are the β-lactam agents (penicillins, cephalosporins, carbapenems). Other important groups include macrolides, tetracyclines, aminoglycosides, glycopeptides and fluoroquinolones. Antibiotics target different bacterial components, including the cell wall (β-lactams, glycopeptides), ribosome (macrolides, aminoglycosides, tetracyclines) and DNA (fluoroquinolones). Antibiotics can be classified as narrow- or broad-spectrum, and whether their action is bactericidal (killing) or bacteristatic (inhibitory). Bactericidal agents are preferred for serious infections, or infections in immunosuppressed patients. Bactericidal activity may be related to the peak antibiotic concentration reached (concentration-dependent killing) or the time that the concentration exceeds the minimum inhibitory concentration (MIC) (time-dependent killing). Some antibiotics exhibit a post-antibiotic effect (PAE).
Routes of administration	Most antibiotics are given orally, but intravenous therapy is used when oral preparations are not available, or when high doses are required for serious or deep-seated infections. Antibiotics can also be used topically for certain superficial infections.
Pharmacokinetics	Antibiotic pharmacokinetics are important when choosing appropriate antibiotic and dosing regimens. Pharmacokinetic factors to consider are absorption, half-life, protein-binding, volume of distribution and tissue concentrations, penetration into difficult sites, metabolism, antibacterial activity of metabolites, excretion, and other drug interactions.
Related topics	Antibiotic susceptibility testing (E3) Antibiotics (2): clinical use and antibiotic resistance (E7)

Introduction The introduction of penicillin and sulfonamides in the 1940s was a major advance in modern medicine. Severe infections (e.g. streptococcal, staphylococcal) could now be treated, often with dramatic cures. Since then a large number of different antibiotics have been discovered, purified, synthesized or chemically modified. Much of **modern medicine** has now come to **rely on antibiotics** either for **prophylaxis** or for **treatment** of bacterial infections.

However, bacteria have by no means been defeated by antibiotics. Resistance to the early antibiotics (e.g. penicillin) is now widespread, and the threat of multiply antibiotic-resistant bacteria is real. **Antibiotic resistance** is now known to be an inevitable consequence of widespread or frequent antibiotic use. This is largely due to the tremendous ability of bacteria to undergo **genetic mutation**, and evolve into antibiotic-**resistant mutants**. Such mutants are then at a **selective advantage** in 'antibiotic-rich' environments (Topic E7).

Classification of antibiotics and mode of action

The classification of antibiotics is shown in *Table 1*. The largest group of antibiotics are the **β-lactam agents**, comprising penicillins, cephalosporins, carbapenems and the monobactam, aztreonam. Other important antibiotic groups include **macrolides, tetracyclines, aminoglycosides, glycopeptides** and **fluoroquinolones**.

Table 1.   Classification of antibiotics

Antibiotic group/class	Examples (and route of administration)	Main clinical use/bacterial activity	Comments
Penicillins* (inhibit synthesis of the bacterial cell wall)	Pencillin (PO, IV, IM)	Streptococcal infections, meningococcal infections	Inactivated by β-lactamase enzymes
	Amoxicillin, ampicillin (PO, IV)	Upper and lower respiratory tract infections. Broader spectrum of activity than penicillin	
	Co-amoxiclav (PO, IV)	Surgical prophylaxis, bites, intra-abdominal sepsis or infections due to β-lactamase-producing bacteria (chest, urinary infections)	A combination of amoxicillin and clavulanic acid – an inhibitor of β-lactamases
	Flucloxacillin (methicillin) (PO, IV)	Staphylococcal infections (but not methicillin-resistant strains, e.g. MRSA)	Stable to β-lactamases.
	Piptazobactam (IV only)	Severe hospital infections (intra-abdominal sepsis, pneumonia, febrile neutropenia). Broad-spectrum activity, including pseudomonads	A combination of piperacillin and tazobactam – another β-lactamase inhibitor
Cephalosporins* (inhibit synthesis of the bacterial cell wall)	Cephradine, cefaclor (mainly PO)	Urinary tract infections, respiratory tract infections	1st generation cephalosporins
	Cefuroxime (mainly IV)	Surgical prophylaxis, pneumonia, intra-abdominal sepsis, pyelonephritis.	2nd generation cephalosporins
	Ceftazidime (IV only)	See 'Piptazobactam'. Active against pseudomonads, but poor activity against staphylococci and anaerobes	3rd generation cephalosporins
	Ceftriaxone, cefotaxime (IV, IM only)	Bacterial meningitis, resistant Gram-negative infections (but not active against pseudomonads)	
Monobactam*	Aztreonam (IV only)	*Pseudomonas* infections	

Table 1.   Classification of antibiotics (continued)

Antibiotic group/class	Examples (and route of administration)	Main clinical use/bacterial activity	Comments
Carbapenems* (inhibit synthesis of the bacterial cell wall)	Imipenem, meropenem, ertapenem (IV only)	See 'Piptazobactam'. Second- or third-line agents for severe hospital infections. Extremely broad-spectrum of activity.	
Glycopeptides (inhibit formation of the bacterial cell wall)	Vancomycin (IV), teicoplanin (IV, IM)	Resistant Gram-positive infections (staphylococci, enterococci). Some strains of enterococci are now resistant (VRE or GRE)	Monitoring of serum levels during therapy may be required
Aminoglycosides (inhibit protein synthesis)	Gentamicin, tobramycin, netilmicin, amikacin (IV, IM)	*Pseudomonas* and other severe Gram-negative infections. Used in combination with cell-wall active agents for staphylococcal sepsis and streptococcal endocarditis	Monitoring of serum levels during therapy is required (nephrotoxic and ototoxic)
	Streptomycin (IM)	Tuberculosis	
Macrolides (inhibit protein synthesis)	Erythromycin, clarithromycin, azithromycin (PO, IV)	Respiratory tract infections, skin-soft tissue infections. Active against atypical pathogens (*Mycoplasma, Legionella, Chlamydia*), *Helicobacter pylori* and *Campylobacter*	Alternative to penicillins in penicillin–allergic patients
Lincosamines (inhibit protein synthesis)	Clindamycin (PO, IV)	Staphylococcal bone and joint infections. Severe cellulitis. Anaerobic infections, including intra-abdominal sepsis	Risk of antibiotic-associated colitis (*C.difficile* infection)
Tetracyclines (inhibit protein synthesis)	Oxytetracycline, doxycycline (PO, TOP)	Chlamydial and rickettsial infections, respiratory tract infections, acne, brucellosis and Lyme disease. Bacterial and chlamydial conjunctivitis	Contraindicated in children and during pregnancy
Oxazolidinones (inhibit protein synthesis)	Linezolid (PO, IV)	Resistant Gram-positive infections (e.g. MRSA, VRE)	
Other protein synthesis inhibitors	Chloramphenicol (PO, IV, TOP)	Bacterial conjunctivitis. Systemic use is now reserved for specific life-threatening infections	Risk of serious aplastic anemia
	Fusidic acid (PO, IV, TOP)	Staphylococcal bone and joint infections. Bacterial conjunctivitis	
	Quinupristin / dalfopristin (IV only)	Resistant Gram-positive infections (e.g. MRSA). Must be given via a central venous catheter due to phlebitis risk	A synergistic mixture of two streptogramin agents
Fluoroquinolones (inhibit DNA gyrase enzyme involved in 'supercoiling' bacterial DNA)	Ciprofloxacin (PO, IV)	Bacterial gastroenteritis (*Salmonella, Campylobacter*), gonorrhea, complicated urinary tract infections, Gram-negative septicemia, typhoid. Only oral antibiotic with *Pseudomonas* activity. Less active against Gram-positive bacteria. Poorly active against anaerobes	Can cause tendonitis and tendon rupture, particularly the Achilles tendon. Risk of tendon rupture is highest in elderly patients and those on steroids. Quinolones are generally not active

Table 1.   Classification of antibiotics (continued)

Antibiotic group/class	Examples (and route of administration)	Main clinical use/bacterial activity	Comments
	Ofloxacin, levofloxacin (PO, IV)	Similar to ciprofloxacin. Levofloxacin is more active against Gram-positive bacteria (e.g. *S.pneumoniae*) but less active against *Pseudomonas*	(and may be a risk factor for) MRSA strains
	Moxifloxacin (PO)	Similar to ciprofloxacin but more active against Gram-positive bacteria	
Rifamycins (inhibit RNA transcription)	Rifampicin (PO, IV)	Tuberculosis. Combination treatment for severe staphylococcal infections	Monitoring of liver function tests required. Numerous drug interactions
Dihydrofolate reductase inhibitors (inhibit folic acid and thus DNA synthesis)	Trimethoprim (PO)	Urinary tract infections.	Contraindicated in 1st trimester of pregnancy (folate antagonist)
	Co-trimoxazole (PO, IV)	Treatment and prophylaxis of *Pneumocystis* pneumonia	A combination of trimethoprim and sulfamethoxazole (sulfonamide)
Nitroimidazole	Metronidazole (PO, PR, IV)	Prophylaxis and treatment of anaerobic infections. Treatment of protozoal infections (giardiasis, amoebiasis, trichomoniasis)	
Polymixins (physicochemical disruption of Gram-negative cell wall)	Colistin (NEB, IV)	Treatment of pseudomonas lower respiratory tract infections in cystic fibrosis (usually via the nebulized route)	Side effects limit systemic use to severe multi-resistant Gram-negative infections

PO, *per orum*; IV, intravenous; IM, intramuscular; TOP, topical; NEB, nebulized; MRSA, methicillin-resistant *Staph. aureus*; VRE, vancomycin-resistant enterococci; GRE, glycopeptide-resistant enterococci.
*Penicillins, cephalosporins, monobactams and carbapenems are all β-lactam antibiotics

Antibiotics are different from most other drugs that are used in humans, as the principal target is the bacterial cell (for efficacy) and not human cells (to minimize toxicity). The bacterial cell has a variety of targets that are not present or are sufficiently different in human cells, including the bacterial cell wall, ribosomes and DNA transcription/translation mechanisms, allowing for the use of drugs with **selective toxicity** for bacterial cells. The primary mechanisms of action of the different antibiotic groups are listed in *Table 1*.

Antibiotics may be classified as either **narrow spectrum** or **broad spectrum**, based on the range of different bacteria against which they are active. Broad-spectrum antibiotics are used in situations where a wide range of bacteria can cause an infection, for polymicrobial infections, and for infections with bacteria resistant to narrow-spectrum antibiotics. 'Broad spectrum' does not, however, mean more powerful. For instance, penicillin (a narrow-spectrum antibiotic) is extremely active against certain important human pathogens.

Some groups of antibiotics are **bactericidal** (i.e. they are capable of killing bacteria). These include β-lactams, aminoglycosides, fluoroquinolones and rifamycins. Other groups of antibiotics (e.g tetracyclines, trimethoprim) are mainly **bacteristatic** (i.e. bacteria are inhibited from multiplying, but are not directly killed). These agents rely on the body's immune system to target and kill the (inhibited) bacteria.

Bactericidal antibiotics are generally preferred for severe or difficult-to-treat bacterial infections (e.g. endocarditis or meningitis). Bacteristatic antibiotics should be avoided for immunosuppressed patients (e.g. those who are profoundly neutropenic).

Some antibiotics (such as aminoglycosides and quinolones) exert a bactericidal action that is related to the peak therapeutic concentration reached (**concentration-dependent killing**). For others (such as β-lactams) bactericidal action is related to the length of time that the therapeutic concentration exceeds the required **minimum inhibitory concentration (MIC)** (**time-dependent killing**), and is not necessarily related to the peak concentration reached.

Some antibiotics also demonstrate a **post-antibiotic effect (PAE)**, where bacterial re-growth is inhibited for several hours after the antibiotic concentration has fallen below the MIC.

These factors have important implications for dosing regimens. As a result, some antibiotics are given by frequent dosing (or even by continuous infusion), especially when their half-life is short, whereas others can be given by more intermittent doses.

Antibiotics are often used in **combination**. This may be to provide a **broader spectrum of activity** (e.g. a cephalosporin plus a macrolide for severe pneumonia to cover both typical and atypical pathogens), or to produce a greater bactericidal effect (**synergistic effect**). The most clinically important synergistic combination of antibiotics is penicillin plus gentamicin, used for the treatment of endocarditis (Topic F13), where the bactericidal effect of penicillin is enhanced in the presence of an aminoglycoside. In contrast, some antibiotics may reduce (**antagonize**) the activity of other antibiotics, and such combinations should be avoided.

Routes of administration

There are several different routes of administration of antibiotics. The main ones are:

- PO (*per orum*) – oral (tablets, capsules, liquid suspension, syrup);
- IV (intravenous) – bolus injection, slow or continuous infusion;
- IM (intramuscular) – by injection;
- PR (*per rectum*) – suppositories;
- topical – creams, ointments, drops, pessaries;
- nebulized – aerosol delivery of antibiotic into the lungs.

Oral antibiotic therapy is usually cheaper and easier to administer, but **IV antibiotics** are often used in hospitals for more severe infections when **higher doses** (than can be given orally) are indicated. Some of the more broad-spectrum antibiotics for systemic use are not available orally, as they are either not absorbed from the gastrointestinal tract or are inactivated in the stomach. These include aminoglycosides, glycopeptides and some β-lactams (piptazobactam, third generation cephalosporins, carbapenems – see *Table 1*). Vancomycin is not systemically absorbed after oral administration (PO vancomycin is used solely as a treatment for *Clostridium difficile*-associated diarrhea). IM injection is an alternative but less desirable parenteral route when IV therapy is not possible (i.e. no IV formulation is available, or no IV access is possible in the patient).

Topical administration of antibiotics is a convenient and effective method of treating a variety of superficial infections. Examples include chloramphenicol **eye drops** for bacterial conjunctivitis, or fusidic acid cream for minor staphylococcal skin infections. The advantage of topical therapy is that very high concentrations of antibiotic can be achieved directly at the site of infection, with minimal risk of untoward systemic side-effects. However, only a minority of infections can be treated this way, and, as with all antibiotic therapy, there is a risk that antibiotic resistance can develop.

Pharmacokinetics   The human pharmacokinetics of different antibiotics, even those within the same group, can vary quite considerably. A detailed description is beyond the scope of this book, but important antibiotic pharmacokinetic factors to consider in choosing appropriate antibiotic and dosing regimens are:

- absorption;
- half-life;
- protein-binding within the bloodstream and extracellular fluid;
- volume of distribution and tissue concentrations;
- penetration into difficult sites (cerebrospinal fluid, bone);
- metabolism (and whether metabolites themselves have antibacterial activity);
- excretion (renal, biliary);
- other drug interactions.

The **absorption** of oral antibiotics is often adversely affected by food or other medications. For instance, **tetracyclines** can be **chelated** by **calcium**-containing preparations or milk, preventing absorption.

The **half-life ($t_{1/2}$)** of an antibiotic is the time taken for the concentration of the drug to halve (usually as measured within the bloodstream), and is highly variable. For instance, benzylpenicillin has a $t_{1/2}$ of ~30 min, and may have to be given every 4 h for severe infections, whereas azithromycin has a $t_{1/2}$ of ~24 h allowing once-daily administration.

Some antibiotics are **highly lipid-soluble** (e.g. **tetracyclines**) or **penetrate inside human cells** (e.g. **macrolides, quinolones**) resulting in high tissue concentrations and a **large volume of distribution (V_D)**. These antibiotics have good efficacy against intracellular pathogens such as *Legionella* or *Chlamydia*. Other antibiotics are more water-soluble and remain confined to the blood and extracellular fluid (e.g. aminoglycosides).

Many antibiotics do not cross the **blood–brain barrier** or penetrate well into cerebrospinal fluid. Very high doses of antibiotics are required to treat **bacterial meningitis** adequately (Topic F3).

Many antibiotics are metabolized by the liver, with the metabolites excreted via the kidneys. Some of these metabolites also have antibacterial activity. The kidneys also excrete most antibiotics directly, and **high concentrations** of antibiotics are usually **found in the urine**, even when systemic concentrations are low. This allows some antibiotics (e.g. cephalexin) to be used to treat urinary tract infections, but not systemic infections caused by the same organism. An important exception is the macrolide group of antibiotics, which do not reach therapeutic concentrations in the urine.

Rifampicin, like many antibiotics, is metabolized by the liver. It is also a potent **inducer of the cytochrome *P*450 metabolic pathway**, and thus alters the metabolism of a large number of other drugs, causing some important **drug interactions** (e.g. rendering the oral contraceptive ineffective).

E7 ANTIBIOTICS (2): CLINICAL USE AND ANTIBIOTIC RESISTANCE

Key Notes

Rational antibiotic use	Overuse of antibiotics is an important driver of antibiotic resistance. Unnecessary antibiotic use is also costly and carries the risk of side-effects. Rational antibiotic use can be promoted through education and audit. Good antibiotic prescribing principles include using narrow-spectrum agents and short durations of treatment when possible, avoiding antibiotics for likely viral infections, and rationalizing antibiotic prescriptions in light of microbiology results.
Antibiotic guidelines	Guidelines are important to facilitate the most appropriate use of antibiotics. Patterns of infections, resistance rates, and availability of antibiotics can vary, and local guidelines should be developed and regularly updated. Guidelines should describe the first-line (empirical) antibiotic regimens for common infections, and offer alternative or second-line regimens for when the preferred regimen cannot be used, or has not worked.
Antibiotic prophylaxis	Prophylaxis is the use of antibiotics to prevent infections from occurring, as opposed to treating established infection. Antibiotic prophylaxis accounts for a significant quantity of the overall antibiotic use within hospitals, but inappropriate prophylactic antibiotic use is common.
Antibiotic resistance	Resistance can be either intrinsic or acquired. Acquired resistance can arise by spontaneous mutation, or by transfer of resistance genes from other bacteria. These can be located on the chromosome, a plasmid or a transposon. Exchange of resistance genes located on a plasmid or transposon occurs more readily. The main mechanisms of antibiotic resistance are inactivation, altered target site, overproduction of the target site, impaired uptake, efflux mechanisms, and bypassing the antibiotic sensitive stage in metabolism.
Antibiotic side-effects	Antibiotics often cause side-effects including skin reactions, gastro-intestinal symptoms, antibiotic-associated diarrhea due to *Clostridium difficile*, and candidiasis. Some antibiotics can cause significant untoward effects including nephrotoxicity (aminoglycosides), tendon damage (quinolones) and liver disturbances (rifampicin).
Related topics	Antibiotic susceptibility testing (E3) Antibiotics (1): classification, mode of action and pharmacology (E6)

Rational antibiotic use

Antibiotics are extremely valuable drugs, but their continued use is threatened by the relentless development and spread of antibiotic resistance. **Overuse of** antibiotics is an important driver of **antibiotic resistance**. Unnecessary antibiotic use is also associated with significant **healthcare costs**, as well as the risk of **side-effects**.

Antibiotic resistance can be particularly problematic in hospitals, where the emergence of antibiotic resistant bacteria is driven by a combination of high volumes of antibiotic use and cross-infection within the hospital environment.

In order to reduce the **selective pressure** for antibiotic resistance, and to preserve the use of these drugs for as long as possible, a sensible and rational approach to antibiotic prescribing should be promoted.

There are a number of general principles that can be applied to reduce the overall use of antibiotics:

- Antibiotics should not be used to treat syndromes that are most likely to be caused by viruses (e.g. sore throat, common cold).
- If a diagnosis is not certain, and the patient not unduly ill, then antibiotics should be withheld pending either investigations or a change in the patient's clinical course.
- Narrow-spectrum antibiotics should be used in preference to broad-spectrum antibiotics when possible.
- The duration of therapy should be as short as possible, except for deep-seated or difficult-to-treat infections (e.g. 3 days for uncomplicated cystitis).
- Within hospitals, all antibiotic prescriptions should be reviewed at 48 h, taking note of the patient's clinical condition and any microbiology results.
- Broad-spectrum antibiotics should be rationalized where possible to narrow-spectrum (targeted) antibiotics on the basis of microbiology investigations.
- Unnecessary antibiotic prophylaxis should be avoided.
- Local antibiotic guidelines should be updated regularly, widely available, and adhered to for the majority of prescriptions.
- When prescribing an antibiotic, full note should be taken of previous microbiological investigations. Expert advice on antibiotic use should be available locally, and consulted in cases of difficulty.

Microbiology investigations, including *in vitro* sensitivity tests, can be a useful guide to appropriate antibiotic therapy. However, there is not always a 100% correlation between laboratory sensitivity tests and clinical response. An organism may be reported as 'resistant' yet a patient may appear to have responded and *vice versa*. This is because it is impossible in the laboratory to control for the many variables that may be present *in vivo*, including the immune response and other defense mechanisms, variations in dosing regimens and pharmacokinetic factors, and the site of infection.

The rational use of antibiotics should be promoted and supported by good education, and by audits of antibiotic use and compliance with antibiotic guidelines.

Antibiotic guidelines

The principal current clinical uses of antibiotics are detailed in the previous chapter (Topic E6, *Table 1*). Common antibiotic regimens are also described in the relevant organ-based chapters of this book (Section F).

It should be recognized, however, that antibiotic guidelines in textbooks can quickly become out of date, as bacterial resistance rates increase, and new antibiotics are introduced into clinical practise. Furthermore, the patterns of antibiotic resistance can vary considerably between different geographical areas, and can vary over time. Even within hospitals, resistance rates can vary signifi-

Table 1.   Some important types of acquired bacterial antibiotic resistance

Mechanism of resistance	Examples	Approximate frequency of resistance*	Comments
Production of β-lactamase enzymes	Penicillin resistance in *Staph. aureus*	>95%	Confers resistance to penicillin and amoxicillin
	Amoxicillin resistance in *Haemophilus influenzae*	10–15%	
	High level penicillin resistance in *Neisseria gonorrhoeae*	1–5%	
	Amoxicillin resistance in *Escherichia coli*	50%	Due to β-lactamases
	Ceftazidime (3rd generation cephalosporin) resistance in *E. coli*	1–5%	Extended spectrum β-lactamases (ESBL) confer resistance to most penicillins and cephalosporins
Altered penicillin-binding protein(s) (PBP)	MRSA (methicillin-resistant *Staph. aureus*	30–40% in hospitals, <5% in the community	Alteration in PBP2 to PBP2′ confers resistance to all β-lactam antibiotics
	Penicillin-resistance in *Strep. pneumoniae* (PRP)	2–10%	Usually confers only low level resistance
	Low level penicillin resistance in *N. gonorrhoea*	30–40%	
Altered target site in peptidoglycan cell wall	GRE (glycopeptide-resistant enterococci)	Generally <5%, but higher in some specialist units (renal, hematology)	Confers resistance to vancomycin ± teicoplanin
Production of aminoglycoside modifying enzymes	Gentamicin resistance in enterobacteria (*E. coli, Klebsiella* etc.)	1–5%	Usually confers resistance to other aminoglycosides as well
Altered target site enzymes (DNA gyrase)	Quinolone resistance in *E. coli*	1–5%	
Multiple resistance mechanisms including: altered cell wall permeability, β-lactamases, efflux pumps	Multiply-resistant Gram-negative bacteria (e.g. *Pseudomonas, Acinetobacter*)	1–5%, but may be more common in some specialist units (ITU, burns)	

*Based on UK data (2005).

cantly between different patient groups and wards. There is also significant national and international variation.

For some infections there are a variety of suitable antibiotic therapies, but the range of available antibiotics varies between different countries. The choice of agents may be guided by local factors such as cost and patient populations, as well as pharmacokinetic factors, side-effect profile, and individual preference amongst clinicians.

For these reasons it is important that healthcare institutions and communities develop and adhere to **locally developed antibiotic guidelines**. These guidelines should be developed in consultation with clinicians, microbiologists, infectious disease experts and pharmacists and take account of local circumstances, including local antibiotic resistance rates. Local guidelines should describe the first-line (empirical) antibiotic regimens for the most common infections, and offer alternative or second-line regimens for when the preferred regimen cannot be used, or has not worked. They should be regularly reviewed and updated.

National and international guidelines are also available for the treatment of specific conditions. General guidance on antibiotic use in the UK can be found in the British National Formulary (Section 5) which is updated bi-annually.

Antibiotic prophylaxis

Antibiotic prophylaxis is the use of antibiotics to prevent infections from occurring, as opposed to treating established infection. Antibiotic prophylaxis (**chemoprophylaxis**) accounts for a significant quantity of the overall antibiotic use within hospitals, but the inappropriate use of prophylaxis is one of the commonest antibiotic abuses.

Antibiotic prophylaxis should be considered when some or all of the following factors are present:

• there is a significant and predictable risk of infection;
• the consequences of infection may be serious;
• the period of highest risk can be ascertained;
• the microbial causes of infection are predictable;
• the antimicrobial sensitivity of the infections is predictable;
• cheap and reasonably safe antimicrobial agents are available.

Inappropriate antibiotic prophylaxis can be associated with:

• increased healthcare cost;
• a risk of adverse effects – side-effects, drug interactions;
• disturbances of the normal human bacterial flora;
• colonization with more antibiotic-resistant bacteria;
• selection of antibiotic resistance.

Examples of appropriate antibiotic prophylaxis include:

(i) co-amoxiclav (one to three doses only) for surgical procedures involving contaminated sites (e.g. colorectal surgery), starting just prior to surgery;
(ii) amoxicillin (one or two doses) for patients having dental surgery who are at risk of endocarditis (see Topic F13);
(iii) rifampicin for close contacts of meningococcal disease (see Topic C10).

Antibiotic resistance

Intrinsic and acquired resistance
Bacterial resistance to antibiotics may be classified as either intrinsic or acquired resistance. **Intrinsic resistance** depends on the natural properties of bacteria and mechanisms of action and other chemical properties of different antibiotics. For example, most Gram-negative bacteria are naturally resistant to vancomycin (vancomycin cannot penetrate the outer membrane of Gram-negative bacteria). Enterococci are generally resistant to cephalosporins, which have very low affinity for enterococcal penicillin-binding proteins.

Acquired resistance can arise by spontaneous mutation, by transfer of resistance genes or genetic elements from other strains of the same bacteria, or even by transfer of resistance genes or genetic elements from other species of bacteria. Bacterial antibiotic resistance genes may be found on either the **chromosome**, on **plasmids**, or on **transposons** (mobile genetic elements found on either chromosome or plasmid). In general, it is easier for bacteria to exchange resistance genes located on a plasmid or transposon.

Mechanisms of acquired antibiotic resistance
There are broadly six different known mechanisms of antibiotic resistance:

(i)   antibiotic inactivation;
(ii)   altered antibiotic target site;
(iii) overproduction of the target site;
(iv)   impaired uptake of the antibiotic;
(v)   enhanced efflux ('out-pumping') of the antibiotic;
(vi)   bypassing the antibiotic sensitive stage in metabolism.

Examples of **plasmid-mediated** resistance mechanisms are:

- β-lactamase enzymes that hydrolyze penicillins;
- extended spectrum β-lactamases (ESBL) that can hydrolyze third generation cephalosporins;
- a range of aminoglycoside-modifying enzymes (AME);
- energy-dependent efflux mechanisms (tetracycline resistance).

An example of a **transposon-encoded** antibiotic resistance mechanism is:

- production of altered peptidoglycan conferring vancomycin resistance in enterococci.

Examples of **chromosomally mediated** resistance mechanisms are:

- altered penicillin-binding protein (PBP) with reduced affinity for β-lactam antibiotics (e.g. MRSA);
- loss of outer membrane protein porins resulting in reduced permeability of antibiotics (e.g. imipenem resistance in *Pseudomonas*);
- mutations in DNA gyrase enzyme (quinolone resistance);
- overproduction of dihydrofolate reductase enzyme (trimethoprim resistance).

Some important types of antibiotic-resistant bacteria, their mechanisms of resistance and approximate frequency in the UK are described in *Table 1*.

Antibiotic side-effects

Antibiotics can have a number of side-effects, in addition to selecting for antibiotic resistance. Skin rashes and gastrointestinal upset (with nausea, vomiting and diarrhea) are common. Diarrhea can be due to direct effects on gut motility (e.g. erythromycin) or due to the significant disturbance of the normal gastrointestinal flora that occurs with most antibiotic use. Antibiotic-associated diarrhea can also be caused by infection with toxigenic strains of *Clostridium difficile* (see Topic C14), a particular problem with hospitalized elderly patients on broad-spectrum antibiotics (cephalosporins, macrolides and clindamycin).

Another common side-effect is oral or vaginal candidiasis (see Topic D1), again caused by the disruption to normal bacterial flora, allowing *Candida* to multiply to large numbers.

There are a wide range of other side-effects of particular antibiotics. Some important examples include:

- aminoglycosides – nephrotoxocity and ototoxicity;
- co-amoxiclav – cholestatic jaundice;
- fluoroquinolones – tendonitis, tendon rupture, risk of seizures (fitting);
- rifampicin – disturbed liver function tests and hepatitis;
- chloramphenicol – aplastic anemia;
- sulfonamides – Stevens–Johnson syndrome.

E8 ANTIFUNGAL DRUGS

Key Notes

Polyenes

Polyenes are broad-spectrum antifungal drugs. They bind to ergosterol and disrupt the fungal cell membrane. Amphotericin B is the principal drug used for invasive fungal infections, but is often poorly tolerated with a number of serious side-effects. However, several lipid-based preparations of amphotericin B are now available that minimize the toxic effects of conventional amphotericin B. These lipid preparations are more efficacious but very expensive. Nystatin is only available for oral or topical use in superficial or mucosal fungal infections.

Azoles

Imidazole and triazole drugs inhibit the biosynthesis of ergosterol. Imidazoles include clotrimazole, miconazole (used as topical preparations for superficial fungal infections) and ketoconazole. Triazoles include fluconazole, itraconazole and voriconazole. Fluconazole is available both orally and parenterally for the treatment of both superficial and invasive candida infections, but is ineffective against *Aspergillus*. Itraconazole and voriconazole have enhanced activity against *Aspergillus* and some fluconazole-resistant *Candida* species.

Allylamines

Terbinafine inhibits the biosynthesis of ergosterol, and is mainly active against dermatophyte infections. It is the treatment of choice for fungal nail infections and is also used for extensive tinea infection where topical treatment is ineffective.

Echinocandins

Caspofungin is the first of a new class of antifungal agents that inhibit the synthesis of glucan in the fungal cell wall. It is available for intravenous treatment of invasive *Aspergillus* of *Candida* infections where other treatments have failed.

Miscellaneous antifungal drugs

Griseofulvin is used for the oral treatment of dermatophyte infections, but has largely been replaced by terbinafine. Flucytosine inhibits fungal DNA synthesis and is sometimes used in combination with other antifungal drugs for the treatment of invasive candidiasis or cryptococcosis. Resistance can emerge during treatment and blood levels must be monitored to minimize the risks of bone marrow suppression.

Related topics

Yeasts (D1)
Molds (D2)

Dimorphic fungi and
    Pneumocystis carinii (D3)

Compared to antibacterial drugs, there are significantly fewer antifungal compounds available. However several new compounds have been developed and introduced into clinical use over the past few years, and this is an evolving area.

Antifungal drugs can be broadly divided into 4 main classes, with a few additional miscellaneous compounds:

(i)   polyenes (e.g. amphotericin, nystatin);
(ii) azoles (e.g. fluconazole, itraconazole, clotrimazole);
(iii) allylamines (e.g. terbinafine);
(iv) echinocandins (e.g.caspofungin);
(v)   miscellaneous antifungal drugs (e.g. flucytosine, griseofulvin).

Polyenes

The main drug in this class is **amphotericin B**, often regarded as the 'gold standard' agent for the treatment of deep-seated fungal infection. Recently, several **lipid preparations** of amphotericin B have been developed to minimize the toxic effects of conventional amphotericin B, and allowing higher doses to be given to maximize efficacy. Nystatin is only used for topical treatment of superficial or mucosal fungal infection.

Polyenes bind directly to **ergosterol**, the principal sterol in the fungal cell membrane, causing disruption to the integrity and function of the membrane. They have a broad-spectrum of antifungal activity including *Candida*, *Cryptococcus*, *Aspergillus*, mucorales and dimorphic fungi. Resistance is rare.

Conventional amphotericin B
This is available for oral, topical or parenteral administration. It is not absorbed after oral administration. Although parenteral therapy with amphotericin B is regarded as the 'gold standard' treatment for deep fungal infections, it is complicated by a number of serious side effects. These include:

• infusion reactions (fever, nausea, phlebitis);
• nephrotoxicity (through renal tubular damage);
• electrolyte disturbances (loss of potassium and magnesium).

Lipid preparations of amphotericin B
Several lipid preparations of amphotericin B are now available:

• liposomal amphotericin B (AmBisome);
• amphotericin B lipid complex (Abelcet);
• amphotericin B colloidal dispersion (Amphocil).

These lipid preparations have fewer adverse reactions compared to conventional amphotericin B, and higher doses can be tolerated. Studies have now shown superior efficacy of these lipid preparations compared to conventional amphotericin B for some types of invasive fungal infection, especially invasive aspergillosis. However, these drugs are very expensive.

Azoles

Antifungal azole compounds can be broadly divided into the **imidazoles** and **triazoles**. These drugs **inhibit the biosynthesis of ergosterol** (the principal sterol in the fungal cell membrane), through inhibition of the 14α-demethylation step (cytochrome *P*450 dependent) in ergosterol formation.

Imidazoles
There are numerous members of this group. Examples include clotrimazole, miconazole and ketoconazole. Clotrimazole and miconazole are commonly used as a topical preparation for superficial candidiasis and dermatophyte infections.

Ketoconazole is available for both topical and oral therapy, and is reasonably well absorbed. However, the use of systemic ketoconazole is complicated by adverse reactions (including hepatotoxicity and alterations in steroid metabolism), and newer triazole drugs are now preferred.

Triazoles
This is an important group of antifungal drugs including fluconazole, itraconazole and, more recently, voriconazole.

Fluconazole is available in both oral and parenteral formulations. It is reasonably well absorbed following oral administration, and has relatively few side-effects. It has a relatively broad spectrum of antifungal activity, but is not active in aspergillosis or mucormycosis. Whilst resistance in *Candida albicans* is rare, some species (e.g. *C. tropicalis*, *C. krusei* and *C. glabrata*) may be resistant. The main clinical uses of fluconazole are:

- treatment of mucosal or cutaneous candida infections;
- treatment and prevention of cryptococcal meningitis;
- prevention of *Candida* infections in immunosuppressed patients;
- treatment of some types of invasive candidiasis.

Itraconazole is more active than fluconazole against some nonalbicans species of *Candida*, and is also active against *Aspergillus* species, dermatophytes and dimorphic fungi. Itraconazole is available for both oral and parenteral administration. It is slightly less well tolerated than fluconazole, has a number of significant drug interactions, and can cause hepatotoxicity. The main clinical uses of oral itraconazole are:

- treatment of dermatophyte infections, pityriasis versicolor and superficial candidiasis;
- prevention and treatment of aspergillosis;
- treatment of histoplasmosis and systemic mycoses caused by other dimorphic fungi.

Voriconazole is a newer triazole with improved activity against *Aspergillus* species and some fluconazole-resistant *Candida* species.

Allylamines

The main drug in this class is **terbinafine**. Allylamines work by disrupting the biosynthesis of ergosterol (the principal sterol in the fungal cell membrane) through inhibition of the enzyme **squalene epoxidase**. Terbinafine is active against dermatophytes and *Candida*, but is ineffective in pityriasis versicolor.
Terbinafine is well absorbed following oral administration and is also available for topical treatment. It is mainly used for the oral treatment of dermatophyte infections, particularly infections of the nails (**onychomycosis**) or other types of tinea infection where topical treatment has failed.

Echinocandins

The only licensed drug in this class is **caspofungin**. This works by inhibiting the synthesis of $\beta(1,3)$-D-glucan, an essential component of the fungal cell wall. Caspofungin has good activity against both *Candida* and *Aspergillus*, but is not active against the agents of mucormycosis. It is only available for parenteral (intravenous) use, and is indicated for severe invasive fungal infections that are unresponsive to amphotericin B or itraconazole, or where these antifungal drugs cannot be tolerated.

Miscellaneous
antifungal drugs

Griseofulvin

Griseofulvin inhibits fungal cell division by: (i) binding to microtubular proteins and inhibiting mitosis; and (ii) inhibiting nucleic acid synthesis. It is active only against the dermatophytes, and is used as an oral treatment for moderate-to-severe dermatophyte infections, including fungal nail infections. Longer treatment courses are required compared to terbinafine.

Flucytosine

5-Fluorocytosine (5-FC) is a synthetic pyrimidine, which inhibits fungal DNA and protein synthesis. It is mainly active against *Candida* and *Cryptococcus*, but some strains are resistant, and resistance can emerge during therapy. It is principally used in combination with other antifungal drugs (e.g. amphotericin B or fluconazole) for the treatment of invasive candidiasis or cryptococcal meningitis. Significant suppression of the bone marrow can occur, and blood levels of the drug must be monitored during treatment.

E9 CONTROL OF HEALTHCARE-ASSOCIATED INFECTIONS

Key Notes

Definitions	Healthcare-associated infection (HAI or 'nosocomial' infection) is an infection occurring >48 h after admission to hospital, or within a defined period after healthcare contact. HAI can be endogenous (from the patient's own microbial flora) or exogenous (transmitted from another patient, member of staff, or the hospital environment). The latter is known as cross-infection. An outbreak is three or more related cases of HAI. Surveillance is an ongoing systematic collection of HAI data with analysis, interpretation and reporting.
Healthcare-associated infections (HAI)	The overall prevalence of HAI is ~10%. In the UK there are estimated to be 100,000 HAI per annum. Good infection prevention measures can reduce the incidence of HAI by 10–20%. Common HAI include urinary tract infections, respiratory infections, surgical site infections, device-related infections and diarrhea. Important 'Alert' organisms include MRSA, Group A streptococci, *Clostridium difficile*, VRE and multiresistant Gram-negative bacteria.
Routes of transmission	Cross-infection within healthcare settings can occur by direct contact (e.g. via the hands of healthcare workers), by indirect contact (e.g. contaminated equipment), by airborne spread (e.g. TB, respiratory viruses, aspergillus), by fecal–oral transmission (e.g. *C. difficile*, norovirus) or by percutaneous inoculation (e.g. blood-borne viruses).
Primary prevention	Endogenous HAI can be prevented by minimizing the use of invasive medical devices (e.g. lines, catheters), by identification of 'at-risk' patients (e.g. MRSA carriers), careful aseptic and surgical techniques, appropriate antimicrobial prophylaxis, and good medical and nursing care. Cross-infection can be reduced by high standards of hand hygiene and cleaning/decontamination (of both medical equipment and the environment). Numerous factors can impact on HAI including bed occupancy rates, staffing levels, volumes of antimicrobial drug use, isolation and sanitary facilities, ventilation systems, and storage facilities. Healthcare workers can also acquire and transmit HAI, and appropriate immunization and monitoring of staff health are important.
Secondary prevention	Additional HAI control measures are taken when an outbreak occurs, or when patients with an Alert organism (e.g. MRSA) are identified. These include isolation of patients or 'cohort' nursing, use of personal protective equipment for staff (gloves, aprons, gowns, masks, and eye protection), identification and screening of contacts, and temporary closure of a ward to admissions or discharges. Surveillance of HAI is

important in order to detect outbreaks, to define baseline infection rates, monitor trends of HAI over time, and to assess the effectiveness of primary prevention measures.

Healthcare-associated infections (HAI) are an important complication of the provision of healthcare. From an epidemiological perspective, hospitals are relatively small geographic areas into which are concentrated those individuals from a particular community who are either suffering from infection, or are at most risk of acquiring one. Whilst there are many benefits of concentrating medical services in such locations, there is an inevitable increase in the risk of transmission of infection in such an environment.

Definitions

Healthcare-associated infection (or hospital-acquired infection, or 'nosocomial' infection)

- An infection that was not present or incubating on admission to hospital, and that occurs as a complication of healthcare.
- This is usually defined as an infection occurring >48 h after admission to hospital, or within a defined period after healthcare contact (i.e. within 3 days of an outpatient procedure, within 10 days of hospital discharge, or a surgical site infection occurring within 30 days of the procedure).

Some infections have a longer incubation period than $\geq$ 48 h (e.g. 2–10 days for Legionnaires' disease) and may potentially be misclassified as healthcare-associated.

Endogenous infection
- An infection acquired from a patient's own microbial flora (i.e. not through cross-infection).

It is important to remember that $\geq$ 50% of HAI is endogenous, and these infections cannot be prevented by measures aimed at reducing cross-infection.

Exogenous infection (or cross-infection)
- HAI that is transmitted to a patient directly or indirectly from another patient, a member of staff, or from the healthcare environment.

Outbreak
- Three or more cases of HAI that are linked by organism, time or place.

Surveillance
- The ongoing systematic collection of HAI data with analysis, interpretation and reporting.

Healthcare-associated infections

In the UK, prevalence studies have shown that at any one time, ~10% of hospital in-patients have an HAI. This equates to 100,000 infections at a cost of > £1 billion. Not all of these infections can be prevented. Indeed, many are the inevitable consequence of undertaking increasingly complex medical and surgical procedures on highly vulnerable patients. However, it is estimated that $\geq$ 10–20% of HAI can be prevented by good infection control and prevention measures.

The commonest types of HAI are:

- urinary tract infection – often related to the use of urinary catheters;
- respiratory infections – following general anesthetic or assisted ventilation;
- surgical site infections;
- diarrhea;
- bacteremia – often associated with indwelling intravenous catheters.

Some microorganisms are a particular concern within the healthcare setting because of their resistance to antibiotics, their transmissibility, their capability to cause serious disease, or a combination of these factors. These are called '**Alert**' or indicator organisms (*Table 1*).

Table 1.   Alert organisms

MRSA (methicillin-resistant *Staphylococcus aureus*)
Group A streptococci
Clostridium difficile
Glycopeptide (vancomycin)-resistant enterococci (GRE or VRE)
Multiresistant Gram-negative bacilli (coliforms, *Pseudomonas, Acinetobacter*)

Routes of transmission

Cross-infection within the healthcare setting can occur via a number of different routes. The commonest route of transmission is via the **hands of healthcare workers**, and this is why regular hand decontamination is vital in the control and prevention of HAI.

Other routes of transmission include:

- indirect contact (e.g. contaminated equipment);
- airborne spread (e.g. tuberculosis, respiratory viruses, *Aspergillus*);
- fecal–oral transmission (e.g. *Clostridium difficile*, norovirus);
- percutaneous inoculation (e.g. blood-borne virus).

Primary prevention

The aim of primary prevention is to prevent HAI from occurring in the first instance.

Endogenous infections
Endogenous HAI can be minimized by a variety of measures, many of which are integrated into good general medical and nursing care. Examples include:

- minimizing the use of invasive medical devices (e.g. lines, catheters);
- identification of 'at risk' patients (e.g. MRSA carriers);
- careful aseptic technique;
- good surgical technique;
- use of appropriate antimicrobial chemoprophylaxis.

Exogenous infections
In addition to the measures outlined above, it is important to reduce the risks of cross-infection at all times, irrespective of whether an Alert organism or outbreak has been detected.

Cross-infection can be reduced by several measures including:

- good standards of **hand decontamination** (hand hygiene);
- good standards of **environmental cleaning**;
- adequate **decontamination of medical equipment**.

Hand hygiene is the single most important factor in preventing cross-infection. Hands should be washed with soap and water and then dried between each patient contact. Alternatively, alcohol-based rinses or hand-rub gel can be used. Other antiseptic soaps can also be used before sterile (aseptic) procedures.

Medical equipment should be adequately **decontaminated** between each patient use. This may be done by **simple cleaning** with detergent and water (e.g. for commodes), by cleaning followed by **chemical disinfection** (e.g. flexible endoscopes), by cleaning followed by **heat disinfection** (e.g. bed-pan washers), or by cleaning followed by **sterilization** (e.g. autoclaving surgical instruments). The physical removal of organic material (cleaning) is the most important component of decontamination. Inadequate cleaning may render disinfection or sterilization processes ineffective.

There are numerous factors that can impact on cross-infection rates including:

- design and layout of wards (e.g. space between beds);
- number of isolation rooms;
- sanitary facilities;
- bed occupancy rates;
- staffing levels (staff:patient ratios);
- provision of appropriate ventilation systems;
- storage facilities.

Many Alert organisms are resistant to multiple antibiotics. High volumes of **antimicrobial use** (particularly broad-spectrum agents) can promote high carriage rates of these organisms, allowing the survival of these organisms within the healthcare setting. The control of antibiotic use is thus another important preventive infection control measure (Topics E6 and E7).

Healthcare workers (HCW) can also acquire and transmit HAI, and appropriate immunization and monitoring of staff health are important. Infections that may be transmitted between HCW and patients include blood-borne viruses, MRSA, tuberculosis, scabies, viral gastroenteritis, chickenpox, and many others. HCW with an acute infection (e.g. infectious diarrhea) should not perform clinical duties until at least 48 h after symptoms have resolved. HCW who are infectious carriers of blood-borne viruses must not perform **exposure prone procedures (EPP)** (e.g. surgery).

Secondary prevention

Alert organisms
A range of additional prevention measures can be used to further reduce the risk of cross-infection when an Alert organism is found. The actual measures employed depend on the organism, its route of transmission, and the healthcare setting. These measures may include:

- strict hand hygiene;
- isolation of patient in a single room;
- wearing of gloves and protective aprons or gowns for patient contact;
- wearing of masks and eye protection;
- dedicated patient equipment (e.g. commode);
- careful disposal of clinical waste;
- careful handling of contaminated equipment, clothing and linen;
- regular cleaning and thorough disinfection of the room after the patient has left.

Outbreaks

The control of an outbreak involves.

- recognition of the outbreak;
- identification of cases, if necessary through additional microbiological screening;
- epidemiological and microbiological investigations (e.g. typing) to help identify the source and route(s) of transmission;
- implementation of outbreak control measures.

Many of the control measures are similar to those taken for patients with Alert organisms. However, patients may have to be **'cohorted'** together if there are insufficient isolation facilities (i.e. patients who are symptomatic or found to be carrying the organism are looked after by dedicated staff in a designated area of the ward, away from other patients).

During an outbreak it may be necessary to restrict admissions to the area, and restrict the transfer of patients (and staff) to other areas, until the outbreak is under control.

Infection control teams

Many hospitals employ infection control teams to give advise on all aspects of primary and secondary infection control and prevention strategies, and to coordinate the investigation and control of outbreaks. These teams usually comprise an Infection Control Doctor and one or more Infection Control Nurses.

Surveillance of HAI

Surveillance of HAI is important in order to detect outbreaks, to define baseline infection rates, monitor trends of HAI over time, and to assess the effectiveness of primary prevention measures.

Surveillance of Alert organisms is usually based on microbiology laboratory data. This may be influenced by the extent of microbiological investigations that are routinely performed in the clinical setting, but the data are relatively straightforward to collect. Surveillance of some of the common types of HAI (e.g. catheter-related UTI, or surgical site infection) is more time-consuming as infections cannot be identified solely from laboratory data. Standardized **definitions** of infection should be used for the purposes of surveillance, in order to allow comparison of data between different wards and hospitals.

In order to calculate **infection rates**, both **numerator data** (i.e. the number of infections) and **denominator data** (i.e. the number of patients at risk, or the number of occupied bed-days) are required.

E10 VACCINES AND IMMUNOPROPHYLAXIS

Key Notes

Passive immunization	This refers to the short-lived protection afforded by injection of preformed antibodies against a variety of bacterial or viral antigens.
Active immunization – vaccination	This generates long-term protection by stimulating a host immune response, including immunological memory, against a variety of bacterial or viral antigens presented to the individual in the form of a vaccine.
Vaccines	There are many forms of vaccine available, each with advantages and disadvantages. Live attenuated organisms generate both humoral and cellular immune responses to a wide range of antigens within the organism under consideration, but they require a secure cold chain, i.e. they must be kept at low temperatures until administration, as otherwise they become nonviable. Inactivated whole organism vaccines are more stable, but less immunogenic, therefore requiring multiple doses, administered by injection. Subunit vaccines consist of only a key part of the organism. These also require multiple doses, and, as with inactivated whole organism vaccines, induce much better humoral than cellular responses. Toxins to be used as subunit vaccines must first be inactivated, e.g. by treatment with formaldehyde, thereby yielding toxoid. Adjuvants are substances which enhance the immune response to the administered vaccine.
Contraindications	There are few genuine contraindications to vaccination: acute illness; a severe allergic reaction to previous vaccination; pregnancy; and live vaccines should be avoided in immunocompromised individuals.
Vaccines given routinely	See table 2.
Vaccines offered to selected high-risk groups	See table 3.

Passive immunization

One way of protecting an individual against an infectious disease is to provide that individual with preformed antibodies (or immune cells, but this is much less practical and rarely done) directed against the infecting agent. This is referred to as **passive immunization**. The protection transferred in this way is immediate, but it only lasts as long as the transferred antibodies last. The antibodies may be human in origin, i.e. derived from blood donors – or may have been purposefully raised in an animal. The latter carries the risk of inducing an immune response to the foreign proteins injected, whereas the former has the potential risk of all material derived from blood donors – possible contamination with blood-borne viruses or even prions. Transplacental transfer of maternal antibodies is an example of passive immunization – this plays an important role in protecting babies from infection in the first 6 months of life, as their own immune system matures.

Active immunization – vaccination

This refers to the process whereby an individual is deliberately exposed to key microbial antigens, in order to induce cellular and/or humoral immune responses to protect the individual against future exposure to the agent from which the antigens were derived. The preparation of antigens is referred to as the **vaccine**. Vaccine-induced protection is not immediate – it takes a finite time for the host system to mount a response to an antigenic challenge. However, active immunization yields much longer-lasting protection than passive immunization, as some immune cells become memory cells, able to be stimulated years later if challenged with the antigen. Vaccination works because such a challenge generates a secondary immune response, rather than a primary one (Topic A2), and secondary responses are faster, of higher affinity, and greater titer than primary ones.

Vaccines

Vaccination is the safest and most cost-effective method of disease prevention; however, developing and introducing protective vaccines against infectious diseases is not an easy task. Several factors determine the development and licensing of vaccines, including the prevalence of disease, seriousness to health or economy, and the availability of safe and protective antigen preparations. There are numerous serious infectious conditions where preventive vaccines would be highly desirable, but protective preparations are not yet available for most of these.

Vaccines may consist of viable but attenuated organisms, nonviable whole cell mixtures or fragments (subunits) of organisms. Only highly protective vaccines against relatively common and serious diseases with proven safety records are introduced into the routine childhood immunization programmes. Partially protective vaccines are made available to certain high-risk groups.

Table 1.   Composition of currently available vaccines.

Type of microbe	Vaccine content	Infection
Viruses	Live attenuated	Polio (Sabin vaccine), measles, mumps, rubella, varicella-zoster, smallpox (vaccinia), yellow fever
	Inactivated	Pollo (Salk vaccine), influenza, hepatitis A, Japanese encephalitis, tick-borne encephalitis, rabies
	Subunit	Hepatitis B, influenza
Bacteria	Live attenuated	BCG, typhoid
	Whole-cell killed	Pertussis, typhoid, plague, Q-fever, cholera
	Single protein (toxoid)	Diphtheria, tetanus
	Multiple proteins	Pertussis, anthrax
	Capsule (carbohydrate)	*Haemophilus influenzae* type b, *Neisseria meningitidis* Group C, typhoid, *Pneumococcus*

Table 1 shows the composition of the currently available vaccines, some of which are given routinely (*Table 2*) and others only when recommended (*Table 3*).

Exposure to live and fully virulent organisms often acts as an effective vaccination (provided the host survives the initial attack). For example, patients who recover from hepatitis A virus infection remain solidly immune to further attacks throughout life.

Live attenuated vaccines stimulate a broad immune response (i.e. both humoral and cellular) to a wide range of microbial antigens, both structural and nonstructural. They should be attenuated to a sufficient degree that they do not of themselves give rise to serious disease. Ideally, they should be administered through the natural route of infection, and thereby stimulate protective immune response in relevant parts of the body (e.g. live attenuated poliovirus vaccine is given by mouth, and stimulates protective immunity in the gut).

Inactivated vaccines, both whole viruses and whole-cell-killed bacteria, are administered by injection. They are not as immunogenic as live vaccines, and therefore multiple doses are necessary in order to achieve protective levels of immunity. However, they are more stable than live vaccines, which require

Table 2.   Routine childhood immunization program in the UK.

Age group	Number of doses	Vaccines
First 6 months of life	3	Diphtheria, pertussis, tetanus, polio, *Haemophilus influenzae* type b (Hib), meningococcus Group C
By 15 months	1	Measles, mumps, rubella (MMR)
By school entry	1 (booster dose)	Diphtheria, polio, MMR, tetanus
By 10–14 years	1	BCG
Before leaving school	1	Polio, tetanus, diphtheria

BCG, Bacille Calmette–Guérin, an attenuated mycobacterium used to vaccinate against tuberculosis.

Table 3.   Other vaccines available to those at risk of infection.

Risk group	Vaccine
Previously unimmunized individuals	Polio, tetanus, diphtheria
Seronegative women before pregnancy	Rubella
Minority groups at high risk of acquiring infection	Hepatitis A, hepatitis B, BCG, varicella, anthrax
Elderly, immunocompromised, patients with organ failure	Influenza, Pneumococcus
Travellers to certain parts of Africa	Yellow fever
Travellers to developing countries	Meningococcus Group A, C, W135, Typhoid, hepatitis A and B, cholera, polio, rabies, Japanese encephalitis, tick-borne encephalitis, and many others depending on the country

BCG, Bacille Calmette–Guérin, an attenuated mycobacterium used to vaccinate against tuberculosis.

maintenance of the **cold chain** prior to administration, i.e. they must be kept at low temperature in order to prevent loss of viability. The immune response to inactivated vaccines is skewed to the humoral, rather than the cellular arm of the immune system.

Subunit vaccines (also nonviable) consist of one or more highly protective components of the organism, selected and purified in the laboratory. These are also administered by injection, and multiple doses are required. The antigens represent key virulence or survival factors of the organism which can generate neutralizing or bactericidal antibodies.

Many vaccines are administered together with an **adjuvant**, something which enhances the local immune response to the vaccine antigens, e.g. aluminium hydroxide. Biologically active toxins, such as diphtheria or tetanus toxins, are rendered into **toxoids** by inactivation with formaldehyde before being introduced to humans, otherwise they may cause serious damage.

Before any vaccine is introduced to routine immunization programmes, rigorous checks are made in the laboratory, in animals, and in a large number of human volunteers. Licensing only occurs once the vaccine is considered safe and free of serious side-effects.

Contraindications

There are very few genuine reasons why certain individuals should *not* be given certain vaccines. These are referred to as **contraindications**:

(i) Acute illness. If an individual is acutely ill at the time when they present for vaccination, the procedure should be postponed until he/she has recovered.

(ii) Allergy to a component of the vaccine. Vaccines may contain a variety of substances other than the relevant antigens. Influenza vaccine is grown in hen's eggs and may contain a trace of egg proteins. Antibiotics are added to some vaccines to prevent bacterial contamination.

(iii) Pregnancy. If at all possible, it is sensible to postpone vaccination until the end of pregnancy, although there will be circumstances where the risk of possible damage to the fetus is outweighed by the risk of not vaccinating, e.g. yellow fever vaccination for someone visiting a yellow fever endemic area.

(iv) Immunosuppression, from whatever cause. Live vaccines in particular should be avoided in immunocompromised patients, as the absence of a fully functioning immune system may allow the attenuated organism to become an opportunistic pathogen.

In addition to the above genuine contraindications, there are, unfortunately, a wide range of false beliefs and myths about vaccination, which collectively mean that large numbers of individuals are being denied the benefits of vaccination.

F1 SKIN AND SOFT TISSUE INFECTIONS

Key Notes

Normal skin flora	The normal skin flora comprises permanent, temporary and transient residents. Permanent residents include coagulase-negative staphylococci (including *Staph.epidermidis*), diphtheroids, micrococci and propionibacteria. Temporary residents may reside for several days or weeks, including *Staph. aureus*. Transient residents such as coliforms or pseudomonads, particularly on the hands, are readily removed by hand washing.
Damaged skin	Damaged or diseased areas of skin are more prone to colonization, and subsequent infection, with potentially pathogenic bacteria such as *Staph. aureus*, hemolytic streptococci, pseudomonads and anaerobes. Enterococci and coliforms are also commonly found on areas of broken skin, particularly on areas below the perineum.
Bacterial skin and soft tissue infections	A wide variety of bacteria can cause skin and soft tissue infections, but *Staph. aureus* and *Strep. pyogenes* are the most common. *Staph. aureus* typically causes spots, boils, carbuncles, impetigo, cellulitis, or wound infections. *Strep. pyogenes* (group A β-hemolytic streptococcus) typically causes impetigo, erysipelas, cellulitis, wound infections or necrotizing fasciitis (rare). *Staph. aureus* is the commonest cause of skin and soft tissue infection in hospitalized patients, often with strains that are methicillin-resistant (MRSA). In addition to direct infection, the skin may be affected by bacterial toxins such as staphylococcal scalded shin syndrome toxin or streptococcal erythrogenic toxin (scarlet fever). β-lactam antibiotics are the mainstay of antibiotic treatment.
Nonbacterial infections of the skin	Viral infections involving the skin include herpesviruses (cold sores, chickenpox), enteroviruses (rashes) and papillomaviruses (warts). Common fungal infections include ringworm (dermatophyte infections or 'tinea') and *Candida* (intertrigo). Skin infestations with ectoparasites include mites (scabies) and lice (headlice, pubic lice).
Related topics	Staphylococci (C1) Streptococci and enterococci (C2)

Normal skin flora   Normal skin has a number of bacteria that permanently colonize the surface, or are found in deeper skin layers or around sebaceous or sweat glands. These **permanent residents** comprise mainly coagulase-negative staphylococci (e.g. *Staphylococcus epidermidis*), micrococci, corynebacteria (diphtheroids) and propionibacteria. Anaerobic streptococci and *Acinetobacter* may also be found.

Temporary residents are bacteria that may colonize areas of skin for several days or weeks but are not present in all individuals. *Staph. aureus* is an important temporary resident. It is estimated that 25–30% of individuals carry *Staph. aureus* in the nose, or other moist body sites, at any one time.

Transient residents are bacteria that can colonize the skin (including hands) for up to several hours, but are usually readily removed by washing. These include *Staph. aureus*, coliforms, pseudomonads and enterococci. This is important from an infection control perspective within hospitals, and emphasizes the need for regular hand decontamination.

Damaged skin

Damaged areas of skin, such as ulcers, burns, wounds, or areas affected by disease (e.g. eczema, psoriasis), are prone to colonization with potentially more pathogenic bacteria than are found on normal skin. Potential pathogens such as *Staph. aureus*, hemolytic streptococci, *Pseudomonas* and anaerobes can found. Enterococci and coliforms are also commonly found on areas of broken skin, particularly on the lower part of the body and legs (areas below the perineum).

Colonization may then lead to infection as the normal defense mechanisms of intact skin are breached.

Bacterial skin and soft tissue infections

A wide variety of bacteria can cause infection of the skin. Some of the main pathogens are listed in *Table 1*, but of these, *Staph. aureus* and *Strep. pyogenes* are the most common by far.

Infections due to Staph. aureus
There is a variety of skin and soft tissue infections caused by *Staph. aureus* (see Topic C1):

- spots, boils, carbuncles (common);
- impetigo (common in children);
- cellulitis;
- infections of traumatic wounds, surgical wounds, and ulcers;
- infections around cannula sites or drains;
- scalded skin syndrome (a staphylococcal toxin-mediated disease).

Staph. aureus is the commonest cause of skin and soft tissue infection, including surgical wounds, for patients in hospital. Many of these infections are now due to methicillin-resistant strains (**MRSA**). Staphylococci are capable of survival on surfaces and within dust, and can cause endemic problems and cross-infection in hospital wards.

Microbiological investigations should include culture of pus swabs, as well as blood cultures for more serious infections (e.g. cellulitis). The mainstay of antibiotic therapy is with flucloxacillin or related antibiotics. Abscesses or carbuncles require surgical drainage of pus. MRSA is resistant to flucloxacillin and other β-lactam antibiotics, and often other classes of antibiotics such as macrolides and quinolones. Antibiotic treatment of MRSA infections should be guided by sensitivity results (e.g. vancomycin, tetracyclines, fusidic acid, rifampicin, trimethoprim). Linezolid is a new antibiotic with good MRSA activity.

Recurrent staphylococcal infections can occur in some patients, particularly those with nasal carriage, or patients with diabetes mellitus.

Infections due to Strep. pyogenes
Strep. pyogenes (β-hemolytic streptococcus group A) also causes a variety of skin and soft tissue infections (see Topic C2):

Table 1. Some bacterial causes of skin/soft tissue infection

Classification	Bacteria	Infection/comment
Gram positive	Staph. aureus	The commonest bacterial pathogens
	Strep. pyogenes	of skin and soft tissue (see text)
	Other haemolytic streptococci	Similar infections to Strep. pyogenes
	Corynebacterium diphtheriae	Cutaneous diphtheria – a cause of skin ulcers in tropical countries
	Bacillus anthracis	Cutaneous anthrax
	Erysipelothrix rhusiopathiae	Erysipeloid in meat or fish handlers
Gram negative	Pseudomonas aeruginosa	Secondary infection of burns or other wounds; common colonizer of leg ulcers; black necrotic ulcers (ecthyma gangrenosum) is a complication of pseudomonas bacteremia in immuno-suppressed patients
	Coliforms	Common colonizer of leg ulcers; may be involved in invasive soft tissue infection along with anaerobes (necrotizing fasciitis)
	Neisseria meningitidis	Meningococcal septicemia with characteristic purpuric rash
	Neisseria gonorrhoeae	Disseminated gonococcal infection can produce pustular lesions on the limbs and body
	Pasteurella multocida	Wound infection and cellulitis following animal bites (cats, dogs)
	Streptobacillus moniliformis	Disseminated pustular lesions can occur in 'rat-bite fever'
Anaerobes	Bacteroides	Necrotising fasciitis (Fournier's
	Fusobacterium	gangrene, Meleney's synergistic
	Anaerobic streptococci	infection) in conjunction with other bacteria
	Clostridium spp	Gas gangrene, tetanus, wound botulism
Higher bacteria	Actinomyces israelii	Soft tissue abscesses
	Nocardia asteroides	
Spirochaetes	Treponema pallidum	Syphilitic ulcers, rash, and gummas
	Borrelia burgdorferi	Lyme disease – erythema migrans
Mycobacteria	Mycobacterium tuberculosis	Lupus vulgaris, cold abscesses
	Atypical mycobacteria	Fish tank granuloma (M. marinum)

- impetigo;
- erysipelas;
- cellulitis;
- wound infections;
- necrotizing fasciitis;
- scarlet fever (a streptococcal toxin-mediated disease).

Microbiological investigations comprise culture of swabs, pus or tissue, and blood cultures for more serious infections (e.g cellulitis, necrotizing fasciitis). Antibiotic treatment is mainly with penicillins or other β-lactam antibiotic.

Impetigo

Impetigo is a superficial skin infection caused by both *Staph. aureus* and *Strep. pyogenes*, sometimes in combination. Vesicular lesions occur, which then crust over.

Cellulitis and erysipelas

These are more serious infections. Cellulitis can occur at any part of the body, but often the lower limb is affected in patients with leg edema. Bacteria can gain entry often from only a minor lesion (e.g. athlete's foot, or insect bite). This is a spreading infection of the skin with marked erythema. Inflammation of the lymphatic vessels and lymph nodes (lymphangitis, lymphadenitis) can occur. Systemic upset is common, and blood cultures may be positive. Erysipelas is similar to cellulitis, producing a characteristic blistering rash on the face.

Cellulitis may also complicate animal bites due to the bacterium *Pasteurella multocida*.

Necrotizing fasciitis

This is a rare but life-threatening infection of deep soft tissue (down to the fascial layers over muscle) that rapidly spreads though tissue planes. Microbiologically, there are two causes of necrotizing fasciitis:

(i)   *Strep. pyogenes;*
(ii)   a synergistic combination of anaerobes, other streptococci and coliforms.

The latter is more common around the lower abdomen or perineum, sometimes complicating surgical procedures. Treatment is by urgent surgical excision of necrotic tissue, and high-dose antibiotic therapy according to the likely causative bacteria, adjusted in light of culture and sensitivity results. Surgery often has to be repeated as the infection extends.

Toxin-mediated skin disease

Staphylococcal scalded skin syndrome (toxic epidermal necrolysis) is due to infection (at any body site) with a *Staph. aureus* strain producing a specific toxin. This causes large areas of skin to peel, as the epidermis is split by fluid.

Scarlet fever is due to infection with **erythrogenic toxin**-producing *Strep. pyogenes*. It causes a florid rash, and characteristic 'strawberry' tongue, together with systemic upset.

Toxic shock syndrome can be associated with both *Staph. aureus* and *Strep. pyogenes* infections. This is a multisystem disease (hypotension, diarrhea, renal failure etc.) with a diffuse red rash a characteristic feature. In the latter stages of recovery large areas of skin may peel away. Toxic shock syndrome is mediated by staphylococcal or streptococcal toxins that act as 'superantigens' (these are antigens that stimulate a polyclonal T-lymphocyte immune response).

Nonbacterial infections of the skin

A large number of nonbacterial microorganisms can cause infection, infestation or clinical manifestations on the skin or in soft-tissue. These are listed in *Table 2*.

Common **viral infections** affecting the skin are herpes simplex and varicella-

Table 2.   Nonbacterial causes of skin infection or infestation

	Organisms	Infections
Viruses	*Herpes simplex* virus	Cold sores, genital lesions, stomatitis, Whitlow
	Varicella-zoster virus	Chickenpox, shingles
	Enteroviruses	Maculopapular rashes, vesicular lesions (hand, foot and mouth disease)
	Measles, rubella, parvovirus	Maculopapular rashes, slapped cheek syndrome (parvovirus)
	Papilloma viruses	Warts
	Molluscum contagiosum	Umbilicated papular lesions (children, HIV-positive patients)
	Other pox viruses	Orf, vaccinia, smallpox
Fungi	*Candida*	Cutaneous thrush (moist/abraded skin areas)
	Dermatophytes	Tinea or ringworm
	Malassezia furfur	Pityriasis versicolor
	Other fungi	Sporotrichosis, mycetoma
Protozoa	*Leishmania*	Cutaneous leishmaniasis
Helminths (worms)	Hookworms (cat, dog)	Cutaneous larva migrans (rare in the UK but occasionally seen in travellers)
	Other (tropical) worm infections	Schistosomiasis, strongyloidiasis, onchocerciasis, guinea worm
Ectoparasites	*Sarcoptes scabiei*	Scabies
	Pediculus humanus capitis, *Pediculus humanus corporis*, *Phthirus pubis*	Headlice, body lice, pubic lice
	Other insects	Bed bugs, fleas, myiasis

zoster viruses (see Topics B5 and B6), enteroviruses (see Topic B15) and papillomaviruses (warts – see Topic B19).

Dermatophyte (ringworm) infections are common fungal skin infections (see Topic D2). There are many different dermatophytes species grouped into three genera:

- *Epidermophyton;*
- *Trichophyton;*
- *Microsporum.*

Infection can occur at many different body sites including the groin (**tinea cruris**), hair (tinea capitis), body (tinea corporis), foot (tinea pedis or athlete's foot) and nails (onychomycosis). Infection produces characteristic skin rashes with dry, flaky skin, damage and discoloration to nails, and sometimes hair loss. Diagnosis is usually made on clinical grounds, but microbiological confirmation can be obtained by microscopy and fungal culture of skin scrapings, hair, or nail clippings. Treatment is generally with **topical antifungal preparations** (e.g. clotrimazole), but systemic treatment with antifungal drugs (griseofulvin, itraconazole, or terbinafine) is used for tinea capitis causing hair loss (often due to *Trichophyton tonsurans*) or for nail infections.

Cutaneous candida infection is also common in warm moist skin areas [nappy (diaper) rash, intertrigo].

Common **ectoparasite** infestations in the UK are scabies and lice.

Scabies is caused by the *Sarcoptes scabiei* mite, which is spread by direct skin contact. Outbreaks of scabies are common in residential institutions, e.g. nursing homes. The female mite burrows into the skin to lay eggs. Burrows are commonly found on the hands (finger webs), wrists and genital areas. An itchy, papular rash (due to an allergic response to the mite) can develop on the body and limbs. Itching is particularly pronounced at night. Chronic infection in debilitated or immunosuppressed patients can cause thickening and crusting of the skin, with minimal itching (**Norwegian encrusted scabies**). This can be diffi-cult to recognise clinically, but is a highly infectious form of scabies. Treatment of scabies (both patient and all direct contacts) involves the topical application (all areas of skin below the neck) of an insecticide (e.g. malathion). Repeated treatment may be necessary.

Headlice (*Pediculus humanus capitis*) are a common problem for young chil-dren and are spread by close contact. The lice feed by biting the scalp skin, causing itching and red papules. The diagnosis can be made by visual detection of either the lice, or eggs (nits) which are firmly adherent to hairs. Treatment is with either insecticidal shampoo and/or physical removal of the lice by regular systematic combing with a fine tooth comb. **Pubic lice** (*Phthirus pubis*) are gener-ally sexually transmitted. **Body lice** (*Pediculus humanus corporis*) are rare, but may be found in vagrants or in refugees. They live in clothing, but feed on the body. They are important vectors of some rickettsial infections (typhus, relapsing fever).

F2 BONE AND JOINT INFECTIONS

Key Notes

Definitions	Septic arthritis – acute infection of a joint. Reactive arthritis – an inflammatory response within a joint (may be triggered by certain infections). Osteomyelitis – infection of bone (either acute or chronic).
Microbiology	The principal pathogens of acute septic arthritis and osteomyelitis are *Staph. aureus* and β-hemolytic streptococci. A variety of Gram-negative bacteria are also encountered. Tuberculosis is an important cause of chronic osteomyelitis. Reactive arthritis can follow gastrointestinal infections, sexually transmitted diseases, and other infections.
Epidemiology	Infections are more common in children, and in adults with certain predisposing conditions that either affect the musculoskeletal system (e.g. rheumatoid arthritis) or increase the risk of bacteremia.
Pathogenesis	Infections of bone and joint usually occur via hematogenous infection, but direct infection (following surgery or trauma) can also occur.
Clinical features	Pain and swelling of joints, or pain and tenderness over affected bones, together with fever and systemic upset occur in acute infections. Vertebral osteomyelitis is an important but rare cause of back pain. Discharging sinuses can occur in chronic osteomyelitis.
Diagnosis	Septic arthritis is diagnosed by microbiological examination of aspirated joint fluid and blood cultures. Both microbiological and radiological investigations are important in osteomyelitis. Blood cultures may identify the causative organism in acute osteomyelitis, but samples of bone or pus are required for chronic osteomyelitis or prosthetic joint infections.
Treatment	High-dose antibiotic therapy for 3–4 weeks (septic arthritis) or 4–6 weeks (acute osteomyelitis), combined with adequate drainage of pus. Chronic osteomyelitis requires many months of treatment. Prosthetic joint infections usually require removal and then replacement of the joint, in addition to antibiotic therapy.
Prevention	Hib immunization has reduced the incidence of Hib bone and joint infections. Ultraclean ventilation and intraoperative antibiotic prophylaxis reduce the risk of prosthetic joint infections.

Definitions
- **Septic arthritis** is an acute infection of a joint.
- **Reactive arthritis** is an inflammatory response within a joint (not infection), that may be triggered by an infection at a distant site.
- **Osteomyelitis** is infection of bone, which may be acute or chronic.

Infections of bone and joint can also be serious complications of orthopedic surgical procedures, particularly prosthetic joint replacements.

Microbiology

The principal causes of **septic arthritis** are *Staphylococcus aureus* (Topic C1) and β-**hemolytic streptococci** (mainly groups A, C and G, but occasionally group B) (Topic C2). *Streptococcus pneumoniae* is a less common cause.

A wide range of other (mainly Gram-negative) bacteria can also occasionally cause septic arthritis including:

- *Neisseria meningitidis* and *Neisseria gonorrhoeae* (Topic C10);
- *Haemophilus influenzae* capsular type b (now less common due to routine Hib immunization) (Topic C9);
- coliforms, *Pseudomonas* and *Salmonella* (Topics C7 and C8);
- *Brucella* (Topic C9).

Arthritis can also be caused by viruses (e.g. rubella, parvovirus), spirochaetes (e.g. syphilis, Lyme disease) and mycobacteria (tuberculosis).

Infections of **prosthetic joint replacements** may involve the main pathogens listed above, but also **less pathogenic skin organisms** (e.g. coagulase-negative staphylococci, diphtheroids and propionibacteria).

Reactive arthritis may follow gastrointestinal infections with *Salmonella*, *Campylobacter* and *Yersinia* (Topic F8), sexually transmitted infections (*Chlamydia*) (Topic F11), meningococcal infections, infections due to *Streptococcus pyogenes* (including rheumatic fever) (Topic C2), mycoplasma infections and viral infections (including rubella, parvovirus and hepatitis B).

Acute osteomyelitis is usually caused by *Staphylococcus aureus*, but hemolytic streptococci and *Streptococcus pneumoniae* are also important. Prior to Hib immunization, *H. influenzae* type b was an important cause in young children. Occasionally Gram-negative bacteria (coliforms, *Salmonella*, *Pseudomonas*) are found.

A wider range of organisms is found in **chronic osteomyelitis**, including staphylococci, streptococci, coliforms, *Salmonella*, *Pseudomonas* and anaerobes. *Mycobacterium tuberculosis* is also an important cause of chronic osteomyelitis, especially in the spine (Topic C6). Miscellaneous chronic infections of bone include *Brucella*, syphilis, actinomycosis, and various fungal infections.

Vertebral osteomyelitis is an important type of osteomyelitis with potentially serious complications (see below). The main causes are *Staph. aureus*, coliforms, *Pseudomonas* and *M. tuberculosis*.

Epidemiology

Septic arthritis is a relatively rare infection that occurs in all ages, but is most common in children. It may be more common in patients whose joints are affected by other diseases (e.g. rheumatoid arthritis) or in those with impaired immunity (e.g. steroid therapy). Patients at risk of bacteremia (Topic F12), including intravenous drug abusers or those in hospital with indwelling venous catheters, are at increased risk of 'seeding' a joint from the bloodstream and developing septic arthritis. Infection of prosthetic joint replacements usually occurs in < 2% of procedures.

Reactive arthritis occurs in only a minority of patients with the infections listed above. Children and young adults are principally affected. However, as some of these infections are quite common (e.g. gastrointestinal infections), reactive arthritis is a relatively common condition.

Acute osteomyelitis due to hematogenous infection (see below) is similar in epidemiology to acute septic arthritis. In both conditions, the number of infections due to *H. influenzae* has been considerably reduced by Hib immunization. Patients with **sickle cell disease** appear to have a particular risk for *Salmonella* osteomyelitis.

Acute osteomyelitis may also occur via direct spread to bone, and thus can complicate either trauma [particularly compound (open) fractures] or surgery.

Pathogenesis

Septic arthritis usually occurs by spread of infection via the **hematogenous** (bloodstream) route. Often there is no apparent primary focus of infection, although sometimes there has been a preceding septic skin lesion. Infection may also spread into a joint from adjacent bone. Occasionally iatrogenic infection may occur following a contaminated injection into the joint (e.g. a steroid injection).

Infection of prosthetic joints can occur at the time of surgery, or later via the hematogenous route. Introduction of normally harmless skin bacteria into the joint during surgery can cause low-grade chronic infection of the prosthesis, associated with the formation of a biofilm, and may not cause clinical problems for many months.

Reactive arthritis is thought to be due to an **immunological response** to an infection at a distant site, either through cross-reactions between joint tissue and the microorganism, or to deposition of immune complexes within the joint. Some patients have a **genetic predisposition** to this condition, notably those of tissue type **HLA-B27** who are predisposed to developing reactive arthritis following infection with *Yersinia enterocolitica*.

Osteomyelitis can occur either by hematogenous spread of infection (as for septic arthritis), or by direct spread of infection into the bone following trauma or surgery. Acute osteomyelitis, if undiagnosed or inadequately treated, can progress to **chronic osteomyelitis**. This is characterized by destruction of bone, with the formation of a **sequestrum** (containing dead bone and adherent microorganisms) and the development of discharging sinuses to the overlying skin.

Clinical features

The clinical features of **septic arthritis** include a short history of fever, pain (particularly on movement), and swelling of the joint and surrounding tissues. Usually only one joint is affected. On examination, the inflamed joint is hot and tender to touch, and there is often a detectable joint effusion. Patients with septic arthritis can also have symptoms and signs of generalized septicemia (Topic F12). Septic arthritis of the hip joint is less easy to diagnose, especially in children, as the joint is much deeper and surrounded by muscles. Children may present with a fever, and an inability to bear weight on the affected leg.

Infections of **prosthetic joints** can present either acutely with fever, pain and swelling (if due to *Staph. aureus* or other pyogenic bacteria), or more chronically with gradual pain and loosening or dislocation of the prosthesis if due to less pathogenic bacteria (e.g. coagulase-negative staphylococci, diphtheroids).

Reactive arthritis causes similar pain and inflammation, but usually affects several joints at once (**polyarthritis**). Patients are generally less unwell, and there may be a history of another infection 1–3 weeks previously (e.g. a history of recent gastrointestinal infection).

Acute osteomyelitis is characterized by pain and tenderness over the affected bone, often with a general systemic upset (bacteremia is also common). **Chronic osteomyelitis** presents with chronic pain in the affected area. Sinuses from the bone to the skin may develop in long-standing osteomyelitis, and these can discharge pus on an intermittent basis.

Vertebral osteomyelitis is one cause of **chronic back pain**. As well as affecting the vertebral bodies, the intervertebral disks may also become infected. This is known as **diskitis**. **Complications** of vertebral osteomyelitis include vertebral collapse and **spinal deformity**, and spread of infection into the spinal canal (**epidural abscess** – Topic F4) or other adjacent tissues (e.g. an abscess in the psoas muscle).

Diagnosis

Clinical diagnosis of acute septic arthritis is generally straightforward, and radiological investigations are not usually required. The main differential diagnoses are acute degenerative osteoarthritis and gout.

Septic arthritis is generally confirmed by microbiological examination of **aspirated joint fluid (or pus)**. This should be obtained using careful aseptic technique, preferably before the start of antibiotic therapy. **Microscopy** is performed for the presence of **white cells** (and also for monosodium urate **crystals** that can occur in gout). A **Gram stain** may rapidly identify the likely causative organism, but is not always positive. The results of **cultures** (and sensitivity tests) may take several days. **Blood cultures** should also be taken prior to the start of antibiotic therapy, since many patients with septic arthritis are bacteremic.

Diagnosis of infection within **prosthetic joints** is more difficult. Clinically and radiologically it may be difficult to distinguish infection from mechanical loosening. Bacteria such as coagulase-negative staphylococci that are found on the skin may also cause infection, and it can often be difficult to distinguish true infection from contamination of samples, even when appropriate samples are collected under full aseptic technique.

Reactive arthritis is usually diagnosed on clinical grounds. There are many potential infective triggers, and the clinical picture should determine the extent of microbiological investigations. Appropriate investigations may include stool samples, investigations for *Chlamydia* and serological tests (e.g. for streptococcal antibodies, Lyme disease, *Yersinia*).

Clinical diagnosis of **acute osteomyelitis** can be more difficult than septic arthritis, and in the early stages of infection the X-ray appearance is unchanged. Other **imaging techniques** [e.g. bone scan, magnetic resonance imaging (MRI)], may be required. Appropriate microbiological investigations include **blood cultures** (before antibiotic therapy is started) and, if necessary, invasive samples of pus or bone obtained by surgery. In chronic osteomyelitis, pus from discharging sinuses can be examined, although deeper samples are always preferred.

If tuberculosis is suspected, then bone or pus samples should be examined for acid–alcohol-fast bacilli (Ziehl–Neelson stain) and appropriate mycobacterial cultures inoculated.

Treatment

Acute septic arthritis is treated with a combination of high-dose (usually intravenous) antibiotics and adequate aspiration of pus from the joint. The joint may require a formal 'washout' within the operating theatre. Antibiotics need to be given in high dose for 3–4 weeks. The choice of antibiotic should be guided by culture and sensitivity results. *Staph. aureus* and hemolytic streptococci are the main pathogens, and flucloxacillin is a suitable empiric choice before microbiological results are known. However, increasing numbers of *Staph. aureus* are now resistant to flucloxacillin [methicillin-resistant *Staph. aureus* (MRSA)], particularly in patients who have been in hospital before, or those who live in nursing homes. In this case treatment with a glycopeptide (e.g. vancomycin) may be required. Adjunctive therapy with a second antibiotic may also be considered (e.g. fusidic acid).

Infections of prosthetic joints are difficult to cure with antibiotics alone, and empiric antibiotic treatment is not advised. The standard approach is to remove the infected prosthesis, taking multiple samples for microbiological evaluation, and then to give 4–6 weeks of antibiotics (guided by culture and sensitivity results) before a new prosthesis is inserted.

Reactive arthritis is treated with anti-inflammatory drugs, rather than antimicrobials.

Acute osteomyelitis is treated similarly to acute septic arthritis. A standard empirical regimen is flucloxacillin and fusidic acid in combination for 4–6 weeks. MRSA is also an increasing concern in bone infections. Other antibiotics with good bone penetration include clindamycin, rifampicin and ciprofloxacin. The latter is useful for Gram-negative osteomyelitis (*Salmonella*, coliforms, *Pseudomonas*). Surgery to remove dead bone and pus is an important adjunct to antibiotic therapy.

Chronic osteomyelitis is difficult to treat effectively. Antibiotics may need to be given for 3–6 months or longer, and surgery is often required. Bone infections due to *M. tuberculosis* require a long course of antituberculous therapy (Topic C6).

Prevention

Aspirations of joint fluid or injections of anti-inflammatory drugs into joints should be done with careful aseptic technique, to avoid introducing bacteria into the joint.

Infections following orthopedic surgery can be minimized by use of appropriate antibiotic prophylaxis as well as ultraclean ventilation (laminar flow) within operating theatres. This is especially important for the insertion of prosthetic joints. Antibiotics (e.g. gentamicin) can also be incorporated into bone cement, or 'beads' within the operative site, and may help to reduce infection in some situations.

Hib immunization has reduced the incidence of bone and joint infections due to *H. influenzae* type b. Rubella immunization (now as part of the MMR vaccine) can prevent rubella-associated arthritis.

F3 MENINGITIS

Key Notes

Microbiology

Enteroviruses are the commonest cause of viral meningitis. The three primary bacterial pathogens are *Neisseria meningitidis, Streptococcus pneumoniae* and *Haemophilus influenzae* capsulate type b (Hib). Group B streptococci and *Escherichia coli* are the main causes of neonatal meningitis.

Epidemiology

Meningitis mostly occurs in children and young adults. Viral meningitis is more common than bacterial meningitis. The incidence of bacterial meningitis due to Hib and meningococcus serogroup C has been significantly reduced by routine immunization programmes. Risk factors for pneumococcal meningitis also include anatomical defects of the skull, and acute or chronic infections of the ear or sinuses.

Pathogenesis

Bacterial meningitis can occur as a result of hematogenous spread of infection to the meninges, usually following recent nasopharyngeal colonization with pathogenic strains. Invasive infection, however, occurs only in the minority of carriers. Nasopharyngeal transmission occurs from person to person by close contact (via respiratory droplets or kissing).

Clinical features

These include headache, fever, neck stiffness and photophobia. Patients with bacterial meningitis are more unwell than with viral meningitis, and the disease can progress rapidly, causing drowsiness, irritability, coma and death. A characteristic petechial or purpuric rash may be present in meningococcal meningitis.

Laboratory diagnosis

The principal laboratory method for diagnosis of meningitis is by examination of cerebrospinal fluid including microscopy, culture and determination of protein and glucose concentrations. Bacterial meningitis is usually associated with increased polymorphs, a raised protein level, and reduced glucose concentration. Viral meningitis is characterized by raised lymphocytes (with some polymorphs), a slightly raised protein, but normal glucose content. Detection methods for bacterial or viral antigens or genome are now available as well as culture. Other important specimens include blood cultures, whole blood for the detection of meningococcal or pneumococcal DNA, and throat swabs.

Treatment

Viral meningitis is usually self-limiting whereas bacterial meningitis is a medical emergency. Patients with suspected meningococcal disease may be given intramuscular penicillin, prior to arrival at hospital. After collecting blood cultures, patients with suspected bacterial meningitis in hospital are given high-dose intravenous antibiotics. Steroids are sometimes used as adjunctive therapy in bacterial meningitis, to reduce the rate of complications (e.g. hearing impairment).

Prevention	Routine primary childhood immunization includes Hib and meningococcus group C, but there is no effective immunization for meningococcus group B. Close contacts of patients with invasive meningococcal infection should be given antibiotic prophylaxis (and immunization if appropriate) to reduce the risk of secondary cases. Infections should be promptly notified to the relevant authorities.
Related topics	Other central nervous system infections (F4) Vaccines and immunoprophylaxis (E10)

Meningitis is infection and/or inflammation of the **meninges** – the membranes overlying the surface of the brain.

Microbiology

A range of microorganisms can cause meningitis including viruses, bacteria, protozoa and fungi (*Table 1*). The **three primary bacterial pathogens** are:

(i)   *Neisseria meningitidis* (meningococcus – see Topic C10);
(ii) *Streptococcus pneumoniae* (pneumococcus – see Topic C2);
(iii) *Haemophilus influenzae* capsulate type b (Hib – see Topic C9).

Enteroviruses (Topic B15) are the commonest viral cause, now that mumps meningitis is prevented by universal immunization.

Epidemiology

Meningitis is most often seen in children and young adults. Viral meningitis is the most common, but usually causes a milder, self-limiting illness. Viral meningitis tends to be seasonal, and outbreaks may occur within closed communities.

Bacterial meningitis is less common but much more serious. There are ~ 2000 cases of bacterial meningitis notified per year in England and Wales.

There is a relationship between age and the likely cause of bacterial meningitis. Meningococcal meningitis occurs in all age groups, but most commonly in children, adolescents, and young adults. The commonest serogroups of *N.meningitidis* that cause meningitis in the UK are groups B and C, but routine immunization against serogroup C strains has now reduced the proportion of cases due to serogroup C. Pneumococcal meningitis is more common in young infants and elderly patients. *H. influenzae* type b (Hib) meningitis used to occur predominantly in children aged < 4 years, but the incidence of this infection has been sharply reduced by Hib immunization. Group B streptococcal meningitis and *Escherichia coli* meningitis are the principal causes of neonatal meningitis (Topic F15), and are rarely seen outside this group.

Risk factors for bacterial meningitis include anatomical defects of the skull (e.g. basal skull fracture following head injury), acute or chronic infections of the ear or sinuses, and some types of immunodeficiency.

Pathogenesis

Most cases of bacterial meningitis occur as a result of **hematogenous** (bloodstream) spread of infection from outside the CNS to the meninges. Meningococci (as well as pneumococci and *H. influenzae*) are often carried asymptomatically in the **nasopharynx**, and spread occurs from person to person by close contact (via respiratory droplets or kissing). From the nasopharynx the bacteria may invade the bloodstream and spread to the meninges, but this only occurs in a small

Table 1.   Organisms causing meningitis

	Organism	Comment
Viruses	Enteroviruses (polio, coxsackie and echo viruses)	Coxsackie and echo viruses are the commonest cause of viral meningitis in the UK (Topic B15)
	Mumps virus	See Topic B13
	Herpesviruses (HSV, VZV)	Rare cause of viral meningitis (Topics B5, B6)
	Arboviruses	Numerous insect-borne viruses can cause meningitis or encephalitis outside of the UK (Topic B17)
Bacteria	*Neisseria meningitidis* *Streptococcus pneumoniae* *Haemophilus influenzae* type B	The three primary pathogens (see text)
	Escherichia coli Group B streptococcus (*Strep.agalactiae*)	The two main causes of neonatal meningitis
	Listeria monocytogenes and *Salmonella* species	Rare causes of meningitis in both neonates and immunosuppressed patients (Topics C5, C7)
	Other bacteria (staphylococci, streptococci, coryneforms, coliforms, pseudomonas, anaerobes)	A wider range of bacteria can cause meningitis following major head trauma or neurosurgery
Mycobacteria	*Mycobacterium tuberculosis*	Important but rare cause of meningitis (Topic C6)
Spirochaetes	*Borrelia burgdorferi* (Lyme disease) and *Leptospira* (leptospirosis)	Rare causes of lymphocytic meningitis (Topic C16)
	Treponema pallidum (syphilis)	Syphilis may affect the CNS in the late stages, as well as in congenital infection (Topic C16)
Fungi	*Cryptococcus neoformans*	Cryptococcal meningitis is mainly seen in HIV patients from Africa
	Candida, *Aspergillus*	Rare causes of meningitis in immunosuppressed patients
Protozoa	*Naegleria* spp.	Free-living amoeba can very occasionally cause meningitis following exposure during swimming

minority of carriers. The usual time from acquisition of nasopharyngeal carriage to infection is 1–3 days.

Direct spread from the nasopharynx to meninges may occur in patients with anatomical defects of the base of the skull (see above). This is particularly relevant to pneumococcal meningitis. Direct spread may also occur from infection at adjacent sites (e.g. middle ear infections, sinusitis). Meningitis can also complicate severe head injuries if there is an open (compound) skull fracture. Postoperative meningitis and other neurosurgical infections are summarized in

Topic F4. Congenital spinal defects (e.g. spina bifida) may also allow direct spread of bacteria from the surface of the skin to the meninges.

Clinical features   The classic features of meningitis include headache, fever, neck stiffness and photophobia (discomfort with bright light). Nausea and vomiting is also common. These features may occur with both viral and bacterial meningitis, but patients with viral meningitis are usually less unwell, and recover spontaneously. Patients with bacterial meningitis are more unwell, and the disease can progress rapidly, causing drowsiness, coma and death (~10% mortality).

Meningococcal meningitis may be accompanied by **meningococcal septicemia**. In this situation, a characteristic **petechial or purpuric (nonblanching) rash** may be present. A rash may also be present in some patients with enteroviral meningitis.

Laboratory diagnosis   The principal laboratory method for diagnosis of meningitis is by examination of **cerebrospinal fluid (CSF)**, which is usually obtained by a lumbar puncture (LP). As bacterial meningitis is a medical emergency, empirical antibiotic therapy should be given before an LP is performed, but this reduces the chances of a positive bacterial culture from CSF. Other important microbiological specimens that may determine the likely etiology include:

- blood cultures (many patients with bacterial meningitis also have bacteremia);
- whole blood for detection of meningococcal or pneumococcal DNA by PCR;
- throat swabs (for detection of meningococcal carriage, or viral culture for enteroviruses);
- viral culture of feces (for the detection of enteroviruses);
- clotted blood for appropriate serological investigations;
- blood glucose (must be done at the same time as the CSF sample).

Every effort should be made to determine the microbial etiology of meningitis, in order to guide the appropriate management of the patient, rationalize antimicrobial therapy, and determine any necessary public health measures.

A summary of the important characteristic CSF changes in different types of meningitis is shown in *Table 2*. **Bacterial meningitis** is usually associated with turbid CSF, containing increased polymorphs, a significantly raised protein level, and reduced ratio of CSF:blood glucose. **Viral meningitis** is characterized by raised lymphocytes (with some polymorphs), a slightly raised protein but normal glucose content. The same appearance is found with a number of other causes of 'lymphocytic' meningitis (*Table 2* footnote). In **tuberculous meningitis**, lymphocytes also predominate, but the protein is usually markedly raised, and the glucose reduced.

Other bacteriological tests on CSF include Gram stain, culture and sensitivity tests, although the sensitivity of these is not particularly high, especially if prior antibiotics have been given. Antigen tests, and more recently molecular tests (PCR), can be applied to CSF samples for the detection of bacterial or viral antigen or genome.

If tuberculous meningitis is suspected, then a Ziehl–Neelson (ZN) stain can be performed, but again the sensitivity is very low. PCR tests are now available, but liquid mycobacterial culture remains the mainstay of diagnosis, although results may take several weeks.

Other virological investigations on CSF include viral culture, PCR (for

Table 2.   Usual CSF changes that occur with different types of meningitis

	Appearance	Polymorphs ($\times 10^6$ l^{-1})	Lymphocytes ($\times 10^6$ l^{-1})	Protein (g l^{-1})	Glucose
Normal CSF (children and adults)	Clear	None	0–5	0.15–0.4	~60% of blood glucose
Normal CSF (neonates)	Clear	0–30	0–5	0.15–1.5	~60% of blood glucose
Bacterial meningitis	Turbid	100–2000*	0–5	0.5–3.0	<40% of blood glucose
Viral meningitis†	Clear or slightly turbid	0–100‡	5–500	0.5–1.0	~60% of blood glucose
Tuberculous meningitis	Clear or slightly turbid	0–100	5–500	1.0–6.0	<50% of blood glucose

*   In some cases there is very little cellular response in bacterial meningitis – a poor prognostic factor.
†   Similar appearances to viral meningitis may occur with partially treated bacterial meningitis,
    spirochaete infections (lyme, leptospirosis), encephalitis, cerebral and spinal bacterial abscesses,
    and certain noninfectious causes of meningitis.
‡   Viral meningitis is typically lymphocytic, but polymorphs may predominate at an early stage.

enteroviruses or herpesviruses) or serological tests for local antibody production.

Cryptococcal meningitis (or other fungal meningitis) may be diagnosed by microscopy (India ink preparation), antigen detection, and culture. Specialized techniques are required for detection of *Naegleria*.

Some CNS infections (e.g. neurosyphilis) are diagnosed by determining the ratio of local specific CSF antibody concentration to the corresponding concentration in peripheral blood, to ascertain if local CNS production of antibody is occurring.

Treatment

Viral meningitis is usually self-limiting and no specific treatment is available for enteroviral meningitis.

Bacterial meningitis requires the prompt administration of high-dose antibiotics. The antibiotics used should be guided by the likely bacterial pathogen, and knowledge of local sensitivity data. In the UK, resistance to penicillin in meningococci is virtually unknown. Pneumococcal resistance to penicillin occurs in 1–10% of infections, but this is mostly low level resistance.

Patients with suspected meningococcal disease should be given intramuscular penicillin, prior to arrival at hospital. After collecting blood cultures, patients with suspected bacterial meningitis in hospital should be given high-dose intravenous antibiotics, usually benzylpenicillin or a third generation cephalosporin (ceftriaxone or cefotaxime). Vancomycin may be added for suspected bacterial meningitis in countries with a high rate of penicillin resistance in pneumococci. In neonates, a combination of benzylpenicillin plus an aminoglycoside, or cefotaxime is usually chosen. Further antibiotic therapy should be guided by the results of investigations and sensitivity tests (if available), together with the clinical response. Steroids may be used as adjunctive therapy in bacterial meningitis, to reduce the rate of complications (e.g. hearing impairment).

Prevention

Primary immunization is now available for *H. influenzae* type b (Hib) and meningococcus group C, and these have been introduced into the routine childhood immunization program. There is no effective immunization for meningococcus group B. Polysaccharide vaccines are available for some other serogroups of meningococcus (e.g. A, W135) for visits to endemic areas.

Conjugate vaccines have been developed to prevent invasive pneumococcal infections (including meningitis) but are only used selectively at present. Immunizations against mumps and polio have made these infections rare in the UK, but there is no immunization available for other enteroviral causes of meningitis.

Close contacts (same household, or kissing contacts) of patients with invasive meningococcal infection, including meningitis, should be given **antibiotic prophylaxis** (and immunization if appropriate) to reduce the risk of secondary cases. Infections should be promptly notified to the relevant authorities. Suitable antibiotics include rifampicin, ciprofloxacin or ceftriaxone. **Outbreaks** of invasive meningococcal disease may occur in closed communities (e.g. schools), and more extensive chemoprophylaxis or immunization strategies may temporarily be required.

Within healthcare settings, workers who are exposed to respiratory secretions (e.g. during resuscitation) may also require antibiotic prophylaxis. Precautions against respiratory droplet spread should be taken for patients with meningococcal meningitis or septicemia, for at least the first 24 h of antibiotic treatment.

Antibiotic prophylaxis should also be given to all household contacts of Hib meningitis, if there are any unimmunised children (other than the index case) aged < 5 years.

Some patients have recurrent pneumococcal meningitis, due to an anatomical skull defect. Long-term prophylaxis with penicillin may reduce the likelihood of such infections.

The incidence of group B streptococcal (GBS) neonatal infections (including meningitis) can be reduced by administering prophylactic antibiotics during labor when there is known maternal vaginal carriage of GBS, or when risk factors are present (preterm delivery, prolonged rupture of membranes or previous infant affected by neonatal GBS infection).

F4 OTHER CENTRAL NERVOUS SYSTEM INFECTIONS

Key Notes

Viral encephalitis	Most cases of encephalitis are due to viruses. *Herpes simplex* encephalitis (HSE) is the commonest, but a wide range of virus infections may result in brain damage. Encephalitis may be caused by direct viral invasion of the brain (e.g. HSV, rabies), or by an immune response to infection which cross-reacts with brain antigens (post-infectious encephalitis). Diagnosis is made by clinical features, including history (travel, exposure to animals) and neuroimaging. Genome amplification from cerebrospinal fluid may identify the specific viral cause. Immediate high-dose intravenous aciclovir is essential in case a patient has HSE. Mumps, measles and rabies encephalitis are all preventable by appropriate vaccination. Risk of arthropod-borne encephalitis may be reduced by appropriate precautions to reduce mosquito and tick bites.
Nonviral causes of encephalitis	Nonviral causes of encephalitis include pyogenic bacteria (secondary to bacterial meningitis or cerebral abscess), atypical organisms (e.g. *Mycoplasma pneumoniae*), and protozoa including *Plasmodium falciparum* (falciparum malaria) and trypanosomes (African sleeping sickness).Toxoplasmosis is an important cause of protozoal encephalitis in immunosuppressed patients, particularly those with HIV.
Cerebral abscesses	Cerebral abscesses are often secondary to infections of adjacent structures (e.g. sinusitis), but they may also occur by hematogenous spread, or following neurosurgery or a penetrating injury. Infections are often polymicrobial with streptococci, *Staph. aureus*, anaerobes and coliforms the principal pathogens. Other causes include tuberculosis (tuberculomas) and opportunistic infections in immunosuppressed patients (e.g. *Aspergillus*). Diagnosis of cerebral abscesses is based on CT or MRI scans, and microbiological examination of aspirated pus. Treatment involves management of raised intracranial pressure, drainage of abscesses where possible, and high-dose antimicrobial chemotherapy.
Other CNS infections	Other CNS infections include various congenital infections (e.g. CMV), epidural abscesses, post-neurosurgical infections (e.g. ventriculitis, shunt infections), bacterial toxin-mediated diseases (tetanus, botulism), Guillain–Barré syndrome and transmissible spongiform encephalopathies (e.g. Creutzfeldt–Jakob disease).
Related topics	Prions (A8) Meningitis (F3)

Encephalitis is inflammation of the brain substance. It is considerably less common than meningitis. Clinical features include fever, headache, drowsiness or even coma, altered personality, confusion, fits, and localized neurological signs.

Viral encephalitis   Viral infections account for the majority of cases of encephalitis. The causes and key features of viral encephalitis are presented in *Table 1*.

Pathogenesis
There are two pathogenic mechanisms whereby viruses may give rise to damage to the brain:

(i)   *Direct invasion of the brain substance.* Viruses may reach the brain via a number of routes. Some viruses travel within nerves (e.g. herpes simplex and rabies viruses); others spread from adjacent structures (e.g. mumps virus may spread from the meninges into the brain substance, giving rise to a meningoencephalitis); finally viruses may be carried into the brain from

Table 1.   *Causes of viral encephalitis*

Virus	Comments	Pathogenesis	Outcome
Herpes simplex viruses	Commonest cause of sporadic viral encephalitis. Any age, including newborn (neonatal herpes)	Direct invasion	High morbidity and mortality
Mumps virus	Meningoencephalitis	Spread from meninges	Most patients recover. Nerve deafness may result
Measles virus	(i) Acute encephalitis 1 in 1000–5000 cases measles	Post-infectious	High morbidity and mortality
	(ii) Subacute sclerosing pan-encephalitis. Presents at age 10–15 1 in 10^6 cases measles	Direct invasion	High mortality
Varicella-zoster virus	(i) With chickenpox (rare)	Post-infectious	
	(ii) With shingles	Probably direct invasion	
Cytomegalovirus	Only in immunosuppressed	Direct invasion	High mortality
Enteroviruses	Usually in immunocompromised host	Direct invasion	High morbidity and mortality
Influenza viruses		Post-infectious	Usually good prognosis
Human immunodeficiency virus		Direct invasion	Dementia is main feature
JC virus	Only in immunosuppressed	Direct invasion	Results in PMLE
Rabies virus	Acquired from animal or bat bite	Direct invasion	Universally fatal
Japanese encephalitis virus	Arthropod-borne, geographically restricted	Direct invasion	
West Nile virus	Arthropod-borne, geographically restricted	Direct invasion	Higher mortality with ↑ age, immunosuppression
Other encephalitis viruses, eg tick-borne, St Louis encephalitis	Arthropod-borne, geographically restricted		

PMLE = progressive multifocal leukoencephalitis.

the bloodstream (e.g. human immunodeficiency virus, carried within CD4-positive T lymphocytes).

(ii) *Induction of an immune response* to infection which cross-reacts with brain antigens, thereby causing inflammation of the brain substance. This mechanism is referred to as **'post-infectious' encephalitis**. The commonest example of this is acute measles encephalitis, which occurs ~ 1–2 weeks after onset of the rash in ~ 1 in 3000 cases of measles. This timing coincides with the production of antimeasles virus antibodies, which appear to be the damaging factor – the virus itself is not present within the brain.

Some viruses utilize both pathogenetic mechanisms – a rare complication of measles is **subacute sclerosing panencephalitis (SSPE)**, which presents years after the initial measles virus infection, and is due to viral infection of the brain substance.

Diagnosis
The precise diagnosis of viral encephalitis can be difficult. Possible clues must be sought in the patient history (e.g. age, past immunizations, travel abroad, animal exposure). Investigations of a patient suspected of having viral encephalitis should include:

(i) **Neuroimaging** (e.g. by CT or MRI scan). This will allow distinction between a diffuse (across all the brain substance) or a focal (one or two foci, or patches, of inflammation) encephalitis. *Herpes simplex* encephalitis (HSE) is invariably focal.

(ii) **Lumbar puncture**. There may be a cellular response in the cerebrospinal fluid (CSF), but this is fairly nonspecific. However, CSF is invaluable in allowing detection of viral genomes (e.g. by PCR).

Treatment
A diagnosis of viral encephalitis is a medical emergency. High-dose intravenous aciclovir must be given as soon as this possibility enters the differential diagnosis, in case the patient has HSE. If subsequent investigations rule out HSE, then the aciclovir can be stopped, but any delay in starting aciclovir in a patient who does have HSE will result in increased brain damage. The management of viral encephalitis other than HSE is expectant, as there are no effective antiviral drugs.

Prevention
The mainstays of prevention are:

(i) *Vaccination*. There are safe and effective vaccines to prevent measles and mumps infection. It is precisely because of the central nervous system complications of these two virus infections that universal vaccination in childhood is recommended. Rabies is also a vaccine-preventable disease. Vaccine can be given pre-exposure to those at occupational or travel risk, or post-exposure in the event of an animal bite in an unimmunized individual.

(ii) *Avoidance of mosquito bites*. *Table 1* lists a number of encephalitis-inducing viruses that are arthropod-borne. Simple precautions (e.g. long sleeves, long trousers) may significantly reduce the risk of infection.

Nonviral causes of encephalitis

Pyogenic bacteria can cause encephalitis as a rare complication of meningitis, or during the development of cerebral (brain) abscess (see below). Encephalitis may occasionally be caused by *Mycoplasma pneumoniae*, *Coxiella burnettii* (Q-fever) and rickettsial infections (Topic C17).

Protozoal infections can cause encephalitis, including *Plasmodium falciparum* (falciparum malaria) and trypanosomes (African sleeping sickness) (Topic D4). Toxoplasmosis (Topic D4) is an important cause of protozoal encephalitis in immunosuppressed patients, particularly those with HIV.

Cerebral abscesses

Abscesses within the skull may be intracerebral (brain abscess), extradural or subdural (empyemas). Abscesses can be single or multiple. These infections are often secondary to infections of adjacent structures (e.g. ear infections, sinusitis, periorbital cellulitis), but they may also occur by hematogenous spread from a distant focus (e.g. endocarditis), or following neurosurgery or a penetrating injury.

Infections are often **polymicrobial**. Streptococci (particularly the *Streptococcus* 'milleri' group and anaerobic streptococci) are frequently associated with abscesses secondary to frontal sinusitis, or following hematogenous spread. Anaerobes (e.g. *Bacteroides*) and coliforms (e.g. *Proteus*) are particularly associated with abscesses secondary to chronic ear infections. *Staph. aureus* is the commonest cause of extradural or subdural empyemas, but other organisms are also found. Tuberculosis can sometimes cause cerebral lesions (tuberculomas). Opportunistic infections can occur in immunosuppressed patients, including reactivation of toxoplasma cysts resulting in multiple cerebral lesions, and fungal abscesses (e.g. *Aspergillus*).

Diagnosis of cerebral abscesses is usually based on CT or MRI scans. Microbiological diagnosis is from culture of aspirated pus and blood cultures.

Treatment involves management of raised intracranial pressure, drainage of abscesses where possible, and high-dose antimicrobial chemotherapy, preferably based on the culture results. Infections at adjacent sites must also be managed, and frequently management involves a collaborative approach between neurosurgeon, ENT surgeon and microbiologist. A combination of third generation cephalosporin (e.g. ceftriaxone) plus metronidazole is often employed as initial antibiotic therapy, and then adjusted according to culture results.

Other CNS infections

Congenital CNS infections
A number of congenital and perinatal infections can affect the CNS including rubella, CMV, HSV, syphilis, *Listeria* and group B hemolytic streptococci (Topic F15).

Epidural abscess
An epidural abscess can occur as a consequence of hematogenous spread (e.g. secondary to bacteremia) (Topic F12), as a complication of vertebral osteomyelitis (Topic F2), or occasionally following an epidural anesthetic. The commonest organisms are *Staph. aureus*, coliforms, *Pseudomonads* and *Mycobacterium tuberculosis*. An epidural abscess may cause irreversible damage to the spinal cord, with paralysis and bladder/bowel dysfunction and constitutes a surgical emergency.

Infections complicating neurosurgical procedures
Post-operative meningitis or ventriculitis is a recognized complication of neurosurgical procedures. Ventriculitis is particularly associated with the use of

external ventricular drains. A wide variety of bacteria can be found, including both conventional pathogens (*Staph. aureus*, coliforms) and opportunistic pathogens (*Staph. epidermidis*, diphtheroids). Opportunistic skin bacteria are also the commonest cause of ventriculo-peritoneal shunt infections (in patients with hydrocephalus). The diagnosis and management of these neurosurgical infections is a specialized topic.

Tetanus

Tetanus is a neuromuscular disorder caused by the exotoxin (**tetanospasmin**) of *Clostridium tetani*. *C.tetani* spores may be found in soil or other organic matter, and may contaminate certain types of wound or injury, with local multiplication of *C.tetani* and systemic absorption of the toxin. The effect of the toxin on the CNS results in repeated spasms of skeletal muscles, which are triggered by a variety of stimuli, and also a disturbance of the autonomic nervous system. Treatment is with **tetanus antitoxin**, as well as debridement of the wound and antibiotic therapy. Despite treatment, mortality is still 10–25% in severe cases. Prevention is by **immunization**, together with appropriate management of tetanus-prone wounds (i.e. wound debridement, antibiotics, a tetanus booster and passive immunoprophylaxis in nonimmune patients).

Botulism

Botulism is another neuromuscular disease caused by the toxins of *Clostridium botulinum*. Three forms of botulism are recognized:

(i)   **food-borne botulism** – following ingestion of pre-formed toxin;
(ii) **wound botulism** – absorption of toxin following multiplication of *C.botulinum* in a contaminated wound;
(iii) **infant botulism** – absorption of toxin following colonization of the gastro-intestinal tract with vegetative *C.botulinum*.

Botulism causes a **descending paralysis**, affecting first the cranial nerves and then skeletal muscles, often resulting in respiratory failure. Treatment is with specific **antitoxin** and supportive intensive care (assisted ventilation). Botulinum toxins are relatively heat-sensitive and appropriate processing (i.e. adequate heat and avoidance of cross-contamination) of canned or bottled food is important in preventing food-borne botulism. Home preservation of vegetables, and eating certain types of raw fish continue to be risk factors.

Guillain–Barré syndrome (GBS)

GBS is an uncommon post-infectious syndrome, possibly of an immunological nature, that results in a bilateral, symmetrical ascending paralysis and also sensory loss. A variety of infections trigger this disease including viruses and bacteria, notably *Campylobacter* and *Mycoplasma pneumoniae*. GBS can also occur rarely following immunization. There is no specific treatment. Supportive measures may include assisted ventilation, but the prognosis is generally favorable.

Transmissible spongiform encephalopathies (TSE)

Prion diseases of humans (Creutzfeldt–Jakob disease, Kuru, see *Table 1* in Topic A8) mostly affect the CNS. Histologically, the most striking feature is vacuolation of the brain substance, giving rise to a spongiform appearance. Clinical presentation is with dementia (global loss of brain function) and motor incoordination. Once symptoms appear, these diseases are relentlessly progressive resulting in death. Diagnosis is by demonstration of spongiform change and detection of abnormal prion protein in brain biopsy or post-mortem material. There is no effective therapy. Prevention is focussed on the possibilities of secondary spread (e.g. through contaminated neurosurgical instruments, cornea or dura mater grafts, or blood donation).

F5 EYE INFECTIONS

Key Notes

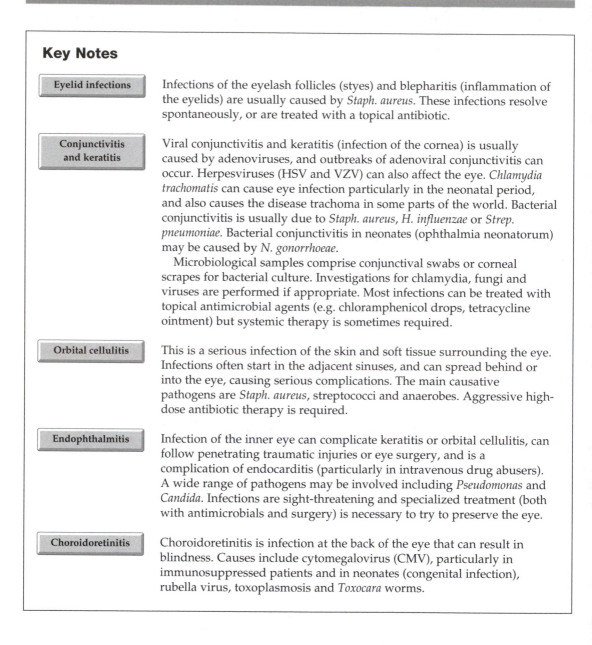

Eyelid infections

Infections of the eyelash follicles (styes) and blepharitis (inflammation of the eyelids) are usually caused by *Staph. aureus*. These infections resolve spontaneously, or are treated with a topical antibiotic.

Conjunctivitis and keratitis

Viral conjunctivitis and keratitis (infection of the cornea) is usually caused by adenoviruses, and outbreaks of adenoviral conjunctivitis can occur. Herpesviruses (HSV and VZV) can also affect the eye. *Chlamydia trachomatis* can cause eye infection particularly in the neonatal period, and also causes the disease trachoma in some parts of the world. Bacterial conjunctivitis is usually due to *Staph. aureus, H. influenzae* or *Strep. pneumoniae*. Bacterial conjunctivitis in neonates (ophthalmia neonatorum) may be caused by *N. gonorrhoeae*.

Microbiological samples comprise conjunctival swabs or corneal scrapes for bacterial culture. Investigations for chlamydia, fungi and viruses are performed if appropriate. Most infections can be treated with topical antimicrobial agents (e.g. chloramphenicol drops, tetracycline ointment) but systemic therapy is sometimes required.

Orbital cellulitis

This is a serious infection of the skin and soft tissue surrounding the eye. Infections often start in the adjacent sinuses, and can spread behind or into the eye, causing serious complications. The main causative pathogens are *Staph. aureus*, streptococci and anaerobes. Aggressive high-dose antibiotic therapy is required.

Endophthalmitis

Infection of the inner eye can complicate keratitis or orbital cellulitis, can follow penetrating traumatic injuries or eye surgery, and is a complication of endocarditis (particularly in intravenous drug abusers). A wide range of pathogens may be involved including *Pseudomonas* and *Candida*. Infections are sight-threatening and specialized treatment (both with antimicrobials and surgery) is necessary to try to preserve the eye.

Choroidoretinitis

Choroidoretinitis is infection at the back of the eye that can result in blindness. Causes include cytomegalovirus (CMV), particularly in immunosuppressed patients and in neonates (congenital infection), rubella virus, toxoplasmosis and *Toxocara* worms.

Infections of the eye include eyelid infections, conjunctivitis, keratitis, orbital cellulitis, endophthalmitis and choroidoretinitis.

Eyelid infections

'Styes' are infections of the eyelash follicles, and blepharitis is inflammation of the eyelids. The main causative pathogen is *Staphylococcus aureus*. These infections may resolve spontaneously, or respond to topical antibiotic treatment.

Conjunctivitis and keratitis

Viral conjunctivitis is usually due to adenoviruses (Topic B12). Outbreaks of infection may occur in closed communities. *Herpes simplex* can cause both conjunctivitis and keratitis (**infection of the cornea**) (Topic B5). Shingles (varicella-zoster virus) can affect the trigeminal nerve and also cause conjunctivitis and keratitis (see Topic B6).

Chlamydia trachomatis is an important pathogen of the eye (see Topic C17). Neonatal conjunctivitis can occur 4–7 days after birth in babies whose mothers have genital tract infection. In adults with *C. trachomatis* genital infection, auto-inoculation (e.g. by finger) of the eye can lead to conjunctivitis. In some tropical countries, repeated eye infections with certain serovars of *C. trachomatis* cause a chronic follicular keratoconjunctivitis. This disease is known as **trachoma** and is associated with over-crowded, unhygienic conditions. Trachoma is an important cause of corneal damage and blindness in some African and Middle East countries.

Bacterial conjunctivitis is usually due to *Staphylococcus aureus*, *Haemophilus influenzae* or *Streptococcus pneumoniae* but other bacteria may be involved. In neonates severe conjunctivitis occurring 1–2 days after birth (**ophthalmia neonatorum**) can be caused by *Neisseria gonorrhoeae* acquired from the mother's genital tract. *Pseudomonas aeruginosa* infection may follow eye surgery, trauma or foreign bodies. It may also complicate extended-wear contact lens use with inadequate cleaning.

Other rarer pathogens include:

- *N. meningitidis* (purulent conjunctivitis);
- *Leptospira* (conjunctivitis) – Weil's disease;
- Fungi (*Aspergillus*, *Candida*, others) – serious invasive eye infections in immunosuppressed patients or following surgery;
- *Acanthamoeba* – free-living amoebae that can cause serious eye infection in contact lens wearers (contaminated lens solutions);
- Worms (*Onchocerca volvulus* – river blindness, *Loa loa*) – found in parts of Africa.

Microbiological investigations include conjunctival swabs for bacterial culture, as well as for chlamydia, fungal and viral investigations if appropriate. Corneal scrapes may also be taken by ophthalmologists. The sensitivity of bacterial culture is highest when agar plates are inoculated directly at the patient's bedside and transported rapidly to an incubator. This is because the inoculum is small, and many of the bacteria are delicate organisms.

Bacterial conjunctivitis can usually be treated with **topical antibiotic drops** or ointment (e.g. chloramphenicol, tetracycline, fusidic acid). Systemic antibiotic therapy is required for ophthalmia neonatorum. Herpes and shingles are treated with aciclovir or other similar antiviral agent. Specialized treatment is required for infections with *Pseudomonas*, fungi and *Acanthamoeba* using appropriate antimicrobial agents.

Orbital cellulitis

Orbital cellulitis is an infection of the skin and other tissues surrounding the eye. Infections often start in the adjacent sinuses. Orbital cellulitis can spread behind or into the eye, causing serious complications (cavernous sinus thrombosis, endophthalmitis). Blood cultures and (if possible) pus should be sent for culture, and aggressive high-dose antibiotic therapy given. The main causative pathogens are *Staph. aureus*, streptococci (including the '*Strep. milleri*' group) and anaerobes. Mixed infections are common.

Endophthalmitis   Infection of the inner eye can occur as a rare complication of bacterial keratitis or orbital cellulitis. It can also follow penetrating traumatic injuries, eye surgery (e.g. cataract surgery), and is a complication of endocarditis (particularly in intravenous drug abusers). A wide range of pathogens may be involved including *Pseudomonas* and *Candida*. Infections are sight-threatening and specialized treatment (both with antimicrobials and surgery) is necessary to try to preserve the eye.

Choroidoretinitis   This is infection at the back of the eye that can result in blindness. Choroidoretinitis can be caused by cytomegalovirus (CMV) (Topic B7), particularly in **immunocompromised patients** (e.g. transplant recipients, HIV infection – Topic F14) and in neonates following **congenital infection** (Topic F15). Varicella-zoster virus (VZV) and toxoplasmosis can also cause choroidoretinitis in immunocompromised patients.

Congenital infections that can affect the retina include rubella virus and toxoplasmosis. Congenital rubella can also cause cataracts.

Toxocara is a helminthic (worm) infection that is sometimes seen in young children. The infection is usually acquired from puppies (*T. canis*) or kittens (*T. catis*) that have not been 'wormed'.

F6 UPPER RESPIRATORY TRACT INFECTIONS

Key Notes

Definitions	Upper respiratory tract infections include all infections of the respiratory tract above the level of the trachea. This includes the nose, throat, mouth, sinuses, ear, larynx and epiglottis. Infections of the trachea and below are called lower respiratory tract infections.
Common cold	A 'cold' is a self-limiting viral infection, usually due to one of the many types of rhinovirus. Other respiratory viruses, including parainfluenza, coronaviruses, RSV, influenza, adenovirus and enteroviruses, may also cause infection. The main features are a nasal discharge and sore throat. These infections can predispose to secondary bacterial infection (e.g. otitis media, lower respiratory tract infection).
Sore throat and tonsillitis	Most sore throats are due to respiratory viruses and do not require antibiotic treatment. The commonest bacterial cause is *Strep. pyogenes*. This is diagnosed by culture (or antigen detection) from a throat swab. Treatment is with penicillin for 10 days. Complications of streptococcal infection include quinsy, scarlet fever, glomerulonephritis and rheumatic fever. Glandular fever is caused by Epstein–Barr virus, but cytomegalovirus and toxoplasmosis cause similar symptoms of sore throat, malaise, and enlarged cervical lymph nodes. Other throat infections include Vincent's angina, oral candidiasis and diphtheria.
Sinusitis	Acute sinusitis is usually a bacterial infection of the sinuses causing pain and headache. The main pathogens are *Strep. pneumoniae*, *Staph. aureus* and *H. influenzae*. Treatment is with antibiotics, but surgical drainage or sinus washout is sometimes required.
Otitis media and otitis externa	Acute otitis media is a bacterial infection of the middle ear that occurs mainly in children, causing ear pain and fever. The main pathogens are *Strep. pneumoniae*, *Strep. pyogenes*, *Staph. aureus* and *H. influenzae*. Infections often resolve spontaneously and antibiotic treatment is not always required. Complications are rare but include mastoiditis, brain abscess and meningitis. Otitis externa is an infection of the external auditory canal causing an ear discharge. Treatment is with topical antimicrobial ear drops.
Acute epiglottitis and croup	Croup occurs in young children and is usually caused by respiratory viruses (parainfluenza, RSV). Bacterial epiglottitis is a more serious condition caused by capsulated *H. influenzae* type b (Hib). Routine Hib immunization has significantly reduced the incidence of this infection.
Related topics	Eye infections (F5) Lower respiratory tract infections (F7)

Definitions

Upper respiratory tract infections (URTI) include infections of the nose (rhinitis), throat (sore throat – pharyngitis, tonsillitis), sinuses (sinusitis), ear (otitis media and otitis externa), larynx (laryngitis) and epiglottis (epiglottitis). Infections below the larynx are referred to as lower respiratory tract infections (LRTI). Some URTI can also cause, or predispose patients to, LRTI.

The majority of URTI are due to **respiratory viruses** and more than one upper respiratory site may be involved. For instance the common cold can cause rhinitis, sore throat, sinusitis and laryngitis, and predispose to acute otitis media.

Common cold

The common cold (**'coryza'**) is a respiratory **viral infection**. Infections are common and occur at all ages, but the peak incidence is in children. The causative viral pathogens are:

- rhinoviruses (many different serotypes) – the commonest cause;
- parainfluenza viruses (1–4);
- coronaviruses;
- respiratory syncytial virus (RSV);
- influenza viruses (A and B);
- adenovirus;
- coxsackie and echoviruses (enteroviruses).

Clinical features include nasal discharge, sore throat, and sometimes systemic symptoms (fever and myalgia). Laryngitis may also occur. Specific investigations and treatment are not required, but patients with coryza are predisposed to developing **secondary bacterial infection** (e.g. sinusitis, otitis media, bronchitis, pneumonia).

Pharyngitis and tonsillitis

The causative pathogens of sore throat and tonsillitis are shown in *Table 1*. The **majority of sore throats are caused by viruses**. The main bacterial cause is *Streptococcus pyogenes*. Clinical features include pain, particularly on swallowing (dysphagia), fever and sometimes painful, enlarged cervical lymph nodes. Young children may have a fever and refuse to eat. On examination, the throat may appear inflamed with enlargement of the tonsils. The tonsils may be ulcerated with purulent discharge apparent on the surface. It is not usually possible to differentiate viral from bacterial infection on clinical grounds.

Vincent's infection, glandular fever, oral candidiasis (thrush) and other infections are briefly described in *Table 1*. A rare cause of serious sore throat infection, usually with marked systemic upset, is **diphtheria** (Topic C4).

Bacteriological investigations are based on examination of **throat swabs** for β-hemolytic streptococci by **culture** (although **rapid antigen tests** for the detection of group A streptococci are also available). If Vincent's infection is suspected, then **microscopy** of a Gram-stained swab may reveal **Vincent's organisms** (*Table 1*). Specialized agar is required for culture of *C. diphtheriae* if diphtheria is suspected clinically. **Viral investigations** are not usually performed on throat swabs from patients with sore throat, but they may be useful for detection of enteroviruses during outbreaks, or from patients with viral meningitis (Topic F3). Throat swabs are also taken from patients with suspected **meningococcal infection** (Topic F3), as the causative pathogenic strain of *Neisseria meningitidis* may be found as part of the **normal flora of the nasopharynx**.

For **glandular fever**, clotted blood may be taken for serological tests (the

Table 1. Organisms causing sore throat and tonsillitis

	Organism (s)	Comment
Viruses (the commonest cause of sore throat)	Rhinoviruses, parainfluenza, influenza, coronaviruses, coxsackie and echoviruses, adenoviruses	Usually cause sore throat as part of the **common cold**. Vesicular lesions may occur on the pharynx and soft palate with some coxsackie virus infections (**herpangina**)
	Epstein–Barr virus (EBV)	**Glandular fever** – sore throat, cervical lymphadenopathy, malaise
	Cytomegalovirus (CMV)	Glandular fever-like illness
	Adenoviruses (certain types)	Pharyngitis and conjunctivitis (pharyngeal–conjunctival fever)
	Herpes simplex	Primary infection can cause stomatitis and lesions within the mouth and throat
	Human immunodeficiency virus	As part of HIV seroconversion illness
Bacteria	*Streptococcus pyogenes* (group A streptococcus)	The commonest bacterial cause of **sore throat** and **tonsillitis**
	Other β-hemolytic streptococci (group C or G)	Can occasionally be recovered from throat swabs from patients with tonsillitis, but their clinical significance is still uncertain
	Borrelia vincenti and anaerobic Gram-negative bacilli (fusiforms) (Vincent's organisms)	A mixed infection that causes painful ulceration of the throat and/or gums (**Vincent's angina**)
	Corynebacterium diphtheriae	**Diphtheria**
	Other corynebacteria	Very rare causes of pharyngitis include *C.ulcerans* and *Arcanobacterium haemolyticum*
	Neisseria gonorrhoeae	Occasional cause of pharyngitis (following oral sex in patients with sexually transmitted disease)
	Treponema pallidum	Both primary and secondary syphilis can affect the throat
Fungi	*Candida albicans*	**Oral candidiasis ('thrush')** causes lesions on the tongue, mouth and throat
Protozoa	*Toxoplasma gondii*	**Toxoplasmosis** can also cause a glandular fever-like illness

Paul–Bunnell test and IgM tests for CMV, EBV and toxoplasma) and appropriate tests for HIV infection if suspected (e.g. PCR). Serological tests for streptococcal antibodies [e.g. anti-streptolysin (ASO)] may be useful for investigation of patients with streptococcal-related glomerulonephritis or rheumatic fever (see below).

Antibiotic treatment for streptococcal sore throat is with penicillin (or erythromycin) for 10 days. Oral cephalosporins may also be used. Amoxicillin (or ampicillin) should be avoided as there is a high instance of rash occurring in patients with glandular fever (which can be difficult to distinguish from a strep-tococcal sore throat on clinical grounds). Antibiotic therapy has only **minimal**

impact on the severity and duration of symptoms of streptococcal sore throat, but reduces the risk of complications.

Complications of *Strep.pyogenes* pharyngitis are:

- peritonsillar abscess ('quinsy') – may require surgical drainage;
- acute otitis media (see below);
- scarlet fever – due to an erythrogenic toxin produced by some strains;
- acute glomerulonephritis and rheumatic fever – caused by an immunological reaction to streptococcal antigens.

Prophylaxis with penicillin is given to children who have had one episode of rheumatic fever to prevent further episodes of streptococcal sore throat, reducing the risk of further attacks of rheumatic fever.

Vincent's infection is treated with either penicillin or metronidazole. Oral thrush is treated with suitable topical or systemic antifungal drugs (e.g. nystatin, fluconazole).

Sinusitis

Acute sinusitis is usually due to *Strep.pneumoniae*, *H.influenzae* and *Staph. aureus*, although many other pathogens may be involved (including *Pseudomonas*, anaerobes and fungi). The clinical features are of pain and tenderness over the affected sinus and associated headache.

Antibiotic treatment is usually given without microbiological investigation, but in some situations microscopy and culture of aspirated pus is required. Suitable antibiotics include amoxicillin–clavulanic acid combination and tetracyclines. Surgical intervention or sinus washouts may be required for complicated infections.

Otitis media and otitis externa

Acute otitis media is bacterial infection of the middle ear that usually occurs in young children. Clinical features include pain in the ear and fever, but also generalized upset, vomiting and diarrhea. The main causative organisms are *Strep. pneumoniae*, *Strep. pyogenes*, *Staph. aureus* and *H. influenzae*. Occasionally *Moraxella catarrhalis* is implicated. Microbiological investigations are not usually performed, unless the eardrum ruptures and pus is discharged or is aspirated via a myringotomy.

There is some debate regarding the routine use of antibiotics to treat acute otitis media, as many infections resolve spontaneously, and antibiotics do not significantly shorten the duration of symptoms. However, antibiotics do reduce the risk of complications (see below), although the risk of these complications is relatively low. Suitable antibiotics, if deemed appropriate, include amoxicillin (or ampicillin), amoxicillin–clavulanic acid combination, or an oral cephalosporin.

Complications of acute otitis media are chronic suppurative otitis media (CSOM), mastoiditis, brain abscess and meningitis (Topics F3 and F4). In addition to the four main pathogens (above), other bacteria may be involved including *Pseudomonas*, *Proteus* and anaerobes (*Bacteroides*).

Otitis externa is an infection of the external auditory canal producing pain and ear discharge. Causative organisms include *Staph. aureus*, *P. aeruginosa*, *Proteus* and fungi (*Aspergillus* and *Candida*). A risk factor for *Pseudomonas* infection is swimming. A swab of the ear discharge should be obtained for microbiological culture. Treatment is with ear drops containing both steroids (to reduce inflammation) and antibiotics (e.g. polymyxin). Adequate aural toilet (regular cleaning of the external auditory canal to remove exudate and debris) is also important, otherwise infection may recur. Repeated courses of topical antibiotics without adequate aural toilet may predispose to infections with antibiotic-resistant bacteria or fungi. Fungal infections are usually treated with nystatin ear drops.

In patients with diabetes or leukemia, *P. aeruginosa* can cause a condition called **malignant otitis externa**. The infection can spread rapidly to involve cartilage and bone. Treatment requires systemic antipseudomonal antibiotic therapy and sometimes surgery.

Acute epiglottitis and croup

Croup occurs in young children. It is usually due to a respiratory viral infection affecting the larynx or epiglottis (parainfluenza viruses, respiratory syncytial virus). Inflammation and swelling can narrow the airway and cause **inspiratory stridor** or 'croup'.

Bacterial epiglottitis is a more serious, potentially life-threatening infection that occurs in children, and occasionally in adults. The commonest cause is capsulated *H.influenzae* **type b**, although infections are now much less common due to routine Hib immunization of infants. Diphtheria is also a serious but rare cause of croup.

Microbiological investigation of epiglottitis is by blood culture. Swabbing the epiglottitis can lead to further swelling and complete obstruction of the airway, and should not be done unless the airway can be protected (e.g. by intubation or tracheostomy).

Some strains of *H. influenzae* produce β-lactamase and the treatment of choice for epiglottitis is with a third generation cephalosporin (e.g. intravenous cetriaxone) until culture and sensitivity results are known.

F7 LOWER RESPIRATORY TRACT INFECTIONS

Key Notes

Acute tracheobronchitis	Acute tracheobronchitis is usually a self-limiting infection caused by respiratory viruses, but secondary bacterial infection can occur, causing a cough with purulent sputum.
Infective exacerbations of chronic obstructive pulmonary disease (COPD)	Infective exacerbations of COPD are usually due to respiratory viruses, but there are also three key bacterial pathogens (*Streptococcus pneumoniae, Haemophilus influenzae* and *Moraxella catarrhalis*). Antibiotic therapy is indicated when two of the following are present: increased breathlessness, increased volume of sputum production, or increased purulence of sputum.
Pneumonia	Pneumonia may be classified as community-acquired or hospital-acquired (including ventilator-associated), each having a different microbial etiology. *Strep. pneumoniae* and 'atypical pathogens' are the commonest causes of community-acquired pneumonia. Investigations and treatment should be guided by the severity of the pneumonia. For severe pneumonia, a wide range of microbiological investigations (including culture, antigen detection and serology) and broad-spectrum empirical antibiotic therapy are employed.
Bronchiectasis and cystic fibrosis	Repeated lower respiratory tract infections cause progressive lung damage in patients with bronchiectasis and cystic fibrosis. Initially, infections are caused by *Strep. pneumoniae, H. influenzae* and *Staph. aureus,* but then *Pseudomonas aeruginosa* can permanently colonize the damaged lung and cause infective exacerbations that are difficult to treat. *Burkholderia cepacia* is another important pathogen.
Empyema	Empyema is a collection of pus within the pleural space, usually as a complication of pneumonia (particularly aspiration pneumonia). Diagnosis is by examination of aspirated pleural pus, and treatment is a combination of drainage and antibiotic therapy.
Lung abscess	Lung abscesses can follow pneumonia, aspiration of a foreign body, or occur in patients with right-sided endocarditis. Treatment is with high-dose antibiotic therapy and surgical drainage of the abscess if necessary.
Other lower respiratory tract infections	These include tuberculosis, whooping cough, pulmonary aspergillosis, *Pneumocystis carinii* pneumonia and other opportunistic infections.

Related topics	Streptococci and enterococci (C2)	Molds (D2)
	Mycobacteria (C6)	Dimorphic fungi and *Pneumocystis carinii* (D3)
	Parvobacteria (C9)	Upper respiratory tract infections (F6)
	Legionella (C11)	Infection in immunocompromised patients (F14)
	Atypical bacteria (C17)	

The **lower respiratory tract**, unlike most of the upper respiratory tract, is **normally sterile**. Microorganisms usually reach the lower respiratory tract via: (i) micro-aspiration of organisms from the nasopharynx; (ii) from inhalation of airborne particles; or (iii) occasionally via the bloodstream (hematogenous spread).

An important **defense mechanism** within the lower respiratory tract is the clearance of inhaled particles by **ciliated respiratory epithelial cells**, which can transport microorganisms upwards from the bronchioles, bronchi and trachea, and out of the lower respiratory tract. Factors which predispose to lower respiratory tract infection include **smoking** and **respiratory virus infection**, both of which can **damage** the ciliated epithelial cells and prevent the mucociliary cleaning of debris and bacteria.

Acute tracheo-bronchitis

Acute tracheobronchitis occurs in a minority of patients with a viral upper respiratory tract infection. Viral causes include parainfluenza, influenza, rhinoviruses and respiratory syncytial virus (Topics B9–11). The clinical features of tracheobronchitis are a dry cough with a feeling of chest 'tightness'. **Secondary bacterial infection** can occur with *Streptococcus pneumoniae*, *Haemophilus influenzae* and sometimes *Staphylococcus aureus*, causing the cough to become productive of purulent sputum.

Microbiological investigations are not usually indicated. Most infections resolve spontaneously, but some patients with secondary bacterial infection may require antibiotic treatment. Suitable antibiotics include amoxicillin, tetracyclines or a macrolide.

Infective exacerbation of chronic obstructive pulmonary disease (COPD)

COPD is a chronic condition of the lungs characterized by a productive cough and breathlessness. This is usually related to long-term cigarette smoking, which damages the normal lower respiratory tract defense mechanisms, and the damaged lungs can become permanently colonized by potentially pathogenic bacteria including *Strep. pneumoniae* and *H. influenzae*.

Infective exacerbations of COPD are common, and a frequent cause of emergency medical admission to hospitals. Patients develop increasing cough and breathlessness. Many of these infective exacerbations are due to viral infection (rhinovirus, parainfluenza, influenza, respiratory syncytial virus, and coronavirus). Bacterial infection is more likely when there is an increase in the volume and purulence of sputum.

The three main bacterial causes of infective COPD exacerbations are:

- *Strep. pneumoniae* (Topic C2);
- *H. influenzae* (Topic C9);
- *Moraxella catarrhalis*.

Staph. aureus and *Pseudomonas aeruginosa* are infrequent but recognized causes.

Most patients with infective exacerbations of COPD do not require microbiological investigation. Bacteriological examination (microscopy, culture and sensitivity) of sputum should, however, be considered when there is either an **increase in volume or purulence of sputum**, especially when patients have failed previous therapy, or are admitted to hospital. Sputum samples should be collected before the start of antibiotic therapy.

The interpretation of positive culture results is complicated by the knowledge that the three main pathogens (see above) can colonize the lower respiratory tract in COPD patients. Sputum specimens taken when the patient is relatively well can yield the same result as those taken during an infective exacerbation, leading to potentially false-positive results.

As many infective exacerbations of COPD are due to viral pathogens, antibiotic therapy is often not indicated. Antibiotic therapy should be considered when at least two of the following clinical features are present:

- increased breathlessness;
- increased volume of sputum production;
- increased purulence of sputum.

Suitable first-line antibiotics include amoxicillin, tetracyclines (e.g. doxycycline) and trimethoprim. Alternative second-line agents (for nonresponse or resistant isolates) are amoxicillin–clavulanic acid combination, fluoroquinolones and oral cephalosporins (e.g. cefaclor, cefixime). The use of macrolides (e.g. erythromycin, clarithromycin) is debated due to uncertainty about activity against *H. influenzae*. Between 10 and 20% of isolates of *H. influenzae* produce β-lactamase and are thus amoxicillin-resistant. *M. catarrhalis* is intrinsically resistant to trimethoprim, and > 50% of strains are also amoxicillin-resistant. In the past, tetracycline resistance rates in all three main COPD pathogens were much higher than currently observed (<5%).

Prevention of infective exacerbations of COPD is through smoking cessation. Annual immunization against influenza is recommended. For some patients with repeated infective exacerbations, prophylactic antibiotics may be used to prevent infections during the winter months.

Pneumonia

Classification
Pneumonia can be classified in several ways. The classification shown in *Table 1* is useful from a microbiological perspective, as there are distinct patterns of organisms associated with each type of pneumonia. **Community-acquired pneumonia** (CAP) is defined as pneumonia presenting within the community, or within 48 h of hospital admission. **Hospital-acquired pneumonia** (HAP) is pneumonia presenting > 48 h after admission to hospital, or within 10 days of hospital discharge. A subset of HAP is **ventilator-associated pneumonia** (VAP) that can occur in patients who are artificially ventilated on intensive care units. **Aspiration pneumonia** can occur in either the community or hospital. Risk factors for aspiration include impaired consciousness (e.g. excess alcohol intake) or impaired swallowing mechanisms (e.g. after a 'stroke').

Epidemiology, etiology, risk factors and pathology
Pneumonia is a relatively common condition, both in the community setting and in hospitals, and is associated with significant morbidity and mortality. There is a wide range of causative organisms (*Table 1*). The incidence of community-acquired pneumonia increases during winter months. One of the risk factors for

Table 1. Classification of pneumonia and microbial etiology

Type of pneumonia	Organism group	Organism(s)	Comments
Community-acquired pneumonia (CAP)	Bacteria	*Streptococcus pneumoniae*	Commonest cause of CAP
		Haemophilus influenzae	Relatively rare (in contrast to COPD exacerbations)
		Staphylococcus aureus	Mainly seen during influenza epidemics
		Klebsiella	Rare cause of severe CAP
	Atypical pathogens	*Legionella*	Legionnaires' disease accounts for 5–10% of CAP
		Mycoplasma pneumoniae	Epidemic seasons occur every 4–5 years
		Coxiella burnettii	Q-fever. Rare cause of CAP
		Chlamydia	*C. psittaci* (psittacosis) is a rare cause of CAP. Role of *C. pneumoniae* in CAP is debated
	Viruses	Influenza, adenovirus, SARS CoV, other respiratory viruses	Seasonal variation. Epidemics may occur in closed communities
Hospital-acquired pneumonia (HAP) (nonventilator associated)	Bacteria	*Streptococcus pneumoniae*	
		Haemophilus influenzae	
		Staphylococcus aureus	MRSA accounts for 30–50% in some hospitals
		Klebsiella and other coliforms (*E.coli, Enterobacter* etc.)	Gram-negative pneumonia is significantly more common in
		Pseudomonas spp.	hospital than the community
	Atypical pathogens	*Legionella*	15–20% of Legionnaires' disease is hospital-acquired
	Viruses	Influenza, SARS	Influenza outbreaks may occur in elderly care units
Ventilator-associated pneumonia (VAP)	Bacteria	*Staphylococcus aureus*	A wide range of bacteria may
		Klebsiella and other coliforms (*E. coli, Enterobacter* etc.)	cause VAP in intensive care units. Many of these organisms are, or
		Pseudomonas aeruginosa and other *Pseudomonas* species	can become, resistant to multiple antibiotics
		Acinetobacter	
	Atypical pathogens	*Legionella*	Rare cause of VAP
Aspiration pneumonia	Bacteria	Streptococci, coliforms anaerobes	Aspiration of normal upper respiratory tract flora

COPD, chronic obstructive pulmonary disease; SARS CoV, severe acute respiratory syndrome coronavirus.

developing bacterial pneumonia is a preceding viral respiratory tract infection. Pneumonia is generally more common in very young or elderly patients, those who are debilitated through other illnesses, conditions associated with aspiration, and in immunosuppressed patients.

The **commonest cause of CAP** is *Strep.pneumoniae* – the 'pneumococcus' (Topic C2). The term **'atypical pathogen'** refers to a group of unrelated organisms that are also important CAP pathogens. Atypical pathogens comprise *Mycoplasma pneumoniae, Legionella, Chlamydia psittaci, C. pneumoniae* **and**

Coxiella burnetii (Topics C11, C17). These intracellular organisms are not readily detected by conventional bacterial culture techniques.

The pathological hallmark of pneumonia is **consolidation** of the lungs. This is where the normal air spaces within the bronchioles and alveoli are replaced by an inflammatory exudate containing organisms, inflammatory cells and fluid.

Clinical features

Patients with pneumonia may have a variety of clinical features including fever, cough, breathlessness, pleuritic chest pain (pain related to breathing) and signs of consolidation on physical examination. Other features such as confusion and diarrhea may occur. Patients may also develop a pleural effusion. Within hospital, diagnosis is usually confirmed by the radiological appearance of consolidation on a chest X-ray.

Microbiological investigations

Microbiological investigation of patients with pneumonia depends on the type of pneumonia, the likely spectrum of pathogens and the severity of infection. A wide range of diagnostic tests is available including conventional bacterial cultures, antigen detection tests and antibody tests (*Table 2*). However, even when all available tests are employed, the etiological cause of pneumonia in an individual often remains unknown.

Antibiotic therapy

Pneumonia is an indication for antibiotic treatment. Most infections are treated with an empirical antibiotic regimen, and often the etiological cause is never established. It should be remembered that atypical pathogens generally do not respond to the cell-wall-active β-lactam antibiotics such as amoxicillin or cephalosporins.

For community treatment of pneumonia, suitable first-line antibiotics include amoxicillin, macrolides or tetracyclines. For hospitalized patients with CAP, a combination of amoxicillin and a macrolide together will cover both typical and atypical pathogens. For severe CAP, high-dose broad-spectrum antibiotics (e.g. second or third generation cephalosporin plus macrolide) are usually given intravenously. An alternative to the above antibiotic combinations is the new fluoroquinolones, which have good activity against both typical and atypical pathogens.

For HAP, empirical treatment for atypical pathogens is usually not required, although the local risk of hospital-acquired Legionnaires' disease should be assessed. Suitable antibiotics include β-lactams (amoxicillin–clavulanic acid combination, second or third generation cephalosporins) or fluoroquinolones. For VAP, empiric treatment should be based on local guidelines that are regularly updated and take full account of the microbiological epidemiology of the particular intensive care unit. Antibiotic therapy should be reviewed with the results of microbiological investigations. Multiresistant Gram-negative bacteria may require treatment with broad-spectrum agents (e.g. piptazobactam, carbapenems or aminoglycosides).

Aspiration pneumonia is usually polymicrobial and may include anaerobic bacteria. Suitable antibiotic therapy includes amoxicillin in combination with metronidazole, or amoxicillin–clavulanic acid combination. Broader-spectrum regimens may be required for hospitalized patients where Gram-negative bacilli or MRSA may also be implicated.

Table 2.   Microbiological investigations for pneumonia

Type of investigation	Specimens	Comments
Conventional bacteriology tests (microscopy, culture and sensitivity tests)	Blood cultures	Low sensitivity but high specificity and an important prognostic marker. Can detect *Strep.pneumoniae*, *H. influenzae*, *Staph. aureus*, coliforms, and *Pseudomonas*, but not viruses or atypical organisms
	Sputum (or other respiratory tract specimen)	Higher sensitivity but less specific than blood cultures
	Pleural aspirate	If pleural effusion or empyema is present
Antigen tests	Urine	*Legionella* and *Strep. pneumoniae* antigen tests available. Good sensitivity and specificity but relatively expensive. Can provide rapid results
	Sputum or other respiratory tract specimen	Direct immunofluorescence (DIF) is available for respiratory viruses, *Chlamydia*, *Mycoplasma pneumoniae* and *Legionella pneumophila* in some laboratories
Antibody tests	Paired serum samples (acute and convalescent)	A variety of serological tests are available to detect antibody rises to viruses (influenza, RSV, pararinfluenza, adenovirus) and atypical pathogens (*Legionella*, *Mycoplasma pneumoniae*, *Chlamydia* and *Coxiella burnetii*)
Specialized culture tests for *Legionella*	Sputum (or other respiratory tract specimen)	Specialized culture techniques and agar are required for *Legionella* which take 3–10 days to grow in the laboratory
Molecular tests (e.g. PCR)	Sputum (or other respiratory tract specimen), urine and blood	Under development for providing a rapid diagnostic service for atypical pathogens

Prevention

Annual immunization against influenza in elderly patients and others with underlying medical conditions can prevent influenza-associated bacterial pneumonia. Pneumococcal immunization is also indicated for patients with pre-existing respiratory disease, and reduces the incidence of pneumococcal pneumonia. Immunization should be supported by antismoking measures.

Within hospitals, various measures can be taken to reduce the likelihood of post-operative or ventilator-associated pneumonia (e.g. adequate analgesia, respiratory physiotherapy, good infection control practise when handling respiratory equipment).

For the prevention of Legionnaires' disease, see Topic C11.

Bronchiectasis and cystic fibrosis

Bronchiectasis is a condition characterized by a cycle of frequent lower respiratory tract infections resulting in progressive destruction of normal lung structure, rendering the patient increasingly susceptible to further infections. **Cystic fibrosis** (CF) is a congenital disease involving the production of highly **viscous secretions** in the lung and elsewhere. This predisposes to recurrent respiratory infections, which cause progressive lung damage.

The microbiological causes of lung infections in these two diseases are similar. *Strep. pneumoniae, H. influenzae* and *Staph. aureus* are commonly found in the early stages, but soon the lungs become infected, and then permanently colonized with one or more strains of *Pseudomonas aeruginosa*. Often these strains are highly **mucoid** when isolated from sputum specimens in the laboratory, and the detection of such mucoid strains may be an important pointer to the underlying diagnosis. Another pathogen of particular concern in cystic fibrosis patients is *Burkholderia cepacia* (a pseudomonas-like pathogen that is often multi-antibiotic resistant). *Aspergillus* is another potential pathogen that can colonize the lungs of these patients.

Diagnosis of an infective exacerbation is usually made on clinical grounds with increasing shortness of breath and cough productive of increasingly purulent sputum. Sputum cultures may give a guide to therapy, but it is more important to analyze the pattern of positive culture and sensitivity results of specimens taken over time, rather than an individual result. This is because, particularly with *Pseudomonas*, there are often several strains in the lungs, and not all are identified from each sputum specimen. There is also considerable variation in the reproducibility of *in vitro* sensitivity tests with mucoid strains of *Pseudomonas*. The detection of new pathogens, particularly the **first detection of *P. aeruginosa* and *B. cepacia*** is, however, of considerable **clinical importance**. Some strains of *B. cepacia* are associated with a very rapid decline in respiratory function, and may also be spread readily from one CF patient to another unless segregation precautions are taken.

Treatment of infective exacerbations is with 10–14 days of high-dose antibiotic therapy. This often involves combination treatment with intravenous **antipseudomonal antibiotics** (e.g. ceftazidime plus tobramycin). Careful monitoring of aminoglycoside blood levels is required. Some strains of *P. aeruginosa*, and most strains of *B. cepacia* are **multiresistant**, and the choice of antibiotics can be very limited.

The frequency of infective exacerbations and progression of lung damage can be reduced by a multifactorial approach and these patients are best managed by specialist multiprofessional teams. **Colistin**, administered regularly by nebulizer, is useful in preventing the frequency of *Pseudomonas* infective exacerbations. Some patients benefit from regular intensive antibiotic treatment courses, without waiting for clinically obvious infective exacerbations to occur.

Outbreaks of *B. cepacia* and multiresistant *P. aeruginosa* are known to occur, particularly when CF patients have close interaction with each other. Most specialized units now **segregate patients** according to their carrier status, both within outpatient clinics and as hospital inpatients.

Empyema

An **empyema** is a collection of **pus** within the **pleural space**. This may be a complication of pneumonia (particularly pneumococcal or aspiration pneumonia), but sometimes no preceding respiratory infection is reported. Sometimes empyema may develop as an infective complication of the insertion of a chest drain for drainage of a pleural effusion or pneumothorax.

The commonest causative bacteria are *Strep. pneumoniae*, the '*Strep. milleri*' group and anaerobic streptococci, *Bacteroides*, *Staph. aureus* and coliforms (particularly *Klebsiella*).

Microbiological diagnosis is made by culture of aspirated pleural pus. Pneumococcal antigen detection may also be used. Blood cultures should also be taken from patients with suspected empyema.

Treatment is by combination of appropriate antibiotic therapy and adequate drainage of the empyema collection. Drainage may be attempted by placement of a chest drain, but surgery is often needed for more extensive or chronic empyema.

Lung abscess

Like empyema, lung abscesses may complicate some types of pneumonia or may follow aspiration or inhalation of a foreign body. The microbiology of lung abscesses is very similar to that of empyema (see above). Tuberculosis is also an important cause of lung cavitation and sometimes abscess formation. Lung abscesses may also occur in patients with tricuspid valve endocarditis (Topic F13).

Microbiological diagnosis is best made from pus aspirated directly from the abscess (obtained by either percutaneous aspiration or by bronchoscopy). Sputum cultures can sometimes give a clue to the etiology. Antibiotic treatment may have to be empirical. Treatment is with several weeks of high-dose antibiotics, and adequate drainage or resection of the abscess may be needed.

Other lower respiratory tract infections

- tuberculosis – Topic C6;
- whooping cough – Topic C9;
- aspergillosis – Topic D2;
- *Pneumocystis carinii* pneumonia – Topic D3;
- opportunistic infections – Topic F14.

F8 GASTROENTERITIS AND FOOD-POISONING

Key Notes

Definitions	Gastroenteritis is the term used to describe acute vomiting and/or diarrhea. Food-poisoning is a food- or water-borne illness causing gastroenteritis or other illness.
Microbiology	The lower gastrointestinal tract is normally colonized by a mixture of microorganisms. A wide range of viruses, bacteria and protozoa can, however, cause gastroenteritis or food-poisoning. Principal bacterial pathogens in the developed world are *Salmonella*, *Shigella*, *Campylobacter*, *Escherichia coli* O157 and *Clostridium difficile*. Other bacteria capable of causing food-poisoning are *C. perfringens*, *Staph. aureus* and *Bacillus cereus*.
Pathogenesis	Transmission of enteric pathogens is principally by fecal–oral spread, from either human (e.g. *Shigella*) or animal (e.g. *Campylobacter*) reservoirs. The infectious dose is lower for some pathogens (e.g. *E.coli* O157) than for others. The acid pH of the stomach is a natural defense mechanism, and drugs that counter gastric acid predispose to gastroenteritis. Symptoms of vomiting and diarrhea are caused by direct invasion of enterocytes, or by the production of enterotoxins. Antibiotics disrupt the normal bowel flora and antibiotic use is the main risk for *C. difficile* infection.
Epidemiology	These are very common infections, especially in children, and particularly in developing countries, where there may be poor sanitation or contaminated water supplies. Foreign travel is a particular risk factor. Some infections show a seasonal variation (e.g. rotaviral gastroenteritis). Outbreaks can occur, either in closed settings (e.g. hospital wards) or on a wider scale (e.g. with food- or water-borne illness).
Clinical features	The principal features are vomiting and diarrhea. Bloody diarrhea indicates a more severe colitis. Systemic upset (e.g. with fever) may occur. Fluid loss can cause dehydration and hypovolemic shock.
Diagnosis	Microbiological diagnosis is only made in a minority of infections, most of which are self-limiting. Diagnosis can be made by microscopy of stool samples (e.g. for protozoa), culture on selective media for specific bacterial pathogens, detection of toxins (e.g. *C. difficile*), or detection of antigen (e.g. rotavirus).
Treatment	Most infections are short-lived, but maintenance of adequate fluid and salt intake is important. Severe fluid loss may require admission to hospital for fluid resuscitation. Antimicrobial drug therapy is indicated only for a minority of infections.

Prevention	General hygiene and public health measures are important in preventing gastrointestinal infections. Preventive measures can be taken to reduce the risk of traveller's diarrhea (e.g. drinking bottled water). Food-poisoning can be avoided by proper handling, cooking and storage of food. Outbreaks of gastroenteritis or food-poisoning should be investigated, and appropriate control measures implemented.
Other gastrointestinal, food- and water-borne infections	These include *Helicobacter pylori*, typhoid fever, listeriosis, toxoplasmosis, botulism, helminthic infections, hepatitis A and polio.

Related topics	Enteroviruses (B15)	Clostridia (anaerobic
	Listeria (C5)	Gram-positive bacteria) (C14)
	Enterobacteriaceae (C7)	Protozoa (D4)
	Camplyobacter and	Helminths and arthropods (D5)
	Helicobacter (C12)	

Definitions

Gastroenteritis is strictly 'inflammation of the stomach and intestine'. The term is used to describe a variety of infections that cause symptoms of **vomiting** and/or **diarrhea**. Many of these infections are transmitted by food or water, but some can also spread directly or indirectly via person-to-person transmission.

Food-poisoning is the term used to describe a variety of illnesses that follow ingestion of either food or water, including gastrointestinal **infections** and **toxin-mediated gastrointestinal illness**. Many of these infections cause gastroenteritis, but some cause other illness (e.g. listeriosis – see Topic C5). For some infections food or water acts simply as the vehicle for transmission to humans, but for others there needs to be substantial multiplication of bacteria (with or without production of enterotoxins) within food before symptoms of food poisoning can occur.

Microbiology

The gastrointestinal tract, especially the large bowel, is heavily colonized by a variety of microorganisms which make up the **normal bowel flora**. These include bacteria, yeasts and protozoa. **Anaerobic bacteria** predominate, but *Escherichia coli*, other coliforms, enterococci and pseudomonads are also found. However, a wide range of other microorganisms can be pathogenic to the gastrointestinal tract causing gastroenteritis. These include viruses, bacteria and protozoa.

Viral causes of gastroenteritis include rotavirus, adenovirus (types 40 and 41), and norovirus (small round structured virus) (Topics B12 and B14). Many different species of bacteria can cause gastroenteritis or food poisoning, and these are summarized in *Table 1*. The major bacterial pathogens in the UK comprise *Salmonella*, *Shigella*, *Campylobacter*, *E. coli* O157 and *Clostridium difficile*. Important protozoal pathogens of the gut are giardia, *Entamoeba histolytica* (amoebic dysentery) and cryptosporidia (Topic D4). Many helminth (worm) infections also involve the gastrointestinal tract (Topic D5).

Pathogenesis

The principal route of transmission of enteric pathogens is via **fecal–oral** spread. This may occur directly during close contact with a symptomatic patient (e.g. hands may become contaminated with subsequent transfer to the mouth), or,

Table 1. Bacterial causes of gastroenteritis and food-poisoning

Bacteria	Route(s) of transmission	Pathogenesis	Comments
Salmonella (excluding *S. typhi* and *S. paratyphi*)	Mainly food-borne (poultry, eggs, meats) Some secondary cases via direct contact (hospitals, nurseries, nursing homes)	Multiple factors including invasion of enterocytes	Common cause of food-poisoning. Both sporadic infections and outbreaks occur. Many different serotypes e.g. *S. enteritidis*, *S. typhimurium*
Shigella	Food or water-borne, or by direct contact with infectious cases	Invasion of enterocytes. *S. dysenteriae* also produces a powerful shiga enterotoxin	*S. sonnei* is the commonest species in the UK causing a mild gastroenteritis. *S. dysenteriae* occurs in tropical countries causing severe bloody diarrhea (dysentery)
Campylobacter	Mainly food-borne (poultry, eggs, meat). Occasionally water-borne	Invasion of enterocytes	Common cause of sporadic food-poisoning. Occasional cause of outbreaks
E.coli O157 (enterohemorrhagic *E. coli* – EHEC)*	Food-borne (meat, milk) and by direct contact with animals or infectious cases	Produces a shiga-like toxin	Rare cause of serious food-poisoning. Can cause bloody diarrhea as well as HUS (hemolytic-uremic syndrome)
Other *E.coli* strains causing diarrhea*	Food- or water-borne	Multiple factors including toxin production, enterocyte invasion or adherence	Enterotoxigenic *E.coli* (ETEC) causes a cholera-like illness – a common cause of 'Traveller's diarrhea'. Enteropathogenic *E. coli* (EPEC) may cause diarrhea in young children Enteroinvasive *E. coli* (EIEC) produces a shigella-like illness
Clostridium difficile (antibiotic-associated diarrhea)	Ingestion of spores, usually by indirect spread from an infectious case	Produces one or more enterotoxins	An important cause of antibiotic-associated diarrhea within hospitals. May cause severe diarrhea and colitis
Clostridium perfringens food-poisoning	Food-borne (meat, gravy)	Ingestion of large numbers of bacteria within food. Release of enterotoxin during sporulation of bacteria within the gut	Causes mainly diarrhea and abdominal pain
Staph. aureus food-poisoning	Food-borne	Ingestion of heat-stable enterotoxin within food	Profuse vomiting occurs 1–6 h after ingestion
Bacillus cereus food-poisoning	Food-borne (reheated rice, meat and vegetable products)	Ingestion of heat-stable enterotoxins within food	Profuse vomiting occurs 1–6 h after ingestion (rice), or diarrhea (10–12 h, meat/vegetables)

Table 1. Bacterial causes of gastroenteritis and food-poisoning (continued)

Bacteria	Route(s) of transmission	Pathogenesis	Comments
Vibrio cholerae (cholera)	Water-borne or occasionally food-borne (seafood e.g. mussels, vegetables)	Powerful enterotoxin producing rapid secretion of fluid into the gut lumen	Cholera is very rare in the UK. It usually occurs in developing countries, or in areas with poor sanitation and hygiene
Vibrio parahaemolyticus	Food-borne (usually seafood)	Uncertain	Rare cause of food-poisoning in the UK. Endemic in South East Asia
Yersinia	Food-borne (pork)		Uncommon infection in the UK. *Y. enterocolitica* causes gastroenteritis. *Y. pseudotuberculosis* causes mesenteric adenitis.
Aeromonas hydrophila	Food- or water-borne		Possible cause of gastroenteritis, especially in children
Listeria monocytogenes	Food-borne (soft cheese, packaged chilled foods, pâté, coleslaw)		Listeriosis is a systemic infection, transmitted by food (Topic C5)

*E. coli is part of the normal bowel flora in humans. However, *E. coli* comprises many different serotypes, and some of these serotypes (e.g. O157), as well as other toxigenic strains, are pathogenic to man.

more commonly, via indirect spread. Indirect transmission is usually via contaminated food or water, but also via fomites (objects) or medical equipment (within a hospital setting). Reservoirs for enteric pathogens include the gastrointestinal tract of both humans (cholera, *Shigella*) and animals (*Salmonella*, *Campylobacter*, *E. coli* O157).

The **infectious dose** varies according to the pathogen. For instance, ingestion of small numbers of salmonellae does not usually result in infection. However, salmonellae can multiply to large numbers within contaminated foods if food hygiene is poor, and the ingestion of large numbers of salmonellae is much more likely to cause symptoms. In contrast, the infectious dose of *E. coli* O157 is much lower.

There are a number of **natural defense mechanisms** against gastrointestinal infection, including the **acid pH of the stomach** and the presence of a rich normal bacterial flora in the lower gastrointestinal tract. **Risk factors** for gastroenteritis include **reduced acid secretion** within the stomach as a result of antacids or other antisecretory drugs, or alteration of the normal gut flora as a result of **antibiotics**. The latter is the main risk factor for *Clostridium difficile* infection (antibiotic-associated diarrhea) as antibiotic-induced alteration of the normal bowel flora allows toxigenic strains of *C. difficile* to multiply to large numbers, if *C. difficile* spores are ingested.

There is a variety of **pathogenic mechanisms** that cause disease, depending on the microorganism. The main pathogenic factors of the major bacterial pathogens are summarized in *Table 1*. Infections can be broadly divided into those that **invade** the cells lining the gut epithelium (**enterocytes**), and those that produce **enterotoxins**, although some bacteria do both. Enterotoxins may be produced within the host (e.g. *Shigella dysenteriae*, *Clostridium difficile*) or they may be pre-formed within food (e.g. food-poisoning due to *Bacillus cereus* or *Staphylococcus aureus*). When pre-formed enterotoxin is ingested within food, the predominant symptom is vomiting occurring within 1–6 h. The incubation period of most other bacterial infections varies from 12 to 48 h.

Many gastrointestinal pathogens remain confined to the gastrointestinal tract. However, some bacteria have the capability of invasion through the intestinal wall and into the bloodstream, causing bacteremia (Topic F12). **Bacteremia** can occur with both *Salmonella* and *Campylobacter* infections, but is not seen with *Shigella*, cholera or *E. coli* O157. *Entamoeba histolytica* (amoebic dysentery) can invade the intestinal wall and be carried via the portal vein to the liver, where a **liver abscess** may develop.

E. coli O157, whilst not causing direct systemic invasion, produces one or more toxins that can act systemically to cause serious complications including hemolytic–uremic syndrome (HUS) in children, and thrombotic thrombocytopenic purpura (TTP) in adults, which carries a high mortality.

Epidemiology

Gastroenteritis and food-poisoning are very **common infections** of worldwide importance. In developing countries they account for major morbidity and mortality, particularly amongst children. Many gastrointestinal infections are endemic in developing countries due to **poor sanitation** and **contaminated water supplies**.

In developed countries, gastroenteritis is generally more common in children and the elderly than in healthy adults, but all age groups can be affected.

Within the community setting, infections may occur sporadically, in small household clusters, or in larger **outbreaks**. The commonest types of food-borne infection are *Salmonella* and *Campylobacter*, and the incidence of these infections has increased sharply over the past 20 years. Both of these bacteria are particularly associated with poultry, and salmonellae (including *S. enteritidis* phage type 4) with eggs. Infections show a seasonal increase during the summer months, as a consequence of warmer ambient temperatures and inadequate cooking of meat and poultry (e.g. during barbeques).

Infections due to *E. coli* O157 are less common, but potentially more serious especially in children and the elderly. Hemolytic–uremic syndrome (HUS) following *E. coli* O157 gastroenteritis is, however, now one of the commonest causes of acute renal failure in young children.

Other common causes of gastroenteritis in children are viruses (particularly rotavirus, adenovirus and norovirus) and *Shigella sonnei*. These can cause outbreaks within nurseries and schools. **Rotaviral gastroenteritis** is a highly **seasonal infection** in the UK, usually peaking around February–March.

Cryptosporidiosis is a water-borne protozoal pathogen. The main reservoir is cattle. Outbreaks may occur following contamination of **drinking water** supplies. Outbreaks related to contamination of **swimming pools** have also been reported.

A significant proportion of cases of gastroenteritis occur in relation to **foreign travel**. Traveller's diarrhea is often due to enterotoxigenic strains of *E. coli*, but *Salmonella*, *Shigella* and *Campylobacter* are also important. Other travel-related infections include cholera (rare), giardiasis, cryptosporidiosis and *E. histolytica* (amoebic dysentery). The increase in both foreign travel and the widespread importation of food products from around the world means that some more 'exotic' gastrointestinal infections are being increasingly encountered in developed countries.

Gastroenteritis is also common within hospitals and nursing homes. The main bacterial pathogen within hospitals is now *Clostridium difficile*. This can become an endemic problem, as well as causing outbreaks. This organism forms spores which are capable of surviving within the hospital environment for many months. Hospital outbreaks of viral gastroenteritis (especially norovirus) are also common, and difficult to control.

Clinical features

The predominant features of gastroenteritis are **diarrhea** and/or **vomiting**. A prodromal 'flu-like illness may precede these symptoms, particularly with campylobacter infection. Abdominal pain and cramps can occur, and patients may be febrile. **Bloody diarrhea** indicates more severe **colitis** (inflammation of the colon).

Vomiting without diarrhea is particularly characteristic of **toxin-mediated food-poisoning** and the early stages of viral gastroenteritis. Profuse watery diarrhea is the hallmark of cholera. Bloody diarrhea (**dysentery**) can occur with *Shigella*, *E. coli* O157, *Salmonella*, *Campylobacter*, *C. difficile* and amoebic dysentery.

Fluid loss due to diarrhea and vomiting may lead to severe **dehydration** and **hypovolemic shock**. Other complications include bacteremia with seeding to other organs (e.g. *Salmonella* to bone – Topic F2), liver abscesses (amoebic dysentery), and renal failure (HUS due to *E. coli* O157).

Some gastrointestinal infections (e.g. *Salmonella*, *Campylobacter* and *Yersinia*) can precipitate a reactive arthritis that occurs 2–3 weeks after the infection in a minority of cases (Topic F2). A rare complication of campylobacter infection is Guillain–Barré syndrome, a serious neurological condition affecting both sensory and motor nerves.

Diagnosis

Many patients do not seek medical help and gastroenteritis is often a short-lived self-limiting disease. For those that seek medical attention, the diagnosis is often made on clinical grounds without laboratory investigation. Investigations should be performed for patients at risk of more serious infection or complications (i.e. the young, elderly or immunosuppressed), those with a recent travel history, or where an outbreak is suspected.

Microbiological investigations are principally based on examination of stool samples. Appropriate investigations should be guided by relevant clinical and epidemiological information which must be stated on the request form. Blood cultures should also be performed if bacteremia is suspected.

In cases of food-poisoning, food (or water) may also be subjected to microbiological examination.

Microscopy

Microscopy of feces can detect protozoal parasites including cysts of cryptosporidium (using a modified Ziehl–Neelsen stain or equivalent), *E. histolytica* and *Giardia*. The ova of many helminths (worms) that infect the gastrointestinal tract can also be detected (Topic D5).

Electron microscopy may be used to detect gastrointestinal viruses, but the sensitivity of this technique is poor, and this approach is being replaced by antigen or genome detection techniques (see below).

Culture

The identification of *Salmonella*, *Shigella*, *E. coli* O157, *Campylobacter* and other bacterial pathogens from stool specimens is by selective culture. A range of highly selective agar plates, together with selective liquid enrichment broths, are available for these different pathogens, which are often present in low numbers compared to the normal flora of the bowel.

Confirmation of the identity of suspected salmonella, shigella and *E.coli* O157 colonies is through **biochemical reactions** and **serotyping** (using O and H antigens). Further laboratory work includes sensitivity testing and referral of the isolates to a reference laboratory for further typing (e.g. phage typing of salmonellae).

Toxin detection

The mainstay of the diagnosis of *C. difficile*-associated diarrhea is by the detection of *C. difficile* toxin(s), using either a **cytotoxic assay** (the 'gold standard' test for toxin B), or one of a number of commercially available kits for the detection of toxins A or B (e.g. by **enzyme immunoassay**). Culture of *C. difficile* from stools can be useful during outbreaks (e.g. for subsequent typing or sensitivity testing), but this will also detect nontoxigenic strains which are not thought to cause gastrointestinal disease.

Antigen detection

Commercially available antigen detection tests are also available for a number of gastrointestinal viruses, including rotavirus, adenovirus and norovirus. Genome detection (e.g. with PCR) is also employed in some laboratories for the detection of norovirus. These techniques are more sensitive than electron microscopy of stool samples.

Treatment

Most types of gastroenteritis and food-poisoning are **self-limiting** and do not require specific antimicrobial treatment. The most important treatment for the more serious infections is to **maintain adequate fluid intake** and electrolyte balance. Life-threatening dehydration and loss of salts can occur, not only with cholera, but also with other bacterial and viral pathogens, and aggressive fluid resuscitation may be required. For mild diarrhea, antimotility drugs may help to control symptoms, but are contraindicated in some infections (e.g. *C. difficile*).

Specific antimicrobial chemotherapy can be given in certain situations, particularly for *Salmonella* and *Campylobacter* infections. Whilst most of these infections resolve spontaneously, antibiotics can possibly shorten the duration of diarrhea, but more importantly prevent the risk of metastatic complications in at-risk patients (e.g. neonates, elderly patients and immunosuppressed patients). The mainstay of therapy for both *Salmonella* and *Campylobacter* infection in adults is with **ciprofloxacin**. An alternative treatment for campylobacter is with erythromycin, particularly as resistance to ciprofloxacin is increasing.

Antibiotics are, however, **contraindicated** in patients with **E. coli O157**, as this may increase the risk of HUS or TTP. Because of this, antibiotics should only be used in gastroenteritis when (i) there is a confirmed microbiological diagnosis, and (ii) risk factors for complicated infection exist.

Prevention

General hygiene and public health measures

Many gastrointestinal infections can be prevented by the provision of clean, uncontaminated **drinking water supplies** and suitable **sanitation** systems for human waste. Lack of these facilities is associated with a high risk of outbreaks of gastrointestinal infections, particularly in the developing world or when major conflicts occur resulting in the displacement of large populations as refugees.

Appropriate **handwashing facilities**, particularly in school toilets, are particularly important in preventing outbreaks of gastroenteritis within closed communities.

Traveller's diarrhea

Various preventative measures can be taken to reduce the risk of gastroenteritis when travelling to less developed countries, for instance drinking (and cleaning teeth in) bottled water only, not eating uncooked vegetables, and peeling fruit. Prophylactic antibiotics may reduce the risk of bacterial gastroenteritis, but their use in this situation is controversial.

Food hygiene

All food should be handled and cooked correctly to avoid the risk of food-poisoning and food-borne infection. The following general principles are advocated:

- food-handlers should wash their hands before handling food, and particularly after visiting the toilet;
- people with gastroenteritis should not handle food until symptoms have resolved for ≥ 48 h;
- raw and cooked food should be stored and handled separately (including utensils or other food-processing equipment);
- food should be adequately cooked (with suitable temperature checking at critical points);
- cooked food should be consumed whilst still hot, or stored at $< 6°C$, to avoid bacterial multiplication;
- cooked food should only be reheated once;
- when reheating food, the food should be 'piping hot';
- frozen food should be thoroughly defrosted before cooking, or additional cooking time used;
- food should not be consumed after the 'use by' date.

High-risk foods include raw poultry (*Salmonella, Campylobacter*), raw shell eggs (*Salmonella*), raw mince or beef (*Salmonella, E. coli*), meat gravy and soup (*Clostridium perfringens*), reheated rice (*Bacillus cereus*), soft cheese, coleslaw and pâté (*Listeria*), and raw shellfish (norovirus, hepatitis A).

Microbiological testing of raw or cooked foods can be undertaken as part of quality assurance within the food industry, and independent checks of the microbiological quality of food products at the point of sale are often undertaken by Environmental Health Officers.

Secondary prevention

'Food-poisoning' is a **Notifiable Disease**, and this includes all food- or water-borne infections causing gastroenteritis. Epidemiological investigation of individual infections, including a detailed food history, can reveal outbreaks associated with a particular food product or food retail outlet, and appropriate control measures can then be implemented.

Transmission of gastrointestinal infections within households or close contacts is common, but can be reduced by careful attention to hand hygiene and general cleaning of sanitary facilities.

Within the healthcare setting, patients with gastroenteritis should be placed in single rooms, with either an en-suite toilet or dedicated commode. Healthcare workers should wear protective gloves and aprons when handling feces, commodes or soiled linen, and adhere to a strict hand hygiene protocol. The control of outbreaks of diarrhea may require a restriction of movement of both patients and staff into and out of the affected area. Outbreaks of viral gastro-enteritis (particularly with norovirus) frequently involve a large number of both patients and staff, and are difficult to contain.

As well as general infection control measures (e.g. increased cleaning), outbreaks of *C. difficile*-associated diarrhea may require a restriction on the use of certain high-risk antibiotics (e.g. broad-spectrum cephalosporins, clindamycin) as part of the control measures.

Other gastrointestinal, food- and water-borne infections

Helicobacter pylori can cause a low-grade chronic inflammatory condition in the stomach, which is now known to be associated with peptic ulcer disease, gastric cancer and gastric lymphoma (Topic C12).

A large number of **helminths** (worms) can be found in the gastrointestinal tract (e.g. hookworm, tapeworm, pinworm, roundworm – see Topic D5).

Whilst not primarily gastrointestinal pathogens, a number of important infections are spread via the fecal–oral route including:

- **hepatitis A and E** (Topic B2);
- enteroviruses, including **polio** (Topic B15);
- **typhoid** and **paratyphoid.**

Typhoid fever and **paratyphoid fever** are systemic diseases caused by *Salmonella typhi* or *Salmonella paratyphi A, B* or *C* respectively. Most infections occur in the developing world, and are related to travel. Humans (chronic carriers) are the main reservoir, in contrast to other salmonellae. After an incubation period (usually 1–3 weeks), clinical features include fever, headache, malaise, 'rose spots' on the trunk, an enlarged spleen and cough. Intestinal complications (hemorrhage, perforation) can occur. Diagnosis in the acute stages is best confirmed by isolation of *S. typhi* or *S. paratyphi* from blood cultures. Prompt antibiotic treatment (e.g. with ciprofloxacin) can reduce overall mortality from 10% to < 1%, but antibiotic-resistant strains are starting to emerge. Typhoid immunization is recommended for travellers to high-risk countries.

Other food-associated nongastrointestinal diseases include **listeriosis** (Topic C5), **toxoplasmosis** (Topic D4), **vCJD** (variant Creutzfeld–Jakob disease – see Topic A8), botulism (caused by *Clostridium botulinum*), and a variety of other toxin-mediated illnesses (e.g. histamine fish poisoning, paralytic shellfish poisoning).

F9 INTRA-ABDOMINAL INFECTIONS

Key Notes

Infections of the liver	Infective hepatitis is usually viral. Non-viral causes of hepatitis include *Coxiella burnettii* (Q-fever), *Leptospira* species (Weil's disease), *Brucella* (brucellosis), and *Mycobacterium tuberculosis.* Liver abscesses can be amoebic (due to *Entamoeba histolytica*) or pyogenic (usually a polymicrobial infection).
Biliary tract infections	Cholecystitis is usually secondary to gallstones, but may be complicated by infection. Empyema of the gallbladder is a serious infection where the gallbladder literally becomes a 'bag of pus'. Cholangitis is usually secondary to obstruction of the biliary tree (e.g. gallstones, malignancy). These infections usually involve coliforms, pseudomonads, enterococci and anaerobes. Treatment is with high-dose broad-spectrum antibiotics and appropriate surgical intervention.
Infections of the spleen	Splenomegaly can occur with a variety of infections including glandular fever (Epstein–Barr virus), brucellosis, visceral leishmaniasis and schistosomiasis (secondary to portal hypertension). Splenic abscesses can occur in patients with infective endocarditis or immunosuppressed patients (e.g. hepatosplenic candidiasis). Pyogenic infections of the spleen are rare.
Infections of the pancreas	Pancreatitis can be due to infection (e.g. mumps virus) as well as non-infective causes (e.g. gallstones, alcohol, and trauma). Secondary bacterial infection can occur in severe pancreatitis. Coliforms, streptococci, staphylococci, pseudomonads and anaerobes are the principal pathogens.
Peritonitis	Spontaneous (primary) bacterial peritonitis is rare, but is more frequently seen in patients with ascites. Peritonitis is more commonly a complication of other intra-abdominal pathology (e.g. appendicitis, carcinoma of the colon), and is usually a polymicrobial infection (e.g. coliforms, anaerobes). Broad-spectrum antibiotic therapy is given as an adjunct to surgery. Peritonitis is also a recognized complication in patients with renal failure on continuous ambulatory peritoneal dialysis (CAPD).
Intra-abdominal collections	Intra-abdominal collections or abscesses are usually found under the diaphragm (subphrenic) or in the pelvis. Infections are often polymicrobial (e.g. coliforms, streptococci and anaerobes) and usually occur as a complication of peritonitis, gastrointestinal or gynecological pathology or surgery. Treatment is by drainage of pus and antibiotic therapy.
Related topics	Viral hepatitis (B2–4)

This topic summarizes the following intra-abdominal infections:

- infections of the liver (non-viral hepatitis, liver abscess);
- biliary tract infections (cholecystitis, empyema of the gallbladder, cholangitis);
- infections of the spleen and pancreas;
- peritonitis;
- intra-abdominal abscesses.

Pelvic inflammatory disease is covered in Topic F11, and pyelonephritis in Topic F10.

Infections of the liver

The two main infective conditions of the liver are hepatitis (inflammation of the liver) and liver abscesses.

Hepatitis is most commonly viral (see Topics B2–4). Nonviral causes of hepatitis include:

- *Coxiella burnettii* (Q fever) (Topic C17);
- *Leptospira* spp (Weil's disease) (Topic C16);
- *Brucella* spp (brucellosis) (Topic C9);
- *Mycobacterium tuberculosis* (Topic C6);
- *Candida* (in immunosuppressed patients) (Topics D1 and F14).

These latter infections often cause a **granulomatous** hepatic inflammation. The spleen (see below) can also be involved.

The clinical features of hepatitis include fever, abdominal pain (right upper quadrant) and tenderness, abnormal liver function tests, and sometimes jaundice. Microbiological investigations include the appropriate serological tests and blood cultures. Sometimes a liver biopsy is required both for microbiological and histological examination. Treatment is directed at the underlying cause.

Liver abscesses can be divided into amoebic liver abscess and pyogenic liver abscess, although these can sometimes be hard to differentiate clinically. A rare cause of liver abscess is hydatid disease (*Echinococcus granulosus* – Topic D5).

Amoebic liver abscess is a complication of gastrointestinal tract infection with *Entamoeba histolytica* (Topics D4 and F8). Diagnosis may be made by travel history, detection of *Entamoeba* cysts in stool specimens, and specialized serological tests for amoebic antibodies in blood. Treatment is with metronidazole.

Pyogenic liver abscess is often polymicrobial. A variety of bacteria may be involved but the most common are:

- streptococci, particularly the *Streptococcus* 'milleri' group and anaerobic streptococci;
- anaerobes (e.g. *Bacteroides* or *Fusobacterium*);
- coliforms (e.g. *E. coli, Klebsiella).

Pyogenic liver abscess can arise as a complication of pathology and/or infection of the gastrointestinal tract (via spread from the portal vein), as a complication of biliary tract sepsis, or as a complication of bacteremia with infection at a distant focus. Sometimes no cause is found. Microbiological investigations include blood cultures and examination of pus (obtained by radiologically guided needle aspiration or open drainage) preferably taken before the commencement of antibiotics. Treatment is by adequate drainage of the pus, together with appropriate high-dose antibiotic therapy for several weeks. Appropriate investigations should be done to look for predisposing causes (e.g. neoplasm of the colon).

Biliary tract infections

The main infections of the biliary tree are cholecystitis, cholangitis and empyema of the gallbladder.

Acute inflammation of the gallbladder (**cholecystitis**) is usually secondary to **gallstones**. It can be either chemically induced or infective in nature. However, chemically induced cholecystitis may also be complicated by infection. Anaerobes (e.g. *Bacteroides*) and coliforms (e.g. *E. coli*) are most frequently involved. Clinical features include abdominal pain (right upper quadrant and radiating through to the back) and tenderness, fever, vomiting and abnormal liver function tests. Blood cultures should be taken before starting broad-spectrum antibiotic therapy (e.g. cephalosporin plus metronidazole). Cholecystitis may lead to empyema of the gallbladder.

Empyema of the gallbladder is a serious life-threatening infection where the gallbladder literally becomes a 'bag of pus', with risk of rupture and peritonitis. Treatment is by emergency cholecystectomy and broad-spectrum antibiotics.

Cholangitis is an infection of the biliary tree, usually secondary to obstruction. Obstruction may be caused by gallstones, malignancy (cholangiocarcinoma or pancreatic carcinoma) and occasionally by worms (including *Ascaris lumbricoides*, *Fasciola hepatica* and *Clonorchis sinensis* – Topic D5). Bacterial infection is usually due to coliforms and anaerobes. Clinical features are similar to acute cholecystitis, but fever, rigors and jaundice are often pronounced. Microbiological investigations should include blood cultures and pus from the biliary tree (if an external biliary drain is inserted). Treatment is with high-dose broad-spectrum antibiotics and removal of the obstruction where possible. Some patients with inoperable malignancy and/or those with biliary stents are prone to repeated attacks of cholangitis, with more antibiotic-resistant organisms appearing over time (e.g. *Pseudomonas*, *Enterococcus*, *Candida*).

Infections of the spleen

A number of infections can involve the spleen. Inflammation of the spleen (splenitis), enlargement of the spleen (splenomegaly) and splenic abscess can all occur. Splenitis is clinically rare. **Splenomegaly** is more common. It can occur in glandular fever (Epstein–Barr virus – Topic B8), and a number of other infections can cause enlargement of both liver and spleen (hepatosplenomegaly) (e.g. brucellosis). A notable cause of gross splenomegaly is visceral leishmaniasis (Topic D4). An indirect infective cause of splenomegaly is schistosomiasis causing liver cirrhosis and portal hypertension (Topic D5).

Splenic abscesses may be single or multiple. Multiple small abscesses or splenic infarcts can occur in some patients with infective endocarditis (Topic F13), or immunosuppressed patients with hepatosplenic candidiasis. Whilst a variety of pyogenic infections can rarely cause splenic abscess, between 30 and 50% of reported infections involve salmonellae.

Infections of the pancreas

Acute inflammation of the pancreas gland (**pancreatitis**) can be due to infection, (e.g. mumps virus), although other causes are more common (gallstones, alcohol, trauma etc.). Secondary bacterial infection can also occur, particularly in severe necrotizing pancreatitis, or when a pseudocyst occurs. Coliforms, streptococci, staphylococci, *Pseudomonas* and anaerobes are the principal pathogens.

Peritonitis

Peritonitis is inflammation of the peritoneal cavity within the abdomen. This is mostly due to pyogenic infection, but chemical peritonitis (e.g. following a leak of bile) can also occur.

Spontaneous bacterial peritonitis is a rare condition in healthy humans, but is more common in patients with **ascites** secondary to liver cirrhosis. *Streptococcus pneumoniae* is an important cause, but other streptococci, coliforms and anaerobes can also be involved.

Peritonitis is more commonly seen as a **secondary complication** to other intra-abdominal pathology including acute appendicitis, diverticulitis, perforation of the bowel or a peptic ulcer, bowel ischemia, and complications of gastrointestinal surgery (e.g. breakdown of an anastomosis). In this situation a **polymicrobial infection** is usually found with bacteria originating from the gastrointestinal tract including coliforms, anaerobes and enterococci. Infections with *Pseudomonas* and *Candida* can also occur in more complicated cases, especially when broad-spectrum antibiotics have already been given.

Microbiological investigations include blood cultures and appropriate specimens of pus (usually obtained at the time of surgery). Treatment is a combination of surgery and broad-spectrum antibiotics (e.g. cephalosporin and metronidazole).

Peritonitis is also a complication in patients with renal failure who are on continuous ambulatory peritoneal dialysis (CAPD) via a peritoneal catheter. Microbiological causes include *Staph. aureus*, coliforms, enterococci, *Pseudomonas* as well as skin organisms (e.g. *Staph. epidermidis* and diphtheroids). The usual portal of entry is via the peritoneal catheter. When this occurs, the peritoneal dialysate turns cloudy and patients develop abdominal pain. Treatment is often with intraperitoneal antibiotics.

Intra-abdominal collections (abscess)

Intra-abdominal collections of pus can occur as complications of other intra-abdominal or gynecological pathology, following peritonitis, or as a complication of surgery. These are one important cause of 'pyrexia of unknown origin' (PUO) as they may cause persistent fever and malaise, but with few other clinical features.

Collections can occur at any site within the abdomen, but **subphrenic** collections (between the diaphragm and liver) or **pelvic collections** are the most common. Infections are often polymicrobial, with anaerobes, coliforms and streptococci (including the *Streptococcus* 'milleri' group). Treatment is principally by open or radiologically guided drainage of pus which should be sent for microbiological culture. Antibiotics, ideally based on the culture results, are given for 3–4 weeks in addition.

F10 URINARY TRACT INFECTIONS

Key Notes

Microbiology	*Escherichia coli* is the principal pathogen. A minority of infections are caused by staphylococci, *Proteus*, *Klebsiella*, enterococci and *Pseudomonas*. Within hospitals more infections are due to *Klebsiella*, other coliforms, *Pseudomonas* and enterococci. Hospital bacteria generally have higher levels of antibiotic resistance.
Pathogenesis	Most infections originate from the patient's bowel flora. Bacteria may colonize the perineum and peri-urethral area, and ascend the urethra to the normally sterile bladder. This is more likely to occur in females (shorter urethra), and with certain strains of *E. coli*.
Epidemiology	Urinary tract infections are common. Infections increase with age and are more common in females. Risk factors include structural abnormalities of the urinary tract, urinary catheters, urological surgery, diabetes and immunosuppression.
Clinical features	The typical symptoms of cystitis are frequency and dysuria, with supra-pubic discomfort. The urine may be cloudy or blood-stained. Routine urine cultures are usually positive. Similar symptoms occur with the 'urethral syndrome', but routine urine cultures are negative. Patients with pyelonephritis are usually unwell with back (loin) pain, high temperature and shaking episodes (rigors). Infections may present differently in children and elderly patients.
Laboratory diagnosis	Mid-stream urine (MSU) samples are analyzed by microscopy, and semi-quantitative culture with appropriate sensitivity tests performed on positive cultures. The presence of pus cells in the urine (pyuria) and $\geq 10^5$ bacteria per ml confirm a diagnosis. Blood cultures should be done if pyelonephritis is suspected.
Treatment	A high fluid intake should be maintained. For uncomplicated cystitis, a short course (e.g. 3 days) of a suitable oral antibiotic is usually sufficient. Empirical antibiotic therapy should be guided by local sensitivity data. More antibiotic-resistant organisms are found in complicated or recurrent infections. Patients who are unwell with pyelonephritis and septicemia require admission to hospital for intravenous high-dose antibiotics and fluids.
Prevention and control	Urinary catheters should be used judiciously, aseptic techniques employed during their insertion and care taken to avoid cross-infection. Noncatheterized patients with recurrent infections may be given prophylactic antibiotics.
Related topics	Genital infections (F11)

Microbiology   *Escherichia coli* is the principal urinary tract pathogen. Other less common causes of urinary tract infection (UTI) are staphylococci (*S. saprophyticus, S. epidermidis, S. aureus*), enterococci (*E. faecalis, E. faecium*), proteus (*P. mirabilis, P. vulgaris*), *Klebsiella*, other coliforms and *Pseudomonas aeruginosa*. Occasionally mixed infections are found.

The frequency with which these pathogens are found varies according to the healthcare setting. In uncomplicated community-acquired infections, *E.coli* predominates (>80% of infections), followed by staphylococci (8–10%) with the remaining pathogens found in only 1–5% of infections. Within hospitals, the proportion of infections due to *E.coli* is reduced (40–60%), with more infections due to *Klebsiella*, other coliforms, *Pseudomonas* and enterococci.

In general, hospital bacteria causing UTI have higher levels of antibiotic resistance, either through **innate resistance** (enterococci, *Pseudomonas*) or **acquired resistance** (*E. coli, Klebsiella*, other coliforms). The local microbiological **epidemiology** of UTI and associated **antibiotic resistance rates** are important to establish within different healthcare settings, to allow for appropriate **empirical antibiotic therapy** of UTI.

Pathogenesis   Most infections originate from the patient's bowel flora (**endogenous infection**). Fecal bacteria, including *E. coli*, may colonize the perineum and peri-urethral area, and ascend the urethra to the normally sterile bladder.

Normal **defense mechanisms** against bladder infection include the **hydrodynamic** effect of the flow of urine, **phagocytosis**, and **IgA antibody**. These defense mechanisms are more likely to be overcome in females than in males as the urethra is much shorter. In some women there is a predisposition to infection, following sexual intercourse, which may be related to increased **transurethral passage of bacteria into the bladder**. Normal defense mechanisms are also breached by instrumentation of the urinary tract (**urinary catheters, surgery**).

Some strains of *E. coli* appear to have a greater ability to **adhere to uroepithelial cells** than other strains, and are more likely to cause UTI. This adhesion to uroepithelial cells is dependent on the antigenic structure of the **bacterial fimbriae** on the cell surface.

Patients who develop UTI within hospital are more likely to have infection with 'hospital bacteria'. This may occur by **endogenous infection** (where the patient's bowel has already become populated with hospital bacteria), or by **exogenous infection** (e.g. by direct transmission of bacteria on the hands of healthcare workers during urinary catheter procedures).

Epidemiology   Urinary tract infections are **common**. The incidence of infections increases with **age**. In general, infections are far more common in **females**. In young men, UTI is relatively rare and warrants investigation for underlying risk factors. In **elderly men** UTI is more common due to enlargement of the prostate gland (**prostatic hypertrophy**). This can cause obstruction to the flow of urine, with incomplete voiding, resulting in **residual urine in the bladder,** a recognized risk factor for UTI.

Other **risk factors** for UTI include:

• structural abnormalities of the urinary tract;
• urinary catheterization;

- urological surgery or other instrumentation of the urinary tract;
- diabetes mellitus;
- immunosuppression.

Clinical features

Cystitis

The typical symptoms of **cystitis** (inflammation of the bladder) are **frequency** (frequent passing of urine) and **dysuria** (pain or stinging on passing urine). There may be a mild **temperature** and **suprapubic discomfort**. The urine may be **cloudy** or **blood-stained (hematuria).**

Urethral syndrome

A proportion of patients may present with the above symptoms suggestive of cystitis, but routine urine cultures are negative. In some cases a **sexually transmitted infection** may be present (e.g. chlamydia or gonorrhea), but often **no pathogen** is identified. This is known as the urethral syndrome. Postulated causes for the urethral syndrome include **fastidious bacteria** that do not grow on conventional agar plates used for urine cultures (e.g. lactobacilli, diphtheroids) or **intracellular bacteria** that are difficult to detect (e.g. *Mycoplasma*, *Ureaplasma*).

Pyelonephritis

Patients with pyelonephritis are usually unwell with **back (loin) pain**, high **temperature** and shaking episodes (**rigors**). Symptoms of cystitis may or may not be present as well. Rigors result from the entry of bacteria from the kidney into the bloodstream (**bacteremia** – Topic F12).

Pyelonephritis can permanently **damage the kidney** causing **fibrosis** (scarring). Repeated attacks of pyelonephritis can, over time, cause the kidneys to stop working (**renal failure**).

UTI in children

UTI should be considered in all children who are unwell, as specific symptoms may be absent. **Bed-wetting** may also be a presenting feature. Proven UTI in children should be adequately treated and investigations performed to look for **anatomical or functional abnormalities** of the urinary tract. Reflux of urine from the bladder into the ureters (**vesico-ureteric reflux**), in association with repeated or **chronic infection**, can cause permanent damage to the growing kidney, and result in premature **renal failure** in early adulthood.

UTI in elderly patients and catheterized patients

Specific clinical features of cystitis or pyelonephritis may be absent in elderly patients, so a UTI should be suspected in all elderly patients who become unwell or develop new confusion. Patients with an indwelling urinary catheter are at increased risk of UTI, but the typical features of frequency and dysuria will be absent.

UTI in pregnancy

UTI is a common complication during pregnancy (Topic F15). Infections may be asymptomatic (**asymptomatic bacteriuria**) so regular screening is important. Antibiotic treatment options are fewer because some antibiotics (e.g. trimethoprim), are contraindicated during pregnancy (potential harm to the developing fetus).

Laboratory
diagnosis

Laboratory confirmation of a diagnosis of UTI is not always essential. Many female patients with uncomplicated cystitis can be treated empirically with urine samples sent from only those who fail to respond. Microbiology investigations should always be done on children, adult males, those with recurrent infections, those with risk factors for a resistant organism (recent hospitalization, residence in nursing home), and those with suspected pyelonephritis.

Specimens
A **mid-stream urine** (**MSU**) is the standard sample for investigation of UTI. The peri-urethral area is cleaned and the first part of the urine sample passed (this clears many of the bacteria that reside at the distal end of the urethra). After temporarily stopping, the next 20–25 ml of urine is collected in a sterile container.

In catheterized patients a **catheter specimen of urine** (**CSU**) can be collected. This must be of freshly produced urine collected directly from the catheter, and not from the collecting bag. **Catheters are frequently colonized** by bacteria and the value of CSU samples is limited. Culture results are often positive in the absence of clinical symptoms or signs and antibiotic treatment should not be given in this situation.

Urine samples are difficult to collect in young children and very elderly patients without contaminating the sample with bacteria from the perineum. In some cases a **suprapubic aspirate** (**SPA**) is required. This is an invasive technique where a needle is passed through the skin directly into the bladder and urine aspirated into a syringe.

Occasionally other samples are obtained from higher up the urinary tract (e.g. ureteric urine). The concentration of bacteria is usually much lower, requiring culture of a greater quantity of urine compared to an MSU.

Blood cultures should also be performed on patients who are unwell, or with suspected **pyelonephritis**.

Urine specimens should be transported rapidly to the laboratory to stop further growth of bacteria within the specimen. Sometimes delay is inevitable. In this case samples should be refrigerated prior to transport (**4°C**). Where regular delays occur, **boric acid containers** or the **dip-slide** method can be used (see Topic E2).

Dipstick analysis
Dipstick analysis of urine can give a rapid indication of a UTI at the bedside. The detection of protein of blood by dipstick is a nonspecific marker of UTI, and strips that detect **leukocyte esterase** (a marker of pyuria) and **nitrates** (produced by many of the bacteria that cause UTI) are more suitable.

Microscopy
Microscopic analysis of urine is performed to look for:

- white blood cells;
- red blood cells;
- epithelial cells (an indication of a contaminated sample);
- casts (a sign of kidney damage).

An abnormal white cell count (**pyuria** – literally pus in the urine) is > 10 white cells per mm^3. In UTI the count is usually much higher than this, but pyuria is not always present in UTI (e.g. in immunosuppressed patients).

An abnormal red cell count (**microscopic hematuria**) is > 5 red cells per mm^3. Hematuria may occur with UTI but there are many other potential causes (e.g. bladder tumor).

The presence of squamous epithelial cells indicates contamination of the sample with cells from the perineum. This may give rise to false-positive culture results.

Urine microscopy can also detect yeast cells, indicative of candida colonization or infection, and occasionally trophozoites of *Trichomonas vaginalis*, the cause of trichomoniasis (Topic F11).

Culture and sensitivity tests

Urine samples are cultured using a **semi-quantitative** method, onto a suitable indicator agar (e.g. CLED agar). Low numbers of organisms (< 10^3 per ml) are ignored as urethral/perineum contaminants, unless the sample is ureteric urine or obtained via an SPA (see above). A count of ≥ 10^5 organisms per ml from an MSU is generally regarded as the significant cut-off, but pure cultures of 10^4 bacteria per ml (especially in the presence of pyuria) are also accepted as indicative of UTI.

Identification of organisms from urine, and subsequent antibiotic sensitivity tests are performed using standard laboratory methods (Topics E2 and E3). In some situations, identification is simply done to the genus level (e.g. 'coliform' bacillus), but it may be necessary to identify to species level (e.g. *Klebsiella oxytoca*), particularly with isolates from inpatients on specialized units, or when multiresistant strains are found.

Sensitivity tests often employ antibiotics at a higher concentration for urine isolates, compared to isolates from other specimens, reflecting the fact that many antibiotics reach the urine in high concentration following renal excretion. Direct sensitivity tests (Topic E3) are sometimes performed on urine samples with pyuria.

Treatment

A **high fluid intake** should be maintained to assist in the physical removal of bacteria from the kidneys and bladder. Most antibiotics are excreted by the kidneys and high concentrations can be found in the urine. Important exceptions to this are macrolides (e.g. erythromycin, clarithromycin) and fusidic acid, which should not be used for treatment of UTI.

Treatment should be guided by laboratory sensitivity tests. However it must be recognized that there is not a 100% correlation between *in vitro* tests and clinical response. Apparently 'resistant' organisms may respond to an antibiotic due to factors such as the high concentration of antibiotic in the urine, physical removal of bacteria by the flow of urine, and the immune response.

Cystitis

For uncomplicated cystitis, a short course (e.g. 3 days) of a suitable oral antibiotic is usually sufficient. Suitable antibiotics include **trimethoprim, cephalexin** and **nitrofurantoin**, but empirical antibiotic therapy should be guided by **local sensitivity data**. Broad-spectrum antibiotics (e.g. fluoroquinolones, co-amoxiclav) should be reserved for complicated or recurrent infections.

In the past **amoxycillin** (or ampicillin) was frequently used, but > **40% of *E. coli*** strains are now **resistant**, making this antibiotic unsuitable for empirical treatment (i.e. before sensitivities are known). Nitrofurantoin is not active against *Proteus* and some *Klebsiella* species. Cephalexin and fluoroquinolones are not active against enterococci.

Pyelonephritis

Patients who are unwell with pyelonephritis may require admission to hospital for intravenous high-dose antibiotics, fluids and other medical support. Broad-spectrum antibiotics are normally given until culture and sensitivity results are available (e.g. intravenous cefuroxime ± gentamicin, or a fluoroquinolone). Treatment should be modified in light of clinical response and microbiology results, and continued for 7–10 days.

Prevention and control

Many UTIs cannot be prevented. However, appropriate investigations should be performed, particularly in children and young men, to exclude anatomical problems of the urinary tract which are a risk for **recurrent infection**.

Antibiotic prophylaxis may be given to certain groups of patients who suffer from recurrent attacks, to reduce the frequency of attacks and prevent long-term kidney damage. Low-dose antibiotics (e.g. once daily at night-time) may be given prophylactically, sometimes on a rotational basis, to overcome resistance problems.

Urinary catheters are a major risk factor for UTI. They are widely used in hospitals, and in elderly patients with urinary incontinence or urinary outflow obstruction. The use of catheters should be kept to a minimum. Prophylactic antibiotics should not be used in this situation due to the rapid emergence of resistant bacteria. Good **aseptic technique** should be used during insertion and care of indwelling urinary catheters to minimize infection risks, and particularly to prevent **cross-infection** between patients. Outbreaks of infections may occur where healthcare workers do not wash hands, or change gloves, when emptying catheter bags sequentially from a number of catheterized patients.

F11 GENITAL INFECTIONS

Key Notes

Microbiology

A wide range of organisms can cause genital infections including viruses, chlamydia, bacteria, fungi, protozoa and ectoparasites. Many of these genital infections are sexually transmitted diseases.

Epidemiology

Common sexually transmitted infections include gonorrhea, chlamydia, genital herpes, genital warts, and trichomoniasis. Syphilis is less common but still occurs. HIV and hepatitis B are of worldwide importance as sexually transmitted diseases, although they do not cause genital manifestation of disease.

Clinical syndromes

Genital infections comprise a range of clinical syndromes including urethritis, vaginal discharge, genital ulceration, pelvic inflammatory disease and uterine infections in women, and urethritis, genital ulceration, prostatitis, epididymitis and orchitis in men.

Chlamydia and gonorrhea both cause urethritis, vaginal discharge, pelvic inflammatory disease, prostatitis and epididymitis. Other common causes of vaginal discharge are candida (thrush), bacterial vaginosis and trichomonas. Important causes of genital ulceration are herpes simplex virus and syphilis (*Treponema pallidum*). Uterine infections may occur in relation to pregnancy. Orchitis is usually due to mumps virus.

Related topics

Human immunodeficiency viruses (B1)
Hepatitis B and D viruses (B3)
Herpes simplex viruses (B5)
Neisseria (C10)

Spirochaetes (C16)
Atypical bacteria (chlamydia) (C17)
Urinary tract infections (F10)

Microbiology

A wide range of organisms can cause genital infections including viruses, chlamydia, bacteria, fungi, protozoa and ectoparasites (*Table 1*). Many of these are **sexually transmitted diseases** (STD). Some important sexually transmitted infections do not cause genital infection, notably human immunodefiency virus (HIV) and hepatitis B virus (HBV). Many of these organisms are described in separate Topics (see *Table 1*).

Epidemiology

Common sexually transmitted infections include gonorrhea, chlamydia, genital herpes, genital warts, and trichomoniasis. Syphilis is less common but still occurs. HIV and hepatitis B are of worldwide importance. Chancroid, granuloma inguinale and lymphogranuloma venereum are rare outside of tropical countries.

Vaginal thrush and bacterial vaginosis are common causes of vaginal discharge but are usually nonsexually transmitted. However recurrent infections may sometimes be due to colonization of the sexual partner.

Table 1.   Genital infections and sexually transmitted diseases

	Disease or syndrome	Organism
Viruses	Genital herpes	Herpes simplex (Topic B5)
	Genital warts	Papilloma viruses (Topic B19)
	Hepatitis B	Hepatitis B virus (Topic B3)
	HIV/AIDS	Human immunodeficiency virus (Topic B1)
	Molluscum contagiosum	Pox virus (Topic B18)
Bacteria	Gonorrhea	*Neisseria gonorrhoeae* (Topic C10)
	Syphilis	*Treponema pallidum* (Topic C16)
	Bacterial vaginosis*	*Gardnerella vaginalis*, anaerobes
	Pelvic inflammatory disease	*Neisseria gonorrhoeae*, coliforms, streptococci, anaerobes (also chlamydia)
	Chancroid	*Haemophilus ducreyi*
	Granuloma inguinale	*Donovania granulomatis*
Chlamydia	Chlamydial urethritis, cervicitis, pelvic inflammatory disease, prostatitis, epididymo-orchitis	*Chlamydia trachomatis* (Topic C17)
	Lymphogranuloma venereum	*Chlamydia trachomatis* (serovars L$_{1-3}$)
Fungi	Vaginal thrush*, penile balanitis	*Candida albicans*, other candida species (Topic D1)
Protozoa	Trichomoniasis	*Trichomonas vaginalis*
Ectoparasites	Genital scabies	*Sarcoptes scabiei*
	Pubic lice	*Phthirus pubis*

*These are frequent causes of vaginal discharge but are not usually sexually transmitted infections.

Clinical syndromes and investigations

Patients with suspected sexually transmitted infection should be referred to specialized clinics for appropriate investigations and treatment. The tracing and treatment of their sexual contacts is also important. Infections with more than one organism are common.

The main pathogens (*Neisseria gonorrhoeae*, *Chlamydia trachomatis*, *Treponema pallidum*, candida and *Herpes simplex*) are described in separate Topics.

Urethritis

The principal causes of urethritis are gonorrhea and chlamydia. Symptoms include **dysuria and frequency** of urine (see Topic F10, UTI), often with a **urethral discharge**. A purulent discharge more frequently occurs in males with gonorrhea, and less so with chlamydia. Infections in **females** can be **asymptomatic**. Urethral swabs for microscopy and gonococcal culture, and a urethral swab or urine for chlamydia tests are the appropriate investigations.

N. gonorrhoeae may also be found at other sites, with or without local symptoms, including the throat and rectum.

Vaginal discharge

Vaginal discharge may be due to cervicitis (gonorrhea, chlamydia), trichomonas vaginitis (*Trichomonas vaginalis*), vaginal thrush (*Candida*) and bacterial vaginosis. **Endocervical swabs** are required for gonorrhea and chlamydia investigations. **High vaginal swabs** are appropriate for trichomonas, thrush, and bacterial vaginosis, but lack sensitivity for chlamydia and gonorrhea.

Cervical carriage of *N. gonorrhoeae* and *C. trachomatis* often occurs without apparent clinical symptoms.

Genital ulcers

Important causes of genital ulceration include syphilis and herpes simplex. Investigations include dark-ground microscopy of ulcer exudate for spirochaetes (syphilis), antibody tests for syphilis, and swabs (in viral transport medium) for herpes simplex culture.

Rare infective causes of genital ulcers in the tropics include chancroid, granuloma inguinale, and lymphogranuloma venereum.

Prostatitis and epididymo-orchitis

Both *N. gonorrhoeae* and *C. trachomatis* can cause prostatitis and epididymo-orchitis. Diagnosis is usually made on clinical grounds, along with the results of urethral swabs and urine investigations.

Acute and chronic prostatitis may also be caused by nonsexually transmitted infections (e.g. *E. coli*, enterococci). Many antibiotics do not penetrate the prostate gland, and the treatment of chronic prostatitis is difficult. Acute epididymitis is sometimes caused by *Pseudomonas*. Chronic epididymitis may occasionally be due to *Mycobacterium tuberculosis*.

The principal cause of orchitis is the mumps virus (see Topic B13).

Pelvic inflammatory disease

Pelvic inflammatory disease (PID) in women includes infection of the Fallopian tubes (**salpingitis**), abscess formation in the ovary and/or Fallopian tubes (**tubo-ovarian abscess**), and local peritonitis. PID is a potentially serious condition, often requiring admission to hospital. Symptoms include lower abdominal pain, backpain, painful intercourse and fever.

Complications can include **infertility** and risk of future **ectopic pregnancy** (where the fetus develops outside of the uterus in the Fallopian tube).

Both gonorrhea and chlamydia are common causes of PID. Sometimes symptoms of lower genital tract infection (e.g. vaginal discharge) are absent. Other nonsexually transmitted causes include mixed infections with aerobes (e.g. *E. coli*, streptococci) and anaerobes (*Bacteroides*, anaerobic streptococci).

Investigations should include routine swabs for gonorrhea and chlamydia (cervical and urethral swabs). Pus within the Fallopian tubes, ovary or peritoneum may be found at laparotomy or laparoscopy and should be sent for culture.

Treatment is with broad-spectrum antibiotics, including agents active against chlamydia in sexually active women, and anaerobes (e.g. oflaxacin and metronidazole in combination).

Bacterial vaginosis

Bacterial vaginosis is due to an overgrowth of bacteria within the vagina, comprising **Gardnerella vaginalis** and anaerobic bacteria. The normal vaginal flora (mainly lactobacilli) is absent and the pH of the vagina is elevated (pH > 5). Bacterial vaginosis produces an offensive (sometimes 'fishy'), nonpurulent vaginal discharge. This is often diagnosed on clinical grounds.

Bacterial vaginosis in pregnancy may be associated with premature labor or abortion.

Diagnosis can be confirmed by microscopic examination of a Gram-stained smear of the discharge (collected by a vaginal swab) for **'clue cells'**. These are vaginal epithelial cells that are coated with many Gram-negative bacteria. Pus cells and lactobacilli are not seen. Sometimes cultures are performed yielding a mixture of *G. vaginalis* and anaerobes.

Treatment is with oral metronidazole or clindamycin vaginal cream.

Trichomonas

Trichomonas vaginalis is a **flagellated protozoon** that is sexually transmitted. It causes an offensive frothy vaginal discharge, with vaginal and vulval soreness and itching. Laboratory diagnosis is by wet microscopy of a vaginal swab or aspirated fluid. Sometimes culture of the organism is attempted. In men the organism can cause balanitis, but is often asymptomatic.

Treatment is with metronidazole. The male sexual partner may need a simultaneous course of treatment.

Ectoparasites

Genital scabies is caused by the mite *Sarcoptes scabiei*. The typical symptom is **itching**. Burrows or papules may be seen. Treatment is with a topical insecticide applied to the whole of the body (excluding the head).

Pubic lice (*Phthirus pubis* or crab louse) are sexually transmitted and also cause itching. **Lice or eggs** are easily visible to the human eye, or with a hand lens. Treatment is with a topical insecticide.

Molluscum contagiosum

This is due to a pox virus and causes small, often umbilicated, white papules. Infection may be sexually transmitted. The lesions usually resolve spontaneously. Widespread, disfiguring molluscum infection can occur in patients with HIV infection.

Tropical sexually transmitted infections

Chancroid is caused by *Haemophilus ducreyi* and produces papules which ulcerate. Local lymph nodes enlarge and may break down to form an abscess.

Granuloma inguinale is caused by *Donovania granulomatis*. Genital papules form that break down into large ulcers.

Lymphogranuloma venereum is caused by particular strains of *Chlamydia trachomatis*, producing painful swollen lymph nodes in the groin that can progress to abscess formation.

The usual treatment for these infections is with tetracycline antibiotics.

Infections of the uterus

Infections of the uterus may occur during pregnancy. **Chorio-amnionitis** is an infection of the placenta and amniotic sac which can occur following systemic infection and seeding via the bloodstream (e.g. **listeria**), or by ascending infection from the vagina, especially if the placental membrane becomes ruptured before labor (**premature rupture of membranes**). This is usually a **mixed infec-**

tion with streptococci, coliforms and anaerobes. This can lead to premature labor, septic abortion, and maternal or fetal death.

Endometritis (infection of the uterine wall) can occur following delivery, usually as a result of ascending infection (**puerperal fever**). *Streptococcus pyogenes* was a common and often fatal cause of puerperal fever in the past, but is now less common. Other streptococci, coliforms and anaerobes are causative pathogens. Treatment is with aggressive antibiotic therapy and removal of any **retained products of conception** (e.g. fragments of placenta).

Pyogenic infection of the uterus (**pyometrium**) can complicate other diseases of the genital tract (e.g. cervical cancer). **Pelvic actinomycosis** can involve the uterus, particularly in women **with intra-uterine contraceptive devices (IUCD)**.

F12 BACTEREMIA AND SEPTICEMIA

Key Notes

Definitions	Bacteremia is the presence of bacteria within the circulating bloodstream. When bacteremia is associated with clinical features of sepsis, the term septicemia is used.
Microbiology and epidemiology	The commonest causes of community-onset bacteremia are *E. coli* and other coliforms (often secondary to UTI), *Staph. aureus* (associated with skin, soft tissue or musculoskeletal infections), *Strep. pneumoniae* (pneumonia) and other streptococci. Meningococcal septicemia and endocarditis are rare but important causes. Healthcare-associated bacteremia has a different microbiological spectrum with more *Staph. aureus* (including MRSA), coagulase-negative staphylococci (related to line infections), a wider variety of Gram-negative pathogens, and higher rates of antibiotic-resistant and fungal pathogens. Bacteremia in hospital is often related to the use of medical devices (e.g. central venous catheters) in vulnerable patients. Polymicrobial bacteremia occurs in ~ 5–10% of episodes.
Clinical features	The clinical features of septicemia include fever, sweats, rigors, tachycardia, tachypnea and hypotension. Specific clinical features may also be present such as a petechial rash (meningococcal disease), blistering cellulitis (*Strep. pyogenes*), urinary tract symptoms, heart murmur or splinter hemorrhages (endocarditis) or pneumonia.
Laboratory diagnosis	Identification of bacteremia is through blood cultures, which should be performed for all patients with suspected serious bacterial infection. As only a few bacteria are present in the bloodstream during bacteremic episodes, laboratory detection relies on using liquid enrichment culture to maximize sensitivity. However, this also increases the chance of false-positive results due to contamination, particularly with skin organisms including *Staph. epidermidis*.
Treatment	The severity of the illness, the likely source of the infection, the likely causative organism(s), and knowledge of local antibiotic sensitivity patterns guide the antibiotic treatment of bacteremia. Antibiotics should be administered promptly in high dose, after taking blood cultures and other urgent specimens. Broad-spectrum antibiotics may be required until culture and sensitivity results are known. Hospitals should have appropriate antibiotic guidelines available, based on their local pattern of infections, patient populations, local resistance rates, and these should be regularly reviewed.

Definitions

Bacteremia is defined as the presence of bacteria within the circulating bloodstream (with or without clinical symptoms or signs). **Septicemia** is a term used to describe bacteremia that is associated with the clinical features of sepsis (see below).

Organisms other than bacteria can cause septicemia, particularly candida or other fungal pathogens. Here the terms **candidemia** or **fungemia** are also used.

Microbiology and epidemiology

A wide range of pathogens can cause bacteremia or septicemia. The pattern of pathogens varies according to whether the infections occur in the community, or in relation to hospital treatment. This is summarized in *Table 1*. In 5–10% of bacteremia, two or more organisms are found (**polymicrobial bacteremia**).

Community-onset bacteremia
Escherichia coli is the commonest cause of community-onset bacteremia, usually as a complication of urinary tract infection (Topic F10) or intra-abdominal sepsis (Topic F9).

Staphylococcus aureus is also frequently found with bacteremia secondary to skin and soft tissue infections (Topic F1), bone and joint infections (Topic F2), and endocarditis (Topic F13). The majority of these infections are with methicillin-sensitive strains, although methicillin-resistant strains (MRSA) are starting to appear.

Streptococcus pneumoniae is an important cause of bacteremia secondary to pneumonia, and sometimes meningitis.

Meningococcal septicemia and meningitis (due to *Neisseria meningitidis*) is an important, but less common cause of community-onset bacteremia.

Table 1.   Causative organisms of bacteremia in community and healthcare-associated infections*

Organisms		Community-onset bacteremia (%)	Healthcare-associated bacteremia (%)
Gram-positive bacteria	*Staph. aureus* – methicillin sensitive	10	16
	Staph. aureus – methicillin resistant (MRSA)	3	10
	Coagulase-negative staphylococci	<1	17
	Strep. pneumoniae	8	1
	Strep. pyogenes	2	<1
	Enterococci	6	11
	Other streptococci	8	1
Gram-negative bacteria	*E. coli*	30	11
	Other coliforms	9	7
	Ps. aeruginosa	2	5
	Salmonella (including *S. typhi*)	1.5	0
	N. meningitidis	1	0
Anaerobes	*Bacteroides*	5	1
	Clostridium	2	<1
Fungi	*Candida*	<1	3

*Data from Nottingham City Hospital, UK (1999–2003).

Endocarditis (Topic F13) is frequently associated with, and often diagnosed by, the identification of bacteremia. Streptococci (particularly the viridans streptococci), enterococci and staphylococci are the commonest organisms.

Healthcare-associated bacteremia

In healthcare-associated bacteremia the proportion of infections with *E. coli* is reduced with more infections due to *Staph. aureus*. A higher proportion of *Staph. aureus* bacteremia is due to **MRSA**.

In contrast to community-onset bacteremia, coagulase-negative staphylococci (including *Staph. epidermidis*) account for a high proportion of bacteremias. These normally harmless skin commensal bacteria are the commonest cause of central-venous catheter infection, which often results in bacteremia.

The proportion of infections with more antibiotic-resistant bacteria (e.g. enterococci and pseudomonads) as well as *Candida*, is increased.

Underlying source of infection

Bacteremia is often a complication of an infection at a specific organ site (e.g. urinary tract infection – Topic F10), although for some infections bacteremia itself is the primary focus (e.g. meningococcal septicemia). With many infections, the underlying source of infection may not be immediately apparent. The identification of the causative organism can then act as a pointer to significant underlying pathology.

As with the pattern of organisms, the underlying source of bacteremia also varies between community and hospital infections. The commonest **community** onset infections associated with bacteremia are:

- urinary tract infections;
- pneumonia;
- biliary sepsis;
- cellulitis;
- endocarditis;
- bone and joint infections.

Within **hospitals** the commonest infections associated with bacteremia are:

- central venous catheter infections;
- febrile neutropenia*;
- urinary tract infection (often catheter-related);
- pneumonia (post-operatively or ventilator-associated);
- skin and soft tissue (surgical wounds);
- intra-abdominal sepsis.

*Bacteremia is common in patients with very low neutrophil counts (e.g. during cancer chemotherapy). Host defenses are severely impaired and spontaneous bacteremia may arise from organisms such as *E. coli* that are naturally found in the large bowel as part of the normal flora.

Risk factors

Risk factors for bacteremia include infection or colonization with bacterial strains of high pathogenicity (virulence), age (neonates or elderly patients), immunosuppression, underlying chronic illnesses (diabetes, malignancy, renal failure) and the presence of indwelling venous catheters (lines) or urinary catheters.

Clinical features

Bacteremia can occur without overt symptoms of septicemia, but usually a patient will have one or more of the following:

- fever;
- sweats;
- rigors (shaking attacks);
- tachycardia (raised heart rate);
- tachypnea (raised respiratory rate);
- hypotension (low blood pressure).

Patients may also have specific clinical features relating to the underlying site of infection or condition. Examples include a petechial rash (meningococcal disease), blistering cellulitis (*Strep. pyogenes*), urinary tract symptoms, heart murmur or splinter hemorrhages (endocarditis) or pneumonia.

Laboratory diagnosis

Identification of bacteremia is by **blood culture**. Blood is normally sterile and the isolation of pathogenic bacteria from the bloodstream is indicative of bacteremia. Blood cultures are one of the most important microbiological specimens and should be obtained from all patients with suspected serious bacterial infection.

Usually only a few bacteria per milliliter of blood are present in the bloodstream during bacteremic episodes. Laboratory detection relies on using liquid enrichment culture to maximize sensitivity, but this also increases the chance of false-positive results due to contamination.

Organisms that contaminate blood cultures are those found on the skin (patient, phlebotomist, or technician), including *Staph. epidermidis* and diphtheroids. Contaminants are a problem because they may mask the true organism, and it is sometimes difficult to distinguish true bacteremia from contamination in blood cultures particularly from patients with central venous catheters.

It is good practise to always obtain a minimum of two sets of blood cultures by separate venepuncture. Good **aseptic technique** is required during venepuncture, inoculation of blood culture bottles, and laboratory processing. This reduces the chances of contamination and helps with interpretation of results. Ideally a minimum of 20 ml of blood should be collected for culture **before antibiotic therapy is started**.

Most diagnostic laboratories now use automated systems for blood cultures. Inoculated cultures are incubated for up to 5 days or more, and automatically monitored for growth continuously or at regular intervals. The commonest surrogate marker used for bacterial growth is the production of CO_2 within the bottle. Organisms from 'positive' bottles are usually identified by conventional bacteriology including Gram stain, culture and antibiotic sensitivity tests.

Treatment

Treatment of bacteremia depends on the severity of the illness, the likely source of the infection, the likely causative organism(s), and knowledge of local antibiotic sensitivity patterns.

Some patients will be severely ill with septicemic shock. These patients require full supportive medical therapy (e.g. fluids and drugs to maintain blood pressure and urine output, and respiratory support with oxygen or assisted ventilation). An obvious source of infection may require surgical intervention. Indwelling urinary catheters or central or peripheral venous catheters may require removal.

Antibiotics should be administered promptly in high dose (usually by the intravenous route) after taking blood cultures and other urgent specimens. Initially broad-spectrum antibiotics may be required until culture and sensitivity results are known. Each hospital should have appropriate antibiotic guidelines available, based on their local pattern of infections, patient populations, local resistance rates and choice of antibiotic drugs. Guidelines should be regularly reviewed. A suggested generic guideline is shown in *Table 2*, but this is not suitable for all situations.

Table 2.   An antibiotic guideline for the treatment of bacteremia or septicemia

Site of infection or likely organism	Community-onset bacteremia	Healthcare-associated bacteremia
Sepsis of unknown origin	Cephalosporin ± aminoglycoside	Cephalosporin + aminoglycoside
Urinary tract infection	Cephalosporin *or* fluoroquinolone	Cephalosporin *or* fluoroquinolone *or* aminopenicillin + aminoglycoside
Pneumonia	Co-amoxiclav* + macrolide *or* cephalosporin + macrolide *or* penicillin + fluoroquinolone	Co-amoxiclav* *or* cephalosporin fluoroquinolone ± glycopeptide
Intra-abdominal infection	Cefuroxime + metronidazole	Aminopenicillin + aminoglycoside + metronidazole *or* Piptazobactam *or* Carbapenem
Suspected *Staph. aureus* infection	Flucloxacillin ± aminoglycoside	Vancomycin ± aminoglycoside
Central venous catheter infection	–	Vancomycin ± aminoglycoside
Meningococcal disease	Penicillin or cephalosporin	–
Febrile neutropenia	–	Antipseudomonal β-lactam ± aminoglycoside

*Co-amoxiclav = amoxicillin–clavulanic acid combination.

F13 ENDOCARDITIS

Key Notes

Microbiology	The common organisms causing endocarditis are viridans streptococci, enterococci and staphylococci. A wide range of other bacteria as well as chlamydia and fungi can occasionally be involved.
Epidemiology	Risk factors for endocarditis include rheumatic heart disease, congenital heart disease, atherosclerotic heart valve disease, poor dentition, intravenous drug abuse and previous heart valve surgery (prosthetic heart valves).
Pathogenesis	Predisposing heart defects can cause damage to the endothelium through turbulent blood flow. Fibrin and platelet deposits build up, which may become infected by bloodborne organisms to form vegetations.
Clinical features	General symptoms of tiredness, weight loss, loss of appetite, and night sweats are common. New or changing heart murmurs occur. Embolic complications include stroke. Immunological reactions can cause kidney damage and vasculitic skin lesions.
Laboratory diagnosis	Repeated culture of the organism from blood cultures, with detailed *in vitro* antibiotic sensitivity tests, including minimum inhibitory concentrations (MIC).
Treatment	High dose intravenous antibiotics according to the organism isolated and sensitivity tests. Synergistic combinations of cell-wall-active agents (e.g. penicillin) plus aminoglycosides (e.g. gentamicin) are used for streptococcal endocarditis.
Prevention	Patients at risk of endocarditis are given prophylactic antibiotics for certain procedures, including dental work that is likely to cause bacteremia.
Other infections of the heart	Myocarditis (inflammation of the heart muscle) can be caused by enteroviruses (coxsackie virus) and *Toxoplasma gondii* (toxoplasmosis). Acute pericarditis can be caused by pyogenic bacteria, including *Staph. aureus*. Chronic pericarditis may be caused by *Mycobacterium tuberculosis* (tuberculous pericarditis).
Related topics	Antibiotic susceptibility testing (E3) Bacteremia and septicemia (F12)

Infective endocarditis is an infection on the **endothelial** (i.e. the inside) surface of the heart, usually associated with the production of **vegetations**. Most infections occur on one or more of the **heart valves**. Previously there was a distinction between acute endocarditis (a rapidly progressive disease) and subacute endocarditis (SBE) (a more indolent but progressive illness). However, with improved methods of diagnosis and earlier antibiotic treatment this distinction is not clinically relevant, and the general term infective endocarditis is now preferred.

Microbiology

The principal pathogens causing infective endocarditis are streptococci, entero-cocci and staphylococci. Most of the streptococcal infections are due to one of the 'viridans' group of streptococci (e.g. *Streptococcus mitis, Strep. salivarius*), but others may be found including group B streptococci (*Strep. agalactiae*) and group G streptococci. *Staphylococcus aureus* is the predominant staphylococcal pathogen (except for prosthetic valve endocarditis – see below). Coagulase-negative staphylococci are rare causes of native valve endocarditis.

Prosthetic valve endocarditis (PVE) may be due to any of the above bacteria, but low-grade pathogens found on the skin, such as coagulase-negative staphy-lococci (e.g. *Staph. epidermidis*) or diphtheroids, may be implanted at the time of surgery. These bacteria can then cause a low-grade but persistent infection of the prosthetic valves, by forming a biofilm on and around the prosthetic material. They are a common cause of PVE presenting within the first 12 months of surgery.

Other rarer causes of endocarditis include:

- the HACEK group of organisms (*Haemophilus, Actinobacillus, Cardiobacterium, Eikenella, Kingella*);
- Gram-negative bacteria (coliforms, pseudomonads);
- fungi (including *Candida*);
- atypical organisms (*Coxiella, Bartonella*).

Epidemiology

Risk factors for endocarditis are congenital or acquired heart diseases, and conditions associated with an increased risk of bacteremia.

There are various **congenital heart conditions** associated with an endo-carditis risk (e.g. ventricular septal defect). Acquired diseases include **rheumatic heart disease** and **atherosclerosis**. Rheumatic heart disease (a complication of group A streptococcus-associated rheumatic fever – Topic F6), was an important cause of heart valve damage in the past, but is much less common today. The principal disease now affecting heart valves is atherosclerotic disease, which is seen mainly in the elderly.

The presence of a **prosthetic heart valve** (from previous cardiac surgery) is a significant risk factor for endocarditis.

Risk factors for bacteremia include poor dentition, dental surgery, some endo-scopic surgical procedures, urinary catheterization, indwelling central venous catheters, and intravenous drug abuse.

Pathogenesis

Predisposing heart defects can cause damage to the endothelial lining of the heart, through turbulent blood flow or high pressure gradients. The **damaged endothelium** can allow **fibrin and platelet deposits** to build up. These may then become infected by blood-borne organisms to form **microbial vegetations**. Most microbial vegetations occur on either the aortic or mitral valves (left-side of the heart) due to the higher pressure gradients compared to the right side. Right-sided endocarditis (mainly tricuspid valve), however, is a particular risk in intravenous drug abusers or hospital patients with infected central venous catheters.

Some particularly pathogenic bacteria (e.g. *Staph. aureus*) are able to cause endocarditis in patients with no clinical evidence of previous heart disease.

Many of the clinical features of endocarditis are due to **immunological responses** to the infection, including high levels of circulating immune complexes (e.g. vasculitic skin lesions, nephritis). Vegetations may also become dislodged and cause **emboli** to other parts of the body (e.g. brain).

Clinical features

Classic 'subacute' endocarditis (SBE) presents with general symptoms of tiredness, weight loss, loss of appetite, and night sweats, developing over several weeks or even months. *Staph. aureus* endocarditis however presents much more acutely, with a febrile, toxic and often confused patient. With generally earlier diagnosis than previously, classic SBE is much less common, and the term 'infective endocarditis' (IE) is preferred.

New or changing **heart murmurs** are a characteristic feature. Embolic complications include cerebrovascular accident (i.e. a stroke). Immunological reactions can cause kidney damage (nephritis, microscopic hematuria) and vasculitic skin/soft tissue lesions (splinter hemorrhages, Janeway lesions, Osler's nodes). In the eye, subconjunctival hemorrhages and retinal lesions (Roth spots) may be found.

Infective endocarditis is a serious infection with 10–15% mortality. Complications can include complete functional destruction of heart valves, leading to acute heart failure and the need for emergency heart valve replacement. Major emboli can cause fatal intracranial infarcts or hemorrhage.

Diagnosis

The diagnosis of endocarditis is usually based on three factors:

(i)   clinical features;
(ii) results of blood cultures or other microbiological investigations;
(iii) results of echocardiography.

A definite diagnosis of infective endocarditis is made on the basis of two major criteria, *or* one major and three minor criteria, *or* five minor criteria (the Duke classification).

Major criteria for the diagnosis of infective endocarditis are:

* positive blood cultures – either a typical endocarditis organism (viridans streptococci, HACEK group), or community-acquired *Staph. aureus* or enterococcus with no obvious primary focus, or a persistently positive blood culture with an organism consistent with endocarditis;
* evidence of definite endocardial involvement on echocardiography;
* new valvular regurgitation (backflow leakage of blood across the valve).

Minor criteria are:

* predisposing factors (heart condition, intravenous drug use);
* fever;
* vascular phenomena (e.g. arterial emboli, conjunctival hemorrhage);
* immunological phenomena (e.g. vasculitic skin lesions, nephritis);
* positive blood cultures (not meeting the above criteria) or serological evidence of infection (e.g. *Coxiella* infection);
* other echocardiography abnormalities.

A definite diagnosis can also be made by microbiological and/or histological examination of vegetations, or intracardiac abscess, if surgery is undertaken.

At least two (preferably three) sets of **blood cultures** should be taken for patients with suspected endocarditis, and the laboratory alerted. The standard incubation time for blood cultures (5–7 days) may need to be extended to 14 days for some organisms. For culture-negative endocarditis, serological tests and/or molecular diagnostic tests (e.g. for *Coxiella*, *Bartonella*) may need to be considered.

Organisms isolated from blood cultures from patients with endocarditis should have careful *in vitro* sensitivity tests performed, including determination of the minimum inhibitory concentrations (MIC) of antibiotics (Topic E3). There should be close liaison between clinicians and microbiologists.

Treatment

The mainstay of treatment is with high-dose intravenous antibiotics. **Bactericidal antibiotics** in high dose are required in order to penetrate and 'sterilize' the vegetations. **Synergistic combinations** of antibiotics are often used for maximum bactericidal activity.

The antibiotics and duration of treatment depend on the causative organisms, and response to treatment (see *Table 1* for a brief summary). **National guidelines for the treatment of endocarditis** are available, and treatment must be discussed with those who are experienced in managing this infection.

Blood cultures should be repeated whilst on antibiotic treatment to insure clearance of the bacteremia, and clinical and other laboratory inflammatory markers should be closely monitored (e.g. C-reactive protein levels). The echocardiogram may need to be repeated to monitor the size of a vegetation.

Surgical intervention to remove the affected valve may be required when valvular damage is severe or a cardiac abscess has developed, when the size of the vegetation is very large, or when the organism is resistant to antibiotics (e.g.

Table 1. *Simplified summary of the intravenous antibiotic therapy of infective endocarditis*

Organism	Usual first line regimen	Alternative regimen
Viridans streptococci (penicillin susceptible)	Benzylpenicillin plus low-dose gentamicin for 2 weeks*	Ceftriaxone or cefotaxime, vancomycin or teicoplanin
Viridans streptococci (reduced susceptibility to penicillin)	Benzylpenicillin plus low-dose gentamicin for 4 weeks	
Enterococci (ampicillin sensitive)	Ampicillin plus low-dose gentamicin for 4–6 weeks	Vancomycin (or teicoplanin) plus low-dose gentamicin
Enterococci (ampicillin resistant)	Vancomycin (or teicoplanin) plus low-dose gentamicin	
Enterococci (vancomycin resistant)	Specialist advice required	
Staph. aureus (methicillin susceptible)	Flucloxacillin for 4–6 weeks (plus gentamcin for 5–7 days)	Vancomycin (or teicoplanin) plus gentamicin
Staph. aureus (methicillin resistant)	Vancomycin (or teicoplanin) plus gentamicin for 4–6 weeks	
Coagulase-negative staphylococci	Vancomycin (or teicoplanin) plus gentamicin for 4–6 weeks	

*Short course (2 week therapy) can be considered for patients with no complicating factors who respond quickly to therapy.

vancomycin-resistant enterococci) or very difficult to treat (e.g. fungal endo-carditis). The treatment of prosthetic valve endocarditis usually requires replacement of the infected valve as well as antibiotics.

Prevention   Patients at **risk of endocarditis** (i.e. those with a known heart valve condition, congenital heart disease, prosthetic heart valve or previous episode of endo-carditis) should be given **prophylactic antibiotics** for procedures that are likely to cause bacteremia. These include:

- dental procedures;
- upper respiratory tract procedures;
- genito-urinary procedures;
- obstetric, gynecological and gastrointestinal procedures.

Details of the prophylactic antibiotics used in the UK are contained in the British National Formulary.

Dental hygiene and care is important for patients at risk of endocarditis.

Antibiotic prophylaxis during cardiac surgery is also an important factor in the prevention of early-onset prosthetic valve infection.

Other infections   **Myocarditis** (inflammation of the heart muscle) can be caused by enteroviruses
of the heart   (coxsackie virus) and sometimes *Toxoplasma gondii* (toxoplasmosis). This may lead to severe acute heart failure.

Acute pericarditis (inflammation of the pericardium) can be caused by pyogenic bacteria, including *Staph. aureus* and *Strep. pneumoniae*. This may arise spontaneously, sometimes as a complication of infection nearby (e.g. pneu-monia), and can also occur following cardiac surgery. Patients are generally unwell with a high fever, central chest pain, a pericardial effusion and general-ized septicemia.

Chronic pericarditis may be caused by *Mycobacterium tuberculosis* (tubercu-lous pericarditis). This causes chronic fibrosis resulting in a constrictive peri-carditis.

F14 INFECTION IN IMMUNOCOMPROMISED PATIENTS

Key Notes

Primary immune deficiencies	Rare diseases can affect neutrophil function (e.g. chronic granulomatous disease), the complement system, or T- or B-lymphocyte function.
Secondary defects in innate defense mechanisms	The skin normally provides an effective barrier to invasion of underlying tissues by microorganisms, but this barrier can be damaged for instance by burns, surgical incisions or indwelling intravenous catheters, allowing a variety of organisms to gain entry. Mucosal defense mechanisms can be impaired by the use of medical devices such as indwelling urinary catheters or endotracheal tubes.
Secondary defects in adaptive immune mechanisms	Many diseases impact on the proper functioning of the normal immune system including malignancy, infections (such as HIV), malnutrition, and metabolic diseases (e.g. diabetes). Immunosuppressive drugs, used for cancer chemotherapy, transplantation and autoimmune disease can cause significant suppression of normal immune responses. These result in an increased risk of infection, particularly with opportunistic pathogens.
Infections in neutropenic patients	The commonest cause of neutropenia is cancer chemotherapy. The risk of infection significantly increases when the circulating neutrophil count falls to $< 0.5 \times 10^9 \, l^{-1}$, and/or the duration of neutropenia is prolonged. Patients with neutropenia who develop a temperature have 'febrile neutropenia'. The principal causes of febrile neutropenia include both Gram-positive and -negative bacteria, and fungi. Microbiological diagnosis is principally by blood cultures, but these are negative in a significant proportion of episodes. Initial treatment is with broad-spectrum antibiotics active against *Pseudomonas aeruginosa,* but other antibiotics or antifungal agents may subsequently be used. Antimicrobial prophylaxis during neutropenia may reduce the frequency of certain infections.
Infections in transplant patients	Recipients of organ transplants are immunocompromised as a result of immunosuppressive drug therapy used to prevent rejection. These drugs principally suppress cell-mediated immunity, leaving patients at risk from a variety of opportunistic infections such as *Pneumocystis carinii* pneumonia (PCP), cytomegalovirus (CMV) and toxoplasmosis. Infections in transplant recipients may be due to reactivation of endogenous dormant foci, or they may be acquired exogenously (e.g. from the environment, from another human, or from the transplanted organ itself). CMV is a particularly problematic pathogen in transplant recipients.

<table>
<tr><td>

Infections in patients with HIV

</td><td>

Primary CMV infection can be transmitted by the organ graft. All high-risk recipients should be carefully monitored for evidence of CMV infection, and antiviral therapy initiated as soon as possible.

Recurrent HSV, VZV and mucosal candida infections are early markers of declining immune function in HIV infection. Life-threatening (and AIDS-defining) infections included *Pneumocystis* pneumonia, tuberculosis and *Mycobacterium avium–intracellulare* (MAI). CNS infections include toxoplasmosis, cryptococcal meningitis and JC virus reactivation. CMV is a multi-system pathogen in HIV-infected individuals.

</td></tr>
</table>

Patients can be immunocompromised by many factors that affect either innate defense mechanisms or the adaptive immune system. These factors may be primary (i.e. a congenitally acquired or inherited disease) or secondary (i.e. acquired after birth through disease, trauma, infection, or as a complication of medical treatment). A broad classification of factors that compromise the immune system is shown in *Table 1*.

Primary immune deficiencies

These are relatively rare diseases. Defects can occur in neutrophil function, the complement system, or in T- or B-lymphocyte function. **Chronic granulomatous disease (CGD)**, one of the best known examples, is due to an impaired ability of neutrophils to kill organisms that have been phagocytosed, rendering patients susceptible to recurrent infections with *Staphylococcus aureus* and other microorganisms (e.g. *Aspergillus*).

Table 1.   Classification of factors that compromise the immune system

	Factors affecting innate defense mechanisms	Factors affecting the adaptive immune mechanisms
Primary factors	• Defects of neutrophil function [e.g. chronic granulomatous disease (CGD)] • Defects in the complement system	• Genetic disease altering B or T lymphocyte function [e.g. severe combined immune deficiency (SCID)]
Secondary factors	• Damage to normal intact skin (e.g. burns, surgical wounds, indwelling venous catheters) • Breach of normal mucosal defenses (e.g. indwelling urinary catheters, endotracheal tubes)	• Malignant disease, particularly of the reticuloendothelial system (e.g. leukemia, lymphoma, myeloma) • Infections, notably HIV • Malnutrition, metabolic diseases (e.g. chronic renal failure, diabetes) • Splenectomy • Immunosuppressive drugs (e.g. cytotoxic chemotherapy, steroids and immunosuppressive agents) used for cancer chemotherapy, organ and bone marrow transplants, and autoimmune diseases

Secondary defects in innate defense mechanisms

Many patients are rendered more susceptible to infection through damage to normal intact skin or by a breach of normal mucosal defenses.

The skin normally provides a very effective barrier to invasion of underlying tissues by microorganisms, but this barrier can be damaged in a variety of ways including:

- burns;
- surgical incisions;
- indwelling intravenous catheters.

Burn wounds can become colonized and infected with a variety of bacteria, particularly *Staph. aureus*, β-hemolytic streptococci and *Pseudomonas aeruginosa*. Invasive infection and septicemia may then follow. Infections with β-hemolytic streptococci can also cause damage to skin grafts, and outbreaks may occur in Burns Units.

Surgical wound infections are a common healthcare associated infection. The principal pathogen is *Staph. aureus*, and many of these are now methicillin-resistant (MRSA). Infections with β-hemolytic streptococci are less common, but often serious. Polymicrobial wound infections with coliforms, enterococci and anaerobes can occur following gastrointestinal or gynecological surgery.

Indwelling intravenous catheters can allow entry of bacteria and other organisms directly into the bloodstream causing bacteremia (Topic F12). Infection can occur with conventional pathogens (e.g. *Staph. aureus*), but also with more **opportunistic pathogens** [e.g. *Staph. epidermidis* – one of the usually harmless residents on the skin (Topic F1)].

A breach of normal **mucosal defenses** can also lead to an increased risk of infection. One of the commonest examples is the increased risk of urinary tract infections in patients with an **indwelling urinary catheter** (Topic F10). There is also a greatly increased risk of lower respiratory tract infection in patients who are ventilated via an **endotracheal tube** (ventilator-associated pneumonia) (Topic F7).

Secondary defects in adaptive immune mechanisms

A wide variety of diseases can impact on the proper functioning of normal immune mechanisms, rendering patients at increased risk of infection (*Table 1*). Different patterns of infections are associated with each of these factors.

Some of these infections also occur in nonimmunocompromised hosts, but they occur with a greater frequency in immunocompromised patients (e.g. the incidence of pneumococcal pneumonia in patients with HIV is significantly higher than in the normal population). Other infections occur with **opportunistic pathogens** – microorganisms that are usually harmless to healthy individuals, but capable of causing infection in immunocompromised hosts (e.g. *Pneumocystis Jiroveci* – Topic D3). The range and pattern of opportunistic infections varies according to the nature of the immunosuppression, with the widest range of opportunistic pathogens found in the most immunosuppressed patients.

Some of the common patterns of infection in selected immunocompromised patients are summarised in the remainder of this topic.

Infections in neutropenic patients

There are a variety of causes of neutropenia, but the commonest cause is **cancer chemotherapy**, particularly for hematological malignancies. The extent and duration of neutropenia vary with the different chemotherapy regimens. The risk of infection is significantly increased when the circulating neutrophil count falls to $< 0.5 \times 10^9 \, l^{-1}$, and is very high with counts $<0.1 \times 10^9 \, l^{-1}$. Infection risks are also related to the duration of neutropenia. For example, the risks of invasive fungal infection (e.g. invasive aspergillosis) particularly increase with neutropenia lasting > 14 days.

Patients with neutropenia who develop a temperature are said to have **febrile neutropenia**. There are many causes of febrile neutropenia, but the principal pathogens are Gram-positive and -negative bacteria and fungi.

Gram-positive bacteria (e.g. *Staphylococcus epidermidis*, diphtheroids) are common causes of bacteremia in neutropenic patients, with the portal of entry often being an indwelling central venous catheter. Enterococci, including strains that are vancomycin-resistant (VRE) are also encountered. **Gram-negative bacteria** (e.g. coliforms, pseudomonads) can cause 'spontaneous' septicemia in neutropenic patients, by bacterial translocation from the gastrointestinal tract into the bloodstream. This arises due to a) chemotherapy-induced mucosal damage in the gastrointestinal tract and b) impaired phagocytosis in the reticuloendothelial system. *Ps. aeruginosa* septicemia is associated with a particularly high mortality in this setting. The principal **fungal pathogens** are *Candida* species (including both *C. albicans* and other species) and *Aspergillus* (Topics D1 and D2). **Anaerobes** are an important cause of peri-anal infection, and *Clostridium tertium* is associated with a particular condition, **neutropenic enterocolitis** (typhilitis). *C. difficile* associated diarrhea is also a frequent complication.

Microbiological diagnosis of febrile neutropenia is principally by **blood cultures**, but cultures are negative in a significant proportion of episodes, and the diagnosis of invasive fungal infections is difficult.

Treatment of febrile neutropenia is with empiric broad-spectrum antibiotics that include activity against *Ps. aeruginosa* and other Gram-negative bacteria (e.g. ceftazidime, piptazobactam or carbapenem, combined with an aminoglycoside). Antibiotic therapy is adjusted when culture and sensitivity results become available (e.g. starting vancomycin if a Gram-positive line-associated bacteremia is identified). Patients may not respond to the first-line regimen and second-line agents, including antifungal drugs (e.g. amphotericin B) may need to be used.

Antimicrobial prophylaxis may reduce the frequency of infections during neutropenic episodes in certain situations. Fluconazole can reduce the incidence of invasive candidiasis. Fluoroquinolone antibiotics are used by some centers to reduce the frequency of Gram-negative sepsis, but this is associated with increased selection of antibiotic-resistant strains.

Infections in transplant patients

Recipients of organ transplants are generally more susceptible to infections as a result of the immunosuppressive drugs (e.g. cyclosporin) that are used to prevent rejection of the transplant [or in the case of bone marrow transplants, to prevent graft-versus-host disease (GVHD)]. These drugs principally suppress **cell-mediated immunity**, leaving patients at risk from a variety of opportunistic infections such as *Pneumocystis* pneumonia (**PCP** – see Topic D3), cytomegalovirus (**CMV** – Topic B7), and **toxoplasmosis** (Topic D4). Transplant patients are also more vulnerable to severe infections with more conventional pathogens, such as varicella-zoster virus (chickenpox, shingles – Topic B6) and *Mycobacterium tuberculosis*.

Bone marrow transplant recipients have the additional risk of being severely neutropenic (often for 2–3 weeks) until engraftment takes place, rendering them susceptible to an even wider range of pathogens (see above).

Infections in transplant recipients may be due to reactivation of endogenous dormant foci as a result of immunosuppression (e.g. toxoplasmosis, tuberculosis), or they may be acquired exogenously (e.g. from the environment, or from another human). Some infections, notably CMV, can be transmitted directly from the transplanted organ. As with blood transfusions, potential **donors** of organs are **screened** for the presence of a range of **blood-borne viruses**, including HIV, HBV, HCV and HTLV-1 in order to minimize the risk of transmission of these agents.

It is important to determine the **CMV 'immunity' status** of both donor and recipient. With solid-organ transplants, the highest risk of CMV disease occurs when a CMV-negative recipient receives an organ from a CMV-positive donor. This can lead to a severe primary CMV infection, as latent CMV in the transmitted organ finds itself in a nonimmune host. The converse is true, however, with bone marrow recipients (i.e. the high-risk individual is a seropositive recipient of a seronegative marrow).

As well as causing a primary infection, CMV can also reactivate in CMV-positive recipients as a consequence of the immunosuppression. Reactivation generally causes less severe disease than primary infections.

Antiviral prophylaxis may be given to reduce the risk of severe disease (e.g. aciclovir to limit the effects of HSV or VZV reactivation, or valganciclovir for CMV prophylaxis). All high-risk recipients should be carefully monitored for evidence of CMV infection, and antiviral therapy initiated as soon as possible.

The diagnosis, treatment and prevention of CMV disease and PCP are covered in Topics B7 and D3 respectively.

Transplant recipients are also at risk of developing EBV-related lymphomas (post-transplant lymphoproliferative disorder – PTLD).

Infections in patients with HIV

HIV infection (Topic B1) causes immunosuppression, principally through depletion of CD4 lymphocytes (T-helper cells). This particularly affects the function of the cell-mediated immune system. There is a wide variety of infections that can occur in patients with HIV, including a number of opportunistic pathogens. Some of these infections are AIDS-defining illnesses (acquired immunodeficiency syndrome).

Some of the important infections relating to HIV are listed in *Table 2*.

Table 2.   *Common infections in patients with HIV infection*

	Organism	Comments
Viruses	HSV, VZV	Herpes simplex may cause frequent severe recurrent oral or genital herpes. Herpes-zoster may be life-threatening in an immuno-suppressed host
	CMV (Topic B7)	CMV retinitis and encephalitis can occur in end-stage HIV infection
	HHV-8 (Topic B8)	HHV-8 is associated with the development of Kaposi's sarcoma, an important AIDS-defining illness
	EBV (Topic B8)	Lymphoproliferative disorders
	JC virus (Topic B19)	Associated with progressive multifocal leukoencephalopathy (PMLE), a cause of progressive dementia in HIV patients
Bacteria	*Streptococcus pneumoniae* (Topic C2)	Increased incidence of pneumococcal pneumonia in HIV patients
	Mycobacterium tuberculosis (Topic C6)	HIV infection is a significant risk factor for tuberculosis
	Mycobacterium avium–intracellulare complex (MAI)	Disseminated infection with MAI can occur in end-stage HIV infection
Fungi	*Candida albicans* (Topic D1)	Recurrent mucosal candidiasis is an early marker of immunosuppression in patients with HIV
	Cryptococcus neoformans (Topic D1)	An important cause of meningitis
	Pneumocystis Jiroveci (Topic D3)	PCP is the principal AIDS-defining infection
Protozoa	*Toxoplasma gondii* (Topic D4)	Reactivation of toxoplasma cysts can cause encephalitis
	Cryptosporidum parvum (Topic D4)	Cryptosporidiosis in HIV patients can cause prolonged, severe diarrhea

F15 INFECTIONS IN PREGNANCY AND NEONATES

Key Notes

Infections in pregnancy	Clinical features and management of infections in a woman of childbearing age may differ depending on whether she is pregnant or not. In addition, infection may pass to the fetus or neonate.
Maternal infections	Varicella is more likely to lead to pneumonia in a pregnant, rather than a nonpregnant, female. Urinary tract infections may be asymptomatic in pregnancy, but if untreated, may result in kidney damage. Antiviral drugs are not licensed for use in pregnancy, but aciclovir and some antiretroviral agents are used in certain circumstances. Many antibiotics are toxic to the developing fetus and should be avoided in a pregnant woman (e.g. tetracyclines, fluoroquinolones, aminoglycosides).
Maternal chronic virus infections	Chronic blood-borne virus infections (HIV, hepatitis B and C viruses) may be transmitted from mother to baby ante-, peri- or post-natally. Vertical transmission of HIV may be prevented by giving antiretroviral therapy to the mother during pregnancy, delivering by Cesarean section, and by avoidance of breastfeeding. Babies of HBV carrier mothers should start a course of vaccine at birth. High-risk babies (of HBeAg-positive mothers) should also receive hepatitis B immunoglobulin. About 4% of HCV-carrier mothers transmit infection to their babies.
Congenital infections	Cytomegalovirus is the commonest congenital infection. Five to 10% of CMV-infected babies are severely damaged at birth. Maternal rubella in the first 12 weeks of pregnancy commonly results in multisystem developmental defects (congenital rubella syndrome). Maternal varicella in the first half of pregnancy may result (1–2%) in congenital varicella. Maternal toxoplasmosis and syphilis infections may also cross the placenta and damage the fetus.
Acute infections in the neonatal period	Life-threatening infections in the neonatal period include septicemia and meningitis, caused by group B streptococci, *E. coli*, or rarely, *Listeria*. Neonatal varicella occurs in neonates whose mothers develop varicella late in pregnancy. Neonatal HSV infection, acquired from an infected birth canal, carries a high mortality. *N. gonorrhoeae* and *C. trachomatis* infections may also be acquired from the maternal birth canal, presenting with acute purulent conjunctivitis, ophthalmia neonatorum.
Related topics	Varicella-zoster virus (B6) Listeria (C5) Cytomegalovirus, HHV-6 Pirochaetes (C16) and -7 (B7) Protozoa (D4) Rubella and parvovirus B19 (B16) Urinary tract infections (F10)

Infections in pregnancy

Women of childbearing age, whether pregnant or not, are susceptible to a wide range of infections. However, physiological changes associated with the pregnant state may alter both disease susceptibility and the clinical manifestations of infection in a pregnant woman. In addition, infectious agents may gain access to the developing fetus, or, if maternal infection is acquired towards the end of pregnancy, to the neonate, with potentially devastating consequences. Recommendations for treatment of infection in a pregnant woman may differ, owing to unacceptable toxicities for the developing fetus. This topic deals with these specific aspects of common infections occurring during pregnancy.

Maternal infections

Varicella (chickenpox) (Topic B6) in an adult is often more severe than in children. There are data to suggest that a pregnant woman with chickenpox is more likely to develop **varicella pneumonia** (the commonest life-threatening complication) than if she were not pregnant. Thus, pregnant women with chickenpox should be assessed carefully for evidence of lung involvement, and consideration given to the need for therapy with intravenous aciclovir.

Urinary tract infections (Topic F10) are more common during pregnancy, and may be asymptomatic. It is thus important to screen urine for **asymptomatic bacteriuria**, particularly in early pregnancy, as left untreated, severe kidney infection can occur (**pyelonephritis**).

Antimicrobial therapy

Management of infection in a pregnant woman must take into account that certain agents are known to be potentially damaging to the developing fetus, and should therefore be avoided. Almost all antiviral agents are not licensed for use in pregnancy, for obvious reasons – most interfere with DNA synthesis. However, there are two important exceptions. One would not hesitate to use aciclovir in a pregnant woman with life-threatening varicella pneumonia. Antiretrovirals are also used in HIV-infected pregnant women for the particular purpose of reducing the risk of mother-to-baby transmission of infection (see below).

A number of antibacterial drugs may be potentially toxic to the developing fetus, and their use should be restricted during pregnancy (particularly in the first trimester). These include trimethoprim, tetracyclines, metronidazole, fluoroquinolones and aminoglycosides.

Maternal chronic virus infections

Mothers who are carriers of a blood-borne virus – i.e. HIV, hepatitis B or C viruses (HBV, HCV) – may transmit the infection to their offspring. Such **vertical transmission** (i.e. mother-to-baby) may occur antenatally through transplacental passage of the virus, perinatally, when the baby is exposed to an infected birth canal and also to maternal blood, or post-natally, through breast milk.

The rate of vertical transmission of **HIV** is ~ 20%, and most infections occur perinatally. This can be reduced to < 1% with appropriate management – HIV carrier mothers should be given antiretroviral therapy during pregnancy, their babies should be delivered by elective Cesarean section, and they should be advised not to breastfeed.

The risk of mother-to-baby transmission of **HBV** is > 90% for HBeAg-positive (Topic B3) mothers, and 30% for anti-HBe-positive mothers. Transmission can be interrupted by giving the baby appropriate prophylaxis at birth – combined active (HBV vaccine) and passive (hepatitis B immunoglobulin) immunization for high-risk babies, active immunization alone for lower-risk babies.

Vertical transmission of **HCV** is unusual – ~ 4% of babies of carrier mothers become infected. Currently, there is no proven form of prophylaxis to reduce this risk.

Note that correct management of HIV- and HBV-infected mothers requires knowledge of who those individuals are. Thus, an essential part of any policy for the prevention of transmission of these infections is universal antenatal screening for evidence of maternal HIV and HBV infection. This is now official policy in the UK.

Congenital infections

A particular anxiety for any pregnant woman with an infectious disease is whether or not this will affect her developing baby. The fetus is in a protected environment within the uterus, but if an organism is present in the maternal bloodstream, it will gain access to, and may cross, the placenta. This may or may not have serious consequences, depending on the stage of gestation, and the nature of the particular organism. A baby born already infected with an organism is said to be **congenitally infected**.

In addition to the specific infections mentioned below which can cross the placenta, there is also a general increased risk of spontaneous abortion (particularly in the early stages of pregnancy) with a range of infections (mainly viral) that cause fever and an inflammatory response (e.g. influenza, measles).

Congenital cytomegalovirus (CMV, see Topic B7) infection is the commonest congenital infection, occurring in 1 in 300 live births. Fortunately, the vast majority (80–85%) of CMV-*infected* babies are not adversely *affected* (i.e. they are normal at birth), and develop normally. Five to 10% of babies are severely damaged, amounting to ~ 300 such babies per year in the UK. They have multiple developmental abnormalities, and many do not survive beyond 1 year. The remaining 5–10% of CMV-infected babies are normal at birth, but are shown to have suffered some damage as they develop (e.g. uni- or bilateral nerve deafness). The outcome of congenital CMV infection is not dependent on either the nature of the maternal infection (i.e. primary or secondary CMV infection), or on the stage in pregnancy at which the fetus becomes infected. There are currently no effective strategies for prevention of congenital CMV infection.

The effects of **congenital rubella** infection are, in contrast to CMV, highly dependent on when the maternal infection occurs. Infection in the first 12 weeks carries a very high risk of the congenital rubella syndrome (CRS), with multiple abnormalities affecting the eyes, central nervous system and the heart. Maternal infection after 18 weeks of pregnancy carries virtually no risk of fetal damage. CRS is preventable by universal childhood rubella vaccination (as part of MMR). This prevents the circulation of virus within the community, and thereby eliminates the risk that a susceptible pregnant woman will come into contact with the virus. Note that maternal **parvovirus B19** (Topic B16) infection, which may be clinically very similar to maternal rubella, does not result in permanent congenital damage. It can, however, cause *hydrops fetalis* and there is an increased risk of intrauterine death/spontaneous miscarriage, but if the pregnancy survives, the baby will be normal.

If a pregnant woman acquires **chickenpox** in the first 20 weeks of pregnancy, there is a 1% risk that her baby will have **congenital varicella**, the main features of which are failure of limb bud development, and areas of severe skin scarring.

Maternal **toxoplasmosis** infection may cross the placenta (Topic D4). The risk of transplacental transmission of this organism increases during the pregnancy, but the earlier this occurs, the greater the risk of severe fetal brain damage.

Maternal **syphilis** (due to *Treponema pallidum* – see Topic C16) can also cause congenital infection. Congenital syphilis affects many parts of the body, and a range of clinical features usually present between the ages of 5 and 15 years. Routine serological screening during pregnancy and treatment with penicillin have made this a very rare condition in the developed world.

Acute infections in the neonatal period

The human immune system is not fully mature at birth. Neonates are protected from infection to a large extent by transplacentally acquired maternal antibodies, but in the absence of such antibodies, otherwise trivial infections may be life-threatening in a neonate.

Bacterial infections can be severe in the early neonatal period, especially in premature and/or low-birthweight neonates. Septicemia and **neonatal meningitis** are principally caused by **group B streptococci** (*Streptococcus agalactiae*) and *Escherichia coli*. These bacteria are usually acquired from the mother's birth canal during delivery, or as a result of premature rupture of membranes complicated by chorioamnionitis. Mothers who are known to be vaginal carriers of group B streptococci (GBS) should be given antibiotics during labor to reduce the risk of early-onset GBS infection. Some countries have developed screening programmes for GBS carriage during pregnancy.

A rare, but potentially serious, bacterial infection during pregnancy is **listeriosis** (*Listeria monocytogenes* – see Topic C5). This is a food-borne illness that can cause either septic abortion, or a chorioamnionitis with early-onset neonatal septicemia or meningitis. *Staphylococcus aureus* is a common cause of minor neonatal skin infections, but can also cause more severe soft tissue infections and sepsis.

Mothers with chickenpox in late pregnancy may infect their fetus/neonate without having passed on protective antibodies. **Neonatal varicella** is a feared disease, with a high mortality. Babies of mothers with varicella should be given varicella-zoster immunoglobulin at birth, and also prophylactic aciclovir for 14 days.

Neonatal herpes infection is also a potentially devastating disease. Infection is acquired from an infected birth canal, in mothers undergoing a primary genital herpes simplex infection late in pregnancy. The virus is disseminated through the bloodstream, and many babies die of herpes encephalitis or through involvement of other internal organs. Survivors may be left with severe brain damage. Fortunately, neonatal herpes is rare.

Ophthalmia neonatorum is an acute purulent conjunctivitis in neonates caused by *Neisseria gonorrhoeae* or *Chlamydia trachomatis* (Topic F5). These infections are acquired from the mother's genital tract, during birth. *C. trachomatis* can also cause neonatal pneumonia.

Premature and low-birthweight neonates on neonatal units are particularly vulnerable to many other infections, particularly as their immature immune system is further compromised by thin, fragile skin, and the presence of invasive medical devices that are required to support their care. In addition to the infections already mentioned, neonatal sepsis with coagulase-negative staphylococci (related to venous catheters), coliforms, pseudomonads and *Candida*, are commonly seen.

INDEX

Bold type is used to indicate the main entry where there are several.